Characterization of Porous Solids and Powders: Surface Area, Pore Size and Density

Particle Technology Series

Series Editor

Professor Brian Scarlett
Technical University of Delft

The Kluwer Particle Technology Series of books is the successor to the Chapman and Hall Powder Technology Series. The aims are the same, the scope is wider. The particles involved may be solid or they may be droplets. The size range may be granular, powder or nano-scale. The accent may be on materials or on equipment, it may be practical or theoretical. Each book can add one brick to a fascinating and vital technology. Such a vast field cannot be covered by a carefully organised tome or encyclopaedia. Any author who has a view or experience to contribute is welcome. The subject of particle technology is set to enjoy its golden times at the start of the new millennium and I expect that the growth of this series of books will reflect that trend.

Characterization of Porous Solids and Powders: Surface Area, Pore Size and Density

by

S. LOWELL

Quantachrome Instruments, Boynton Beach, U.S.A.

JOAN E. SHIELDS

C.W. Post Campus of Long Island University, U.S.A.

MARTIN A. THOMAS

Quantachrome Instruments, Boynton Beach, U.S.A.

and

MATTHIAS THOMMES

Quantachrome Instruments, Boynton Beach, U.S.A.

KLUWER ACADEMIC PUBLISHERS

DORDRECHT / BOSTON / LONDON

A C.I.P. Catalogue record for this book is available from the Library of Congress.

ISBN 1-4020-2302-2 (HB)
ISBN 1-4020-2303-0 (e-book)

Published by Kluwer Academic Publishers,
P.O. Box 17, 3300 AA Dordrecht, The Netherlands.

Sold and distributed in North, Central and South America
by Kluwer Academic Publishers,
101 Philip Drive, Norwell, MA 02061, U.S.A.

In all other countries, sold and distributed
by Kluwer Academic Publishers,
P.O. Box 322, 3300 AH Dordrecht, The Netherlands.

Printed on acid-free paper

Printed in the Netherlands.

Preface

The growth of interest in newly developed porous materials has prompted the writing of this book for those who have the need to make meaningful measurements without the benefit of years of experience. One might consider this new book as the 4th edition of "Powder Surface Area and Porosity" (Lowell & Shields), but for this new edition we set out to incorporate recent developments in the understanding of fluids in many types of porous materials, not just powders. Based on this, we felt that it would be prudent to change the title to "Characterization of Porous Solids and Powders: Surface Area, Porosity and Density".

This book gives a unique overview of principles associated with the characterization of solids with regard to their surface area, pore size, pore volume and density. It covers methods based on gas adsorption (both physi- and chemisorption), mercury porosimetry and pycnometry. Not only are the theoretical and experimental basics of these techniques presented in detail but also, in light of the tremendous progress made in recent years in materials science and nanotechnology, the most recent developments are described. In particular, the application of classical theories and methods for pore size analysis are contrasted with the most advanced microscopic theories based on statistical mechanics (e.g. Density Functional Theory and Molecular Simulation).

The characterization of heterogeneous catalysts is more prominent than in earlier editions; the sections on mercury porosimetry and particularly chemisorption have been updated and greatly expanded.

The book will appeal both to students and to scientists in industry who are in need of accurate and comprehensive pore and surface area characterization of their materials, and those who have the need to learn quickly the rudiments of the measurements. This book therefore retains the successful style of the earlier series in that the first half of the book is devoted to theoretical concepts, while the second half presents experimental details, including instrument design factors.

Thanks are due to Robert Swinson for providing much of the new artwork, and to Scott Lowell for his keen-eyed proofreading. Two of us (M.A.T and M.T.) express our sincerest appreciation to our spouses and families for their patient endurance during the preparation of this book.

Boynton Beach, Florida, November 2003.

Contents

5   Surface Area from the Langmuir and BET Theories

6   Other Surface Area Methods

10   Mercury Porosimetry: Non-Wetting Liquid Penetration

11   Pore Size and Surface Characteristics of Porous Solids by Mercury Porosimetry

12 Chemisorption: Site Specific Gas Adsorption

PART 2 EXPERIMENTAL

13 Physical Adsorption Measurements – Preliminaries

14 Vacuum Volumetric Measurements (Manometry)

15   Dynamic Flow Method

16   Volumetric Chemisorption: Catalyst Characterization by Static Methods

17 Dynamic Chemisorption: Catalyst Characterization By Flow Techniques

18 Mercury Porosimetry: Intra and Inter-Particle Characterization

19   Density Measurement

Index

1 Introduction

1.1 REAL SURFACES

There is a convenient mathematical idealization which asserts that a cube of edge length, ℓ cm, possesses a surface area of $6\ell^2$ cm^2 and that a sphere of radius r cm exhibits $4\ell r^2$ cm^2 of surface. In reality, however, mathematical, perfect or ideal geometric forms are unattainable since under microscopic examinations all real surfaces exhibit flaws. For example, if a 'super microscope' were available one would observe surface roughness due not only to voids, pores, steps, and other surface imperfections but also due to the atomic or molecular orbitals at the surface. These surface irregularities will always create a real surface area greater than the corresponding theoretical area.

1.2 FACTORS AFFECTING SURFACE AREA

When a cube, real or imaginary, of one meter edge length is subdivided into smaller cubes each one micrometer (10^{-6} meter) in length there will be formed 10^{18} particles, each exposing an area of 6 x 10^{12} square meters (m^2). Thus, the total area of all the particles is 6 x 10^6 m^2. This million-fold increase in exposed area is typical of the large surface areas exhibited by fine powders when compared to undivided material. Whenever matter is divided into smaller particles, new surfaces must be produced with a corresponding increase in surface area.

In addition to particle size, the particle shape contributes to the surface area of the powder. Of all geometric forms, a sphere exhibits the minimum area-to-volume ratio while a chain of atoms, bonded only along the chain axis, will give the maximum area-to-volume ratio. All particulate matter possesses geometry and, therefore, surface areas between these two extremes. The dependence of surface area on particle shape is readily shown by considering two particles of the same composition and of equal mass, M, one particle a cube of edge length ℓ and the other spherical with radius r. Since the particle density, ρ, is independent of particle shape[†] one can write

---

[†] For sufficiently small particles the density can vary slightly with changes in the area to volume ratio. This is especially true if particles are ground to size and atoms near the surface are disturbed from their equilibrium position.

$$M_{cube} = M_{sphere} \tag{1.1}$$

$$(V\rho)_{cube} = (V\rho)_{sphere} \tag{1.2}$$

$$\ell_{cube}^3 = \tfrac{4}{3}\pi r_{sphere}^3 \tag{1.3}$$

$$\frac{S_{cube}\ell_{cube}}{6} = S_{sphere}\frac{r_{sphere}}{3} \tag{1.4}$$

$$\frac{S_{cube}}{S_{sphere}} = \frac{2r_{sphere}}{\ell_{cube}} \tag{1.5}$$

Thus, for particles of equal weight, the cubic area, S_{cube}, will exceed the spherical area, S_{sphere}, by a factor of $2\,r/\ell$.

The range of specific surface area[‡] can vary widely depending upon the particle's size, shape, and porosity.[§] The influence of pores can often overwhelm the size and external shape factors. For example, a powder consisting of spherical particles exhibits a total surface area, S_t, as described by equation (1.6):

$$S_t = 4\pi(r_1^2 N_1 + r_2^2 N_2 + \ldots + r_i^2 N_i) = 4\pi\sum_{i=1} r_i^2 N_i \tag{1.6}$$

where r_i and N_i are the average radii and numbers of particles respectively in the size range i. The volume of the same powder sample is

$$V = \tfrac{4}{3}\pi(r_1^3 N_1 + r_2^3 N_2 + \ldots + r_i^3 N_i) = \tfrac{4}{3}\pi\sum_{i=1} r_i^3 N_i \tag{1.7}$$

Replacing V in equation (1.7) by the ratio of mass to density, M/ρ, and dividing equation (1.6) by (1.7) gives the specific surface area

$$S = \frac{S_t}{M} = \frac{3\sum_i N_i r_1^2}{\rho\sum_i N_i r_1^3} \tag{1.8}$$

For spheres of uniform radius equation (1.8) becomes

[‡] The area exposed by one gram of powder is called the 'specific surface area'.
[§] Porosity is defined here as surface features that are deeper than they are wide.

$$S = \frac{3}{\rho r} \qquad (1.9)$$

Thus, powders consisting of spherical particles of 0.1 micrometer (μm) radius with densities near 3 g cm^{-3} will exhibit surface areas about 10^5 cm^{2}g^{-1} (10 m^{2}g^{-1}). Similar particles with radii of 1.0 μm would exhibit a tenfold decrease in surface area. However, if the same 1.0 μm radius particles contained extensive porosity they could exhibit specific surface areas well in excess of 1,000 m^{2}g^{-1}. This clearly indicates the significant contribution that pores can make to the surface area.

1.3 SURFACE AREA FROM PARTICLE SIZE DISTRIBUTIONS

Although particles can assume all regular geometric shapes, and in most instances highly irregular shapes, most particle size measurements are based on the so-called 'equivalent spherical diameter'. This is the diameter of a sphere that would behave in the same manner as the test particle being measured in the same instrument. For example, the electrical sensing zone method [1] is a commonly used technique for determining particle sizes. Its principle is based on the momentary increase in the resistance of an electrolyte solution that results when a particle passes through a narrow aperture between two electrodes. The resistance change is registered in the electronics as a rapid pulse. The pulse height is proportional to the particle volume and therefore, the particles are sized as equivalent spheres.

Stokes' law [2] is another concept around which several instruments are designed to give particle size or size distributions. Stokes' law is used to determine the settling velocity of particles in a fluid medium as a function of their size. Equation (1.10) is a useful form of Stokes' law

$$D = \sqrt{\frac{18\eta v}{(\rho_s - \rho_t)g}} \qquad (1.10)$$

where D is the particle diameter, η is the coefficient of viscosity, v is the settling velocity, g is the gravitational constant, and ρ_s and ρ_r are the densities of the solid and the fluid, respectively. Allen [3] gives an excellent discussion of the various experimental methods associated with sedimentation size analysis. Regardless of the experimental method employed, nonspherical particles will be measured as larger or smaller equivalent spheres depending on whether the particles settle faster or more slowly than spheres of the same mass. Modifications of Stokes' law have been used in centrifugal devices to enhance the settling rates but are subject to the same limitations of yielding only the equivalent spherical diameter.

Optical devices, based upon particle attenuation of a light beam or measurement of scattering angles and intensity, also give equivalent spherical diameters.

Permeametric methods, discussed in chapter 6, are often used to determine average particle size. The method is based upon the impedance offered to the fluid flow by a packed bed of powder. Again, equivalent spherical diameter is the calculated size.

Sieving is another technique that sizes particles according to their smallest dimension but gives no information on particle shape.

Electron microscopy techniques can be used to estimate particle shape. A limitation is that only relatively few particles can be viewed.

Attempts to measure surface area based on any of the above methods will give results significantly less than the true value, in some cases by factors of 10^3 or greater depending upon particle shape, surface irregularities and porosity. At best, surface areas calculated from particle size will establish the lower limit by the implicit assumptions of sphericity or some other regular geometric shape, and by ignoring the highly irregular nature of real surfaces.

1.4 REFERENCES

1. ISO 13319 (2000) *Determination of particle size distributions – Electrical sensing zone method,* International Organization for Standardization, Geneva, Switzerland.
2. Orr Jr. C. and DallaValle J.M. (1959) *Fine Particle Measurement*, Macmillan, New York.
3. Allen T. (1981) *Particle Size Measurement*, Chapman and Hall, London.

2 Gas Adsorption

2.1 INTRODUCTION

Gas adsorption is one of many experimental methods available for the surface and pore size characterization of porous materials. These include small angle x-ray and neutron scattering (SAXS and SANS), mercury porosimetry, electron microscopy (scanning and transmission), thermoporometry, NMR-methods, and others. Each method has a limited length scale of applicability for pore size analysis. An overview of different methods for pore size characterization and their application range was recently given by IUPAC [1]. Among these methods gas adsorption is the most popular one because it allows assessment of a wide range of pore sizes (from 0.35 nm up to > 100 nm), including the complete range of micro- and mesopores and even macropores. In addition, gas adsorption techniques are convenient to use and are not that cost intensive as compared to some of the other methods. A combination of mercury porosimetry and gas adsorption techniques allows even performing a pore size analysis over a range from ca. 0.35 nm up to ca. 400 μm.

Adsorption can be understood as the enrichment of one or more components in an interfacial layer; in gas adsorption we consider the gas/solid interface. The solid is called the adsorbent and the gas, which is capable of being adsorbed, is called the adsorptive. The fluid in the adsorbed state is called adsorbate [2].

Invariably the amount adsorbed on a solid surface will depend upon the absolute temperature T, the pressure P, and the interaction potential E between the vapor (adsorbate) and the surface (adsorbent). Therefore, at some equilibrium pressure and temperature the weight W of gas adsorbed on a unit weight of adsorbent is given by

$$W = F(P, T, E) \tag{2.1}$$

Usually the quantity adsorbed is measured at constant temperature and equation (2.1) reduces to

$$W = F(P, E) \tag{2.2}$$

A plot of W versus P, at constant T, is referred to as the sorption isotherm of a particular gas-solid interface.

2.2 PHYSICAL AND CHEMICAL ADSORPTION

Depending upon the strength of the interaction, all adsorption processes can be divided into the two categories of chemical and physical adsorption. The former, also called irreversible adsorption or chemisorption, is characterized mainly by large interaction potentials, which lead to high heats of adsorption often approaching the value of chemical bonds. This fact, coupled with other spectroscopic, electron spin resonance, and magnetic susceptibility measurements confirms that chemisorption involves true chemical bonding of the gas or vapor with the surface. Because chemisorption occurs through chemical bonding it is often found to occur at temperatures above the critical temperature of the adsorbate. Strong bonding to the surface is necessary in the presence of higher thermal energies, if adsorption is to occur at all. In addition, chemisorption is usually associated with an activation energy, as is true for most chemical reactions. Furthermore, chemisorption is necessarily restricted to, at most, a single layer of chemically bound adsorbate on the surface. Another important feature of chemisorption is that the adsorbed molecules are more localized on the surface when compared to physical adsorption. Because of the formation of a chemical bond between an adsorbate molecule and a specific site on the surface the adsorbate is less free to migrate about the surface. This fact often enables the number of active sites on catalysts to be determined by simply measuring the quantity of chemisorbed gas.

The second category, reversible or physical adsorption, is a general phenomenon, which occurs whenever an absorbable gas (the adsorptive) is brought in contact with the surface of the solid adsorbent. Physisorption exhibits characteristics that make it most suitable for surface area-determinations as indicated by the following:

1. Physical adsorption is accompanied by low heats of adsorption with no violent or disruptive structural changes occurring to the surface during the adsorption measurement.
2. Unlike chemisorption, physical adsorption may lead to surface coverage by more than one layer of adsorbate.
3. Pores can be filled completely by the adsorptive for pore volume measurements. Such pore condensation phenomena can be used also to calculate the pore size and its distribution.
4. Physical adsorption equilibrium is achieved rapidly since no activation energy is required as is generally true in chemisorption. An exception here is adsorption in small pores where diffusion can limit the adsorption rate.
5. Physical adsorption is fully reversible, enabling both the adsorption and desorption processes to be studied.

6. Physically adsorbed molecules are not restricted to specific sites and are free to cover the entire surface. For this reason, surface areas rather than number of sites can be calculated.

2.3 PHYSICAL ADSORPTION FORCES

Upon adsorption, the entropy change of the adsorbate, ΔS_a, is necessarily negative since, due to the loss of at least one degree of translational freedom, the condensed state is more ordered than the gaseous state. A reasonable assumption for physical adsorption is that the entropy of the adsorbent remains essentially constant and certainly does not increase by more than the adsorbate's entropy decreases. Therefore, ΔS for the entire system is necessarily negative. The spontaneity of the adsorption process requires that the Gibbs free energy, ΔG, also be a negative quantity. Based upon the entropy and free energy changes, the enthalpy change, ΔH, accompanying physical adsorption is always negative, indicating an exothermic process, as shown by

$$\Delta H = \Delta G + T\Delta S \tag{2.3}$$

The so-called van der Waals' forces [3] are most important for the occurrence of physisorption:

1. Dispersion forces: these forces are present regardless of the nature of other interactions and often account for the major part of the adsorptive-adsorbent potential.
2. Ion-dipole: an ionic solid and electrically neutral but polar adsorbate.
3. Ion-induced dipole: a polar solid and polarizable adsorbate.
4. Dipole-dipole: a polar solid and polar adsorbate.
5. Quadrupole interactions: symmetrical molecules, such as nitrogen and carbon dioxide, possess no dipole moment but do have a quadrupole (e.g. $^-O\!-\!C^{++}\!-\!O^-$), which can lead to interactions with polar surfaces.

The nature of dispersion forces was first recognized in 1930 by London [4] who postulated that the electron motion in an atom or molecule would lead to a rapidly oscillating dipole moment. At any instant, the lack of symmetry of the electron distribution about the nuclei imparts a transient dipole moment to an atom or molecule, which vanishes when averaged over a longer time interval. When in close proximity, the rapidly oscillating dipoles of neighboring molecules couple into phase with each other leading to a net attracting potential. This phenomenon is associated with the molecular dispersion of light due to the light's electromagnetic field interaction with the oscillating dipole.

It is evident from the above that adsorption forces are similar in nature and origin to the forces that lead to liquefaction of vapors and that the same intermolecular interactions are responsible for both phenomena.

2.4 PHYSICAL ADSORPTION ON A PLANAR SURFACE

The London-van der Waals' interaction energy $U_s(z)$ of a gas molecule with a planar surface is given by

$$U_s(z) = C_1 z^{-9} - C_2 z^{-3} \qquad (2.4)$$

where C_1 and C_2 are constants, and z is the distance of the gas molecule from the surface. The first term describes the repulsive forces that occur when the location of the molecule is too close to the surface (so-called Born repulsion). The second term represents the attractive fluid-wall interactions. The interaction potential $U_s(z)$ exhibits a minimum relatively close to the surface of the adsorbent and tends to zero for large distances from the surface (see Fig. 2.1). The (attractive) interaction energy at the minimum of the gas-solid potential is typically ten times greater than the thermal energy $k_b T$, where k_b is the Boltzmann constant and T is the absolute temperature. As a consequence, gas molecules will accumulate in the vicinity of the surface (see Fig.2.1).

At sufficiently low temperatures (typically around the boiling temperature of the adsorptive) a dense monolayer of molecules is formed at pressures P far below the saturation pressure P_0 and a multilayer adsorbed film of increasing thickness and liquid like density builds up on strongly adsorbing substrates as P_0 is approached. In this low-temperature region the adsorption of gases can be analyzed in terms of a two-phase model, in which an adsorbed phase coexists with the bulk phase [5]:

$$n^g = n_{ads} + n_{bulk} \qquad (2.5)$$

$$V^g = V_{ads} + V_{bulk} \qquad (2.6)$$

where n^g is the amount of gas and V^g the overall macroscopic volume accessible to the gas molecules; n_{ads} represents the amount and V_{ads} the volume of the adsorbed phase.

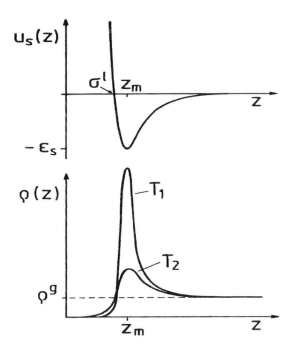

Figure 2.1 Gas-solid interaction potential $U_s(z)$ (upper graph) and density profile $\rho(z)$ of an ideal adsorbed gas at a flat, homogeneous surface for two temperatures $T_2 > T_1$. From [5].

However, at higher temperatures the model of an adsorbed phase becomes progressively unrealistic because (i) the tendency of molecules to accumulate near the surface of the adsorbent becomes less pronounced and (ii) due to the weaker physisorption at elevated temperatures higher pressures have to be applied in order to reach significant surface coverage [5]. As a consequence the density of the bulk gas phase is no longer negligible relative to the density near the surface and a clear separation between adsorbed phase and bulk gas phase is not possible, i.e. the profile of the local density $\rho(z)$ exhibits a smooth transition from the surface into the bulk gas. For this situation the definition of the adsorbed amount, i.e., the adsorption space, becomes problematic and other concepts have to be applied. One possibility is to express the adsorbed amount in terms of the surface excess, a concept that was first introduced by Gibbs [6].

The surface excess (not the adsorbed amount) is the quantity that is actually determined when using the volumetric or gravimetric technique (see chapters 13 and 14) to measure adsorption isotherms. Here a known amount of gas, n^g, is contained in a volume V^g in contact with the solid adsorbent. The experimentally determined adsorbed amount represents the excess amount over and above the amount that would be present if the density of the gas remained constant and equal to that of the bulk fluid up to the surface [7]. Accordingly, experimentally the surface excess n^σ instead of the adsorbed amount n^a is determined

$$n^\sigma = n^g - \rho^g V^g \tag{2.7}$$

where ρ^g is the (molar) density of the bulk gas at the experimental temperature and pressure. Combining equations (2.5) – (2.7) leads to

$$n^\sigma = n_{ads} - \rho^g V_{ads} = (\rho_{ads} - \rho^g) V_{ads} \tag{2.8}$$

where the term $\rho^g V_{ads}$ represents the amount of gas in a volume equal to V_{ads} somewhere in the bulk phase and $\rho_{ads} = n_{ads}/V_{ads}$ is the mean density of the adsorbed phase.

At sufficiently low temperatures and pressures, the gas density is negligibly small against the density near the surface ($\rho^g \ll \rho_{ads}$) and thus the surface excess n^σ corresponds to the adsorbed amount n_{ads}, i.e. $n^\sigma \approx n_{ads}$. This is the typical situation encountered for nitrogen and argon adsorption at their boiling temperatures (77.35 K and 87.27 K, respectively), which is used for the surface and pore size characterization of solids and finely divided matter.

In contrast, adsorption in significant amounts occurs close to the critical point and in particular above the critical temperature (supercritical adsorption) only at higher pressures [5]. The bulk gas density is here so high that it cannot be neglected anymore, and as indicated before adsorption data are therefore usually given in terms of the surface excess.

2.4 REFERENCES

1. Rouquerol J., Avnir D., Fairbridge C.W., Everett D.H., Haynes J.H., Pernicone N., Ramsay J.D.F., Sing K.S.W. and Unger K.K. (1994) *Pure Appl. Chem.* **66**, 1739.
2. Sing K.S.W., Everett D.H., Haul R.A.W., Moscou L., Pierotti R.A., Rouquerol J. and Siemieniewska T. (1985) *Pure Appl. Chem.* **57**, 603.
3. Israelachvili J.N. (1985) *Intermolecular and Surface Forces,* Academic Press, London.
4. London F. (1930) *Z. Phys.* **63**, 245.
5. Findenegg G.H. and Thommes M. (1997) In *Physical Adsorption: Experiment, Theory and Application* (Fraissard J. and Conner W.C., eds), Kluwer, Dordrecht.
6. Gibbs J.W. (1957) *The Collected Works, Vol.1,* Yale University Press, New Haven, p219.
7. Everett D.H. (1972) *Pure Appl. Chem.* **31**, 579.

3 Adsorption Isotherms

3.1 PORE SIZE AND ADSORPTION POTENTIAL

The shape of sorption isotherms of pure fluids on planar surfaces and porous materials depends on the interplay between the strength of fluid-wall and fluid-fluid interactions as well as the effects of confined pore space on the state and thermodynamic stability of fluids confined to narrow pores. The International Union of Pure and Applied Chemistry [1] proposed to classify pores by their internal pore width (the pore width defined as the diameter in case of a cylindrical pore and as the distance between opposite walls in case of a slit pore), i.e., *Micropore:* pore of internal width less than 2 nm; *Mesopore:* pore of internal width between 2 and 50 nm; *Macropore:* pore of internal width greater than 50 nm.

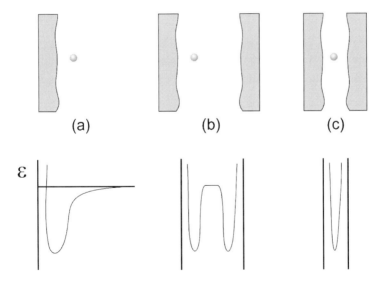

Figure 3.1 Schematic illustration of adsorption potential, ε, on (a) planar, nonporous surface; (b) mesopore; (c) micropore.

The sorption behavior in macropores is distinct from that of mesopores and micropores. Whereas macropores are so wide that they can be considered as nearly flat surfaces (see Fig. 3.1a) the sorption behavior in micropores is dominated almost entirely by the interactions between fluid molecules and the pore walls; in fact the adsorption potentials of the opposite pore walls are overlapping. Hence the adsorption in micropores (i.e., micropore filling) is distinct from the adsorption phenomena occurring in mesopores. As illustrated in Fig. 3b, the pore potential in mesopores is not dominant anymore in the core of the pores. Hence, the adsorption behavior in mesopores does not depend only on the fluid-wall attraction, but also on the attractive interactions between fluid molecules, which may lead to the occurrence of capillary (pore) condensation. Pore condensation represents a phenomenon whereby gas condenses to a liquid-like phase in pores at a pressure less than the saturation pressure P_0 of the bulk fluid. It represents an example of a shifted bulk transition under the influence of the attractive fluid-wall interactions.

3.2. CLASSIFICATION OF ADSORPTION ISOTHERMS

Based upon an extensive literature survey, performed by Brunauer, Demming, Demming and Teller (BDDT)[2], the IUPAC published in 1985 a classification of six sorption isotherms [1], which reflects the situation discussed above in connection with figure 3.1. The appropriate IUPAC classification is shown in Fig. 3.2. Each of these six isotherms and the conditions leading to its occurrence are now discussed according to Sing *et al* [1].

The reversible type I isotherm is concave to the P/P_0 axis and the adsorbed amount approaches a limiting value as $P/P_0 \rightarrow 1$. Type I isotherms are obtained when adsorption is limited to, at most, only a few molecular layers. This condition is encountered in chemisorption, where the asymptotic approach to a limiting quantity indicates that all of the surface sites are occupied. In the case of physical adsorption, sorption isotherms obtained on microporous materials are often of type I. Micropore filling and therefore high uptakes are observed at relatively low pressures, because of the narrow pore width and the high adsorption potential. The limiting uptake is being governed by the accessible micropore volume rather than by the internal surface area.

Type II sorption isotherms are typically obtained in case of non-porous or macroporous adsorbent, where unrestricted monolayer-multilayer adsorption can occur. The inflection point or knee of the isotherm is called point B. This point indicates the stage at which monolayer coverage is complete and multilayer adsorption begins to occur.

The reversible type III isotherm is convex to the P/P_0 axis over its entire range and therefore does not exhibit a point B. This indicates that the attractive adsorbate-adsorbent interactions are relatively weak and that the adsorbate-adsorbate interactions play an important role. Isotherms of this type are not common, but an example is nitrogen adsorption on polyethylene or the adsorption of water vapor on the clean basal plane of graphite.

Type IV isotherms are typical for mesoporous materials. The most characteristic feature of the type IV isotherm is the hysteresis loop, which is associated with the occurrence of pore condensation. The limiting uptake over a range of high P/P_0 results in a plateau of the isotherm, which indicates complete pore filling. The initial part of the type IV can be attributed to monolayer-multilayer adsorption as in case of the type II isotherm.

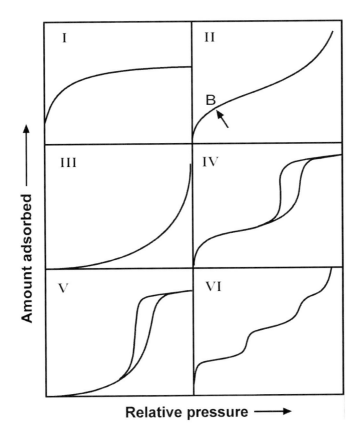

Figure 3.2 IUPAC classification of sorption isotherms. From [1].

Type V isotherms show pore condensation and hysteresis. However, in contrast to type IV the initial part of this sorption isotherm is related to adsorption isotherms of type III, indicating relatively weak attractive interactions between the adsorbent and the adsorbate.

The type VI isotherm is a special case, which represents stepwise multilayer adsorption on a uniform, non-porous surface [3], particularly by spherically symmetrical, non-polar adsorptives. The sharpness of the steps depends on the homogeneity of the adsorbent surface, the adsorptive and the temperature. Type VI isotherms were for example obtained with argon [4] and krypton [5] on graphitized carbons at liquid nitrogen temperature.

3.3 REFERENCES

1. Sing K.S.W., Everett D.H., Haul R.A.W., Moscou L., Pierotti R.A., Rouquerol J. and Siemieniewska T. (1985) *Pure Appl. Chem.* **57,** 603.
2. Brunauer S., Deming L.S., Deming W.S. and Teller E. (1940) *J. Am. Chem. Soc.* **62**, 1723.
3. Hill. T.L. (1955) *J. Phys. Chem.* **59**, 1065.
4. Polley M.H., Schaeffer W.D. and Smith W.R. (1953) *J. Phys. Chem.* **57**, 469.
5. Greenhalgh E. (1967) *J. Phys. Chem.* **71**, 1151.

4 Adsorption Mechanism

4.1 LANGMUIR AND BET THEORIES (KINETIC ISOTHERMS)

The success of kinetic theories directed toward the measurements of surface areas depends upon their ability to predict the number of adsorbate molecules required to cover the solid with a single molecular layer. Equally important is the cross-sectional area of each molecule or the effective area covered by each adsorbed molecule on the surface. The surface area then, is the product of the number of molecules in a completed monolayer and the effective cross-sectional area of an adsorbate molecule. The number of molecules required for the completion of a monolayer will be considered in this chapter and aspects of the adsorbate cross-sectional area will be discussed in chapter 5.

4.1.1 The Langmuir Isotherm

The asymptotic approach of the quantity adsorbed toward a limiting value indicates that type I isotherms are limited to, at most, a few molecular layers. In the case of chemisorption, only one layer can be bonded to the surface and, therefore, true chemisorption always exhibits a type I isotherm. Although it is possible to calculate the number of molecules in the monolayer from the type I chemisorption isotherm, some serious difficulty is encountered when attempts are made to apply the cross-sectional adsorbate area. This difficulty arises because chemisorption tightly binds and localizes the adsorbate to a specific surface site so that the spacing between adsorbed molecules will depend upon the adsorbent surface structures as well as the size of the adsorbed molecules or atoms. In those cases where the surface sites are widely separated, the calculated surface area will be smaller than the actual values because the number of molecules in the monolayer will be less than the maximum number which the surface can accommodate. Nevertheless, it will be instructive to consider the type I isotherm in preparation for the more rigorous requirements of the other five types.

Using a kinetic approach, Langmuir [1] was able to describe the type I isotherm with the assumption that adsorption was limited to a monolayer. According to the kinetic theory of gases, the number of molecules N striking unit area of surface per second is given by

$$N = \frac{\overline{N}P}{\sqrt{2\pi\overline{M}RT}} \tag{4.1}$$

where $\overline{N}$ is Avogadro's number, P is the adsorbate pressure, $\overline{M}$ is the adsorbate molecular weight, R is the gas constant and T is the absolute temperature. If θ_0 is the fraction of the surface unoccupied (i.e., with no adsorbed molecules) then the number of collisions with bare or uncovered surface per unit area of surface each second is

$$\frac{dN}{dt} = kP\theta_0 \tag{4.2}$$

where k is $\overline{N}/\sqrt{2\pi\overline{M}RT}$. The number of molecules striking and adhering to each unit area of surface is

$$N_{ads} = kP\theta_0 A_1 \tag{4.3}$$

where A_1 is the condensation coefficient and represents the probability of a molecule's being adsorbed upon collision with the surface.

The rate at which adsorbed molecules leave each unit area of surface is given by

$$N_{des} = N_m \theta_1 v_1 e^{-E/RT} \tag{4.4}$$

where N_m is the number of adsorbate molecules in a completed monolayer of unit area, θ_1 is the fraction of the surface occupied by the adsorbed molecules, E is the energy of adsorption and v_1 is the vibrational frequency of the adsorbate normal to the surface when adsorbed. The product $N_m \theta_1$ is the number of molecules adsorbed per unit area. Multiplication by v_1 converts this number of molecules to the maximum rate at which they can leave the surface. The term $e^{-E/RT}$ represents the probability that an adsorbed molecule possesses adequate energy to overcome the net attractive potential of the surface. Thus, equation (4.4) contains all the parameters required to describe the rate at which molecules leave each unit area of surface.

At equilibrium the rates of adsorption and desorption are equal. Thus equating (4.3) and (4.4):

$$N_m \theta_1 v_1 e^{-E/RT} = kP\theta_0 A_1 \tag{4.5}$$

Recognizing that $\theta_0 = 1 - \theta_1$, one obtains

$$N_m \theta_1 v_1 e^{-E/RT} = kPA_1 - \theta_1 kPA_1 \tag{4.6}$$

then

$$\theta_1 = \frac{kPA_1}{N_m v_1 e^{-E/RT} + kPA_1} \tag{4.7}$$

Allowing

$$K = \frac{kA_1}{N_m v_1 e^{-E/RT}} \tag{4.8}$$

Substitution of equation (4.8) into (4.7) gives

$$\theta_1 = \frac{KP}{1 + KP} \tag{4.9}$$

The assumption implicit in equation (4.8) is that the adsorption energy E is constant, which implies an energetically uniform surface. Up to and including one layer of coverage one can write

$$\theta_1 = \frac{N}{N_m} = \frac{W}{W_m} \tag{4.10}$$

where N and N_m are the number of molecules in the incomplete and complete monolayer, respectively, and W/W_m is the weight adsorbed relative to the weight adsorbed in a completed monolayer. Substituting W/W_m for θ_1 in equation (4.9) yields

$$\frac{W}{W_m} = \frac{KP}{1 + KP} \tag{4.11}$$

Equation (4.11) is the Langmuir equation for Type I isotherms. Rearrangement of equation (4.11) gives

$$\frac{P}{W} = \frac{1}{KW_m} + \frac{P}{W_m} \tag{4.12}$$

A plot of P/W versus P will give a straight line of slope $1/W_m$ and intercept $1/KW_m$ from which both K and W_m can be calculated. Having established W_m, the sample surface area S_t can then be calculated from equation (4.13):

$$S_t = N_m A_x = \frac{W_m \overline{N} A_x}{\overline{M}} \qquad (4.13)$$

where A_x and $\overline{M}$ are the cross-sectional area and the molecular weight of the adsorbate, respectively, and $\overline{N}$ is Avogadro's number.

Although the Langmuir equation describes type I and sometimes chemisorption isotherms, it fails to be adequately general to treat physical adsorption and the type II-type V isotherms. In addition, surface area measurements obtained from type I isotherms are subject to uncertainties, regardless of whether chemisorption or physical adsorption is occurring. In chemisorption, localization of the adsorbate molecules leaves the value of A_x seriously in question, since the adsorbate will adsorb only at active surface sites, leaving an unspecified area around each chemisorbed molecule. When applied to physical adsorption, the type I isotherm is associated with the pore filling of micropores with no clearly defined region of monolayer coverage.

4.1.2 The Brunauer, Emmett and Teller (BET) Theory [2]

During the process of physical adsorption, at very low relative pressure, the first sites to be covered are the more energetic ones. Those sites with higher energy on a chemically pure surface reside within narrow pores where the pore walls provide overlapping potentials. Other high-energy sites lie between the horizontal and vertical edges of surface steps where the adsorbate can interact with surface atoms in two planes. In general, wherever the adsorbate is afforded the opportunity to interact with overlapping potentials, or an increased number of surface atoms, there will be a higher energy site. On surfaces consisting of heteroatoms, such as organic solids or impure materials, there will be variations in adsorption potential depending upon the nature of the atoms of functional groups exposed at the surface.

That the more energetic sites are covered first as the pressure is increased does not imply that no adsorption occurs on sites of lower potential. Rather, it implies that the average residence time of a physically adsorbed molecule is longer on the higher-energy sites. Accordingly, as the adsorbate pressure is allowed to increase, the surface becomes progressively coated and the probability increases that a gas molecule will strike and be adsorbed on a previously bound molecule. Clearly then, prior to complete surface coverage the formation of second and higher adsorbed layers will commence. In reality, there exists no pressure at which the surface is covered with exactly a completed physically adsorbed monolayer. The effectiveness of the Brunauer, Emmett and Teller (BET) theory is that it enables an experimental determination of the number of molecules required to form a monolayer despite the fact that exactly one monomolecular layer is never actually formed.

Brunauer, Emmett, and Teller, in 1938, extended Langmuir's kinetic theory to multilayer adsorption. The BET theory assumes that the uppermost molecules in adsorbed stacks are in dynamic equilibrium with the vapor. This means that where the surface is covered with only one layer of adsorbate, an equilibrium exists between that layer and the vapor; where two layers are adsorbed, the upper layer is in equilibrium with the vapor, and so forth. Since the equilibrium is dynamic, the actual location of the surface sites covered by one, two or more layers may vary but the number of molecules in each layer will remain constant.

Using the Langmuir theory and equation (4.5) as a starting point to describe the equilibrium between the vapor and the adsorbate in the first layer,

$$N_m \theta_1 v_1 e^{-E_1/RT} = kP\theta_0 A_1 \qquad \text{(cf. 4.5)}$$

By analogy, for the fraction of surface covered by only two layers one may write

$$N_m \theta_2 v_2 e^{-E_2/RT} = kP\theta_1 A_2 \qquad (4.14)$$

In general, for the nth layer one obtains

$$N_m \theta_n v_n e^{-E_n/RT} = kP\theta_{n-1} A_n \qquad (4.15)$$

The BET theory assumes that the terms v, E, and A remain constant for the second and higher layers. This assumption is justifiable only on the grounds that the second and higher layers are all equivalent to the liquid state. This undoubtedly approaches reality as the layers proceed away from the surface but is somewhat questionable for the layers nearer the surface because of polarizing forces. Nevertheless, using this assumption one can write a series of equations, using L as the heat of liquefaction

$$N_m \theta_1 v_1 e^{-E_1/RT} = kP\theta_0 A_1 \qquad \text{(cf. 4.5)}$$

$$N_m \theta_2 v e^{-L/RT} = kP\theta_1 A \qquad (4.16a)$$

$$N_m \theta_3 v e^{-L/RT} = kP\theta_2 A \tag{4.16b}$$

and, in general, for the second and higher layers

$$N_m \theta_n v e^{-L/RT} = kP\theta_{n-1} A \tag{4.16c}$$

From these equations, it follows that

$$\frac{\theta_1}{\theta_0} = \frac{kPA_1}{N_m v_1 e^{-E_1/RT}} = \alpha \tag{4.17a}$$

$$\frac{\theta_2}{\theta_1} = \frac{kPA}{N_m v e^{-L/RT}} = \beta \tag{4.17b}$$

$$\frac{\theta_3}{\theta_2} = \frac{kPA}{N_m v e^{-L/RT}} = \beta \tag{4.17c}$$

$$\frac{\theta_n}{\theta_{n-1}} = \frac{kPA}{N_m v e^{-L/RT}} = \beta \tag{4.17d}$$

then

$$\theta_1 = \alpha\theta_0 \tag{4.18a}$$

$$\theta_2 = \beta\theta_1 = \alpha\beta\theta_0 \tag{4.18b}$$

$$\theta_3 = \beta\theta_2 = \alpha\beta^2\theta_0 \tag{4.18c}$$

$$\theta_n = \beta\theta_{n-1} = \alpha\beta^{n-1}\theta_0 \tag{4.18d}$$

The total number of molecules adsorbed at equilibrium is

$$N = N_m\theta_1 + 2N_m\theta_2 + ... + nN_m\theta_n = N_m(\theta_1 + 2\theta_2 + ... + n\theta_n) \tag{4.19}$$

Substituting for $\theta_1, \theta_2, ...$ from equations (4.18 a-d) gives

$$\frac{N}{N_m} = \alpha\theta_0 + 2\alpha\beta\theta_0 + 3\alpha\beta^2\theta_0 + \ldots + n\alpha\beta^{n-1}\theta_0 \tag{4.20a}$$

$$= \alpha\theta_0(1 + 2\beta + 3\beta^2 + \ldots + n\beta^{n-1}) \tag{4.20b}$$

Since both α and β are assumed to be constants, one can write

$$\alpha = C\beta \tag{4.21}$$

This defines C by using equations (4.17a) and (4.17b-d) as

$$\frac{A_1 V_2}{A_2 V_1} e^{(E_1 - L)/RT} = C \tag{4.22}$$

Substituting $C\beta$ for α in equation (4.20) yields

$$\frac{N}{N_m} = C\theta_0(\beta + 2\beta^2 + 3\beta^3 + \ldots + n\beta^n) \tag{4.23}$$

The preceding summation is just $\beta/(1-\beta)^2$. Therefore,

$$\frac{N}{N_m} = \frac{C\theta_0\beta}{(1-\beta)^2} \tag{4.24}$$

Necessarily

$$1 = \theta_0 + \theta_1 + \theta_2 + \ldots + \theta_n \tag{4.25}$$

Then

$$\theta_0 = 1 - (\theta_1 + \theta_2 + \ldots + \theta_n) = 1 - \sum_{n=1}^{\infty} \theta_n \tag{4.26}$$

Substituting equation (4.26) into (4.24) gives

$$\frac{N}{N_m} = \frac{C\beta}{(1-\beta)^2}\left(1 - \sum_{n=1}^{\infty} \theta_n\right) \tag{4.27}$$

Replacing θ_n in equation (4.27) with $\alpha\beta^{n-1}\theta_0$ from equation (4.18d) yields

$$\frac{N}{N_m} = \frac{C\beta}{(1-\beta)^2}\left(1 - \alpha\theta_0 \sum_{n=1}^{\infty} \beta^{n-1}\right) \tag{4.28}$$

and introducing $C\beta$ from equation (4.21) in place of α gives

$$\frac{N}{N_m} = \frac{C\beta}{(1-\beta)^2}\left(1 - C\theta_0 \sum_{n=1}^{\infty} \beta^{n}\right) \tag{4.29}$$

The summation in equation (4.29) is

$$\sum_{n=1}^{\infty} \beta^{n} = \beta + \beta^2 + \ldots + \beta^n = \frac{\beta}{1-\beta} \tag{4.30}$$

Then

$$\frac{N}{N_m} = \frac{C\beta}{(1-\beta)^2}\left(1 - C\theta_0 \frac{\beta}{1-\beta}\right) \tag{4.31}$$

From equation (4.24) we have

$$\frac{C\beta}{(1-\beta)^2} = \frac{N}{N_m}\frac{1}{\theta_0} \tag{cf. 4.24}$$

Then equation (4.31) becomes

$$1 = \frac{1}{\theta_0}\left(1 - C\theta_0 \frac{\beta}{1-\beta}\right) \tag{4.32}$$

and

$$\theta_0 = \frac{1}{1 + C\beta/(1-\beta)} \tag{4.33}$$

Introducing θ_0 from equation (4.33) into (4.24) yields

$$\frac{N}{N_m} = \frac{C\beta}{(1-\beta)(1-\beta+C\beta)} \qquad (4.34)$$

When β equals unity, N/N_m becomes infinite. This can physically occur when adsorbate condenses on the surface or when $P/P_0 = 1$.

Rewriting equation (4.17d) for $P = P_0$, gives

$$1 = \frac{kAP_0}{N_m v e^{-L/RT}} \qquad (4.35)$$

but

$$1 = \frac{kAP}{N_m v e^{-L/RT}} \qquad (\text{cf. } 4.17d)$$

then

$$\beta = \frac{P}{P_0} \qquad (4.36)$$

Introducing this value for β into (4.34) gives

$$\frac{N}{N_m} = \frac{C(P/P_0)}{(1-P/P_0)[1-P/P_0+C(P/P_0)]} \qquad (4.37$$

Recalling that $N/N_m = W/W_m$ (equation 4.10) and rearranging equation (4.37) gives the BET equation in final form,

$$\frac{1}{W[P/P_0-1]} = \frac{1}{W_m C} + \frac{C-1}{W_m C}\left(\frac{P}{P_0}\right) \qquad (4.38)$$

If adsorption occurs in pores limiting the number of layers then the summation in equation (4.37) is limited to n and the BET equation takes the form

$$\frac{W}{W_m} = \frac{C}{[P/P_0-1]} \frac{\left[1-(n+1)(P/P_0)^n+n(P/P_0)^{n+1}\right]}{\left[1+(C-1)P/P_0-C(P/P_0)^{n+1}\right]} \qquad (4.39)$$

Equation (4.39) reduces to (4.38) with $n = \infty$ and to the Langmuir equation with $n = 1$.

The application of the BET (essentially equation (4.38)) and the Langmuir approach (essentially equation (4.11)) for the determination of the specific surface area will be discussed in chapter 5.

4.2 THE FRENKEL-HALSEY-HILL (FHH) THEORY OF MULTILAYER ADSORPTION

Physisorption at temperatures below the critical temperature T_c and in the complete wetting regime leads to the development of multilayer adsorption by approaching the saturation pressure P_0. The BET theory describes adsorption of the first two or three layers in a satisfying way, but fails to assess correctly the range of the adsorption isotherm, which is associated with the development of thick multilayer films.

Beyond a film thickness of two or three molecular layers, the effect of surface structure is largely smoothed out and close to the saturation pressure the adsorbed layer has a thickness, which allows to consider the adsorbed film as a slab of liquid. It is assumed that here the adsorbed film has the same properties (i.e., density etc.) as the bulk liquid would have at this temperature. This is the basic assumption of the slab approximation, which was first proposed by Frenkel [3] and was later also derived independently by Halsey [4] and Hill [5]. The only modification to its free energy of the adsorbed liquid slab arises from the interaction with the solid, i.e., the adsorption forces (dispersion forces). The interaction energy $U_s(z)$ of a gas molecule at distance z from a solid surface is approximately given as

$$U_s(z) \approx C_{sf}\rho_s/z^3 \tag{4.40}$$

where C_{sf} is a measure for the strength of attractive fluid-wall interactions and ρ_s represents the solid density.

Within the spirit of the FHH approach, the chemical potential difference $\Delta\mu = \mu_a - \mu_0$ between an adsorbed, liquid-like film (μ_a) of thickness $z = l$ and the value (μ_0) at gas-liquid coexistence of the bulk fluid is given by

$$\Delta\mu = \mu_a - \mu_0 = -RT \ln(P/P_0) = u(z) = -\alpha\, l^{-3} \tag{4.41}$$

The more general equation (4.41) is known as the Frenkel-Halsey-Hill (FHH) equation:

$$\Delta\mu \; = \; \mu_a - \mu_0 = - RT \ln(P/P_0) \; = \; -\alpha \, l^{-m} \tag{4.42}$$

where α is an empirical parameter, characteristic for the gas-solid interaction. For non-retarded van-der-Waals' interactions (i.e., dispersion forces), one expects $m = 3$ (as expected from equation (4.41)). According to equation 4.42, the FHH-equation predicts that the thickness of a film l ($z = l$) adsorbed on a solid surface is expected to increase without limit ($l \to \infty$) for $\mu_a \to \mu_0$, i.e. by approaching $P/P_0 = 1$.

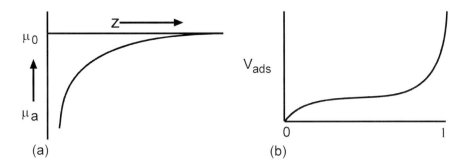

Figure 4.1 (a) Chemical potential difference of an adsorbed film as function of distance z (i.e. film thickness l) from the adsorbent surface. The film thickness diverges by approaching μ_0, which corresponds to a relative pressure $P/P_0 = 1$. (b) Corresponding adsorption isotherm revealing the diverging of the film thickness (and V_{ads} for $P/P_0 \to 1$).

In the case of low temperature adsorption (e.g. adsorption of nitrogen and argon at their boiling temperatures) the adsorption can be analyzed in terms of a two-phase model in which a clearly defined adsorbed phase coexists with a bulk gas phase of low density (see chapter 2). In this case the thickness of the adsorbed liquid-like multilayer, l, can be related to the volume V_{liq} of the adsorbed phase, viz

$$l = V_{liq}/S \tag{4.43}$$

where S is the total surface area. Inserting this expression into the equation (4.42) gives

$$\ln(P/P_0) = -\alpha \, (V_{liq}/S)^{-m} \tag{4.44}$$

The validity of the FHH equation can be tested by plotting $\log\log(P/P_0)$ against $\log(V_{liq}/S)$ (the classical FHH plot). In the multilayer region of the sorption isotherm a straight line should be obtained; the slope is indicative of Frenkel-Halsey-Hill exponent m. Experimental values usually found for m are often significantly smaller than the theoretical values of 3, i.e., values of m = 2.5-2.7 are found even for strongly attractive adsorbents like graphite [6], as well as for samples with oxidic surfaces like silica, alumina, rutile etc. [7]. The deviations from the theoretical value m = 3 were often attributed to interparticle condensation (in case of powders [6,8]), which overlaps with multilayer adsorption, as well as to surface roughness and fractality of the adsorbent surface (see chapter 7). In addition it was found that the relative pressure range over which a linear FHH plot is achieved seems to depend on the nature of the adsorbent-adsorbate interaction [7]. Please note also that the FHH theory is only applicable in the regime of high relative pressures, where the assumption that the adsorbate can be considered as slab of liquid with bulk-like properties can be indeed justified. Accordingly, when the FHH theory is applied to the low or middle range of isotherms the values obtained for m etc. can only be considered as empirical.

The temperature dependence of the FHH-equation was tested by Findenegg and co-workers [6,8,9] over a large temperature up to the critical temperature T_c. In the region of higher temperatures and pressures the relative pressure P/P_0 has to be replaced by the ratio of appropriate fugacities f/f_0. The correspondent FHH equation can then be written in the form: $\ln(f/f_0) = -\alpha l^{-m}$ [10] and it could be concluded that the simple FHH-equation remains indeed applicable up to nearly the critical point.

4.3 ADSORPTION IN MICROPOROUS MATERIALS

4.3.1 Introduction
According to IUPAC [11] pores are classified as macropores for pore widths greater than 500 Å, mesopores for the pore range 20 to 500 Å and micropores for pore widths less than 20 Å. Because of the intense potential fields in very narrow pores (overlapping fields from opposite pore walls), the mechanism of pore filling is different as in mesopores. Mesopores fill via pore condensation (see chapter 4.4), which represents a first order gas-liquid phase transition. In contrast the filling of micropores reflects in most cases a continuous process. The micropore range is subdivided into those smaller than about 7 Å (ultramicropores) and those in the range from 7 to 20 Å (supermicropores). The filling of ultramicropores (pore width smaller < 7 Å) occurs at very low relative pressures and is entirely governed by the enhanced gas-solid interactions. However, in addition to the strong adsorption potential a cooperative mechanism may play a role in the pore filling process of so-called supermicropores [12]. The relative pressure where micropore filling occurs is dependent on a number of factors including the size and nature of the molecules of the adsorptive, the pore shape and the effective pore width. The pore filling capacity depends essentially on the accessibility of the pores for the probe molecules, which is

determined by the size of the molecule and the chosen experimental conditions.

In an ideal case microporous materials exhibit type I isotherms (see IUPAC classification, i.e., Fig. 3.2 in chapter 3). However, many microporous adsorbents (e.g., active carbons) contain pores over a wide range of pore sizes, including micro- and mesopores. Accordingly, the observed adsorption isotherm reveals features from both type I and type IV isotherms. An example is shown in Fig. 4.2, which shows the nitrogen isotherm (at ~77 K) on a disordered active carbon sample. The observed hysteresis loop is indicative of mesoporosity, whereas the type I behavior is clearly visible in the lower relative pressure range. Another example is shown in Fig. 4.3, which shows the adsorption isotherm at 87 K (i.e., liquid argon temperature) in a faujasite zeolite. In order to reveal details of the adsorption isotherm (in particular in the range of the low relative pressures where micropore filling occurs), the isotherm is favorably represented in a semi-logarithmic scale of the relative pressure. The strong increase of the adsorbed amount close to saturation pressure results from pore condensation into large meso- and macropores.

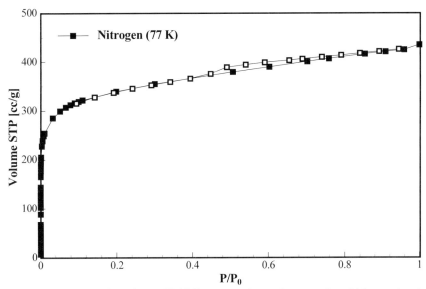

Figure 4.2 Nitrogen adsorption at 77.35 K on an active carbon sample, which contains, in addition to its microporosity, some mesoporosity indicated by the occurrence of hysteresis and the fact that the adsorption isotherm does not reveal a truly horizontal plateau at relative pressures > 0.1; the observed slope being associated with the filling of mesopores.

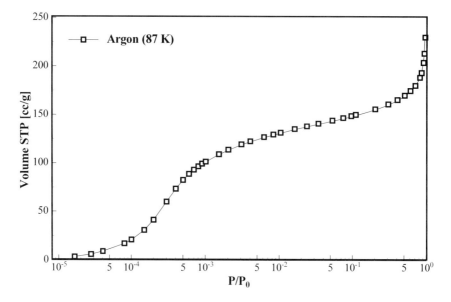

Figure 4.3 Semi-logarithmic isotherm plot of argon at 87K on a faujasite zeolite which clearly resolves the micropore filling in the low relative pressure range. The steep increase close to the saturation pressure represents the pore filling of large meso- and macro-pores.

In order to interpret sorption isotherms measured on microporous materials various methods and theories have been developed. The so-called 'classical methods' are based on macroscopic, thermodynamic assumption, i.e., they assume that the adsorbed pore fluid is liquid-like and that it reveals essentially the same properties as a bulk liquid at the same temperature.

Such classical approaches are, for instance, the theories for micropore characterization by Polanyi [13], Dubinin [14-16], Stoeckli [17] including the more recent approaches by Horvath-Kawazoe (HK) [18] and related methods [19]. In contrast to these macroscopic approaches, methods like the Density Functional Theory (DFT) [20] or methods of molecular simulation (e.g., Monte Carlo simulation methods (MC), Molecular Dynamics methods (MD)) [21,22] provide not only a microscopic model of adsorption but lead also to a better assessment of the thermodynamic properties of the pore fluid. These theories, which are based on statistical mechanics, connect macroscopic properties to the molecular behavior allowing a much more realistic description of micropore filling, which is the prerequisite for an accurate and comprehensive pore size analysis.

In chapter 4.3.2 we will discuss some aspects of macroscopic, classical theories for adsorption in microporous materials. In chapter 4.3.3, we focus on some aspects of the microscopic methods (e.g., Density Functional Theory (DFT) and Molecular Simulation), which are meanwhile frequently used to describe adsorption in micropores.

4.3.2 Aspects of Classical, Thermodynamic Theories for Adsorption in Micropores: Extensions of Polanyi's Theory

Polanyi's potential theory of adsorption [13] views the area immediately above an adsorbent's surface as containing equipotential lines that follow the contour of the surface potential. When a molecule is adsorbed, it is considered trapped between the surface and the limiting potential plane at which the 'adsorption potential' has fallen to zero. Fig. 4.4 illustrates these equipotential planes. In the diagram, Y represents a pore and X depicts some surface impurity.

According to the potential theory, the volume $\tilde{V}$, defined by the adsorbent's surface and the equipotential plane, E_n, can contain adsorbate in three different conditions, depending upon temperature. Above the critical temperature, the adsorbate cannot be liquefied and the gas in the adsorption volume, $\tilde{V}$, simply becomes more dense near the surface.

At temperatures just below the critical temperature, the adsorbate is viewed as a liquid near the surface and a vapor of decreasing density away from the surface.

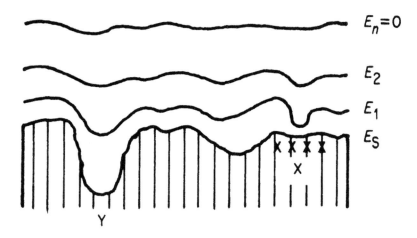

Figure 4.4 Polanyi's potential planes.

At temperatures much less than the critical temperature ($T \leq 0.8T_c$), the adsorption volume is considered to contain only liquid. Under the latter conditions one can write

$$\tilde{V} = \frac{W}{\rho} \tag{4.45}$$

where W and ρ are the adsorbate weight and density, respectively.

The potential theory asserts that when the adsorbate is in the liquid state, the adsorption potential is given by

$$E = RT \ln \frac{P_0}{P} \tag{4.46}$$

According to the preceding equation, E is the isothermal work required to compress the vapor from its equilibrium pressure, P, to the saturated pressure, P_0, of the liquid in the adsorption volume.

Using equations (4.45) and (4.46), both $\widetilde{V}$ and E can be calculated from an experimental isotherm. Therefore,

$$\widetilde{V} = F(E) \tag{4.47}$$

Plots of $\widetilde{V}$ versus E take the form shown in Fig.4.5 and are called 'characteristic curves'. If two adsorbates fill the same adsorption volume, as shown by the vertical dotted line in Fig. 4.5, their adsorption potentials, E and E_0, will differ only because of differences in their molecular properties. Consequently, the ratio of adsorption potentials is assumed by Dubinin [14,15] to be constant. Dubinin calls E/E_0 the 'affinity coefficient', which, for an adsorbate pair, is a measure of their relative affinities for a surface. Using the adsorption for one vapor, say E_0, as a reference value, the ratio of potentials can be written as

$$\frac{E}{E_0} = \beta \tag{4.48}$$

Substitution into equation (4.47) then gives, for the reference vapor,

$$\widetilde{V} = F\left(\frac{E}{\beta}\right) \tag{4.49}$$

Using benzene as the reference or standard vapor ($\beta = 1$), Dubinin and Timofeev [14a] were able to calculate values of β for other adsorbents. The characteristic curves, shown in Fig. 4.5, appear similar to the positive side of a Gaussian curve.

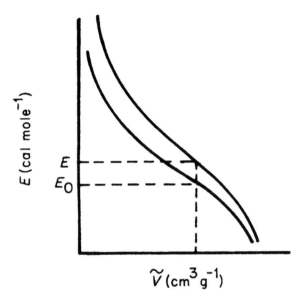

Figure 4.5 Characteristic curves for two vapors.

These similarities led Dubinin and Radushkevich ("DR") [14b] to postulate that the fraction of the adsorption volume, $\widetilde{V}$, occupied by liquid adsorbate at various values of adsorption potentials, E, can be expressed as a Gaussian function. Thus,

$$\widetilde{V} = \widetilde{V}_0 \exp\left(-KE_0^2\right) \tag{4.50}$$

where K is a constant, determined by the shape of the pore size distribution, and $\widetilde{V}_0$ is the total adsorption volume or the microporous volume. Substituting the value for E_0 from equation (4.48) gives

$$\widetilde{V} = \widetilde{V}_0 \exp\left(-K(E/\beta)^2\right) \tag{4.51}$$

Equation (4.51) is applicable to micropores, rather than larger pores, because the overlapping potential from the walls of pores only slightly larger than an adsorbate molecule will considerably enhance the adsorption potential. Substituting for E in equation (4.51) using equation (4.46) yields

$$\widetilde{V} = \widetilde{V}_0 \exp\left[-K\left(\frac{RT}{\beta}\ln\frac{P_0}{P}\right)^2\right] \tag{4.52}$$

which can be rewritten as

$$\log W = \log(\widetilde{V}_0\rho) - 2.303K\left[\frac{RT}{\beta}\log\left(\frac{P_0}{P}\right)\right]^2 \qquad (4.53)$$

where W and ρ are the weight adsorbed and the liquid adsorbate density, respectively. Simplifying equation (4.53) yields

$$\log W = \log(\widetilde{V}_0\rho) - k\left[\log\left(\frac{P_0}{P}\right)\right]^2 \qquad (4.54)$$

where

$$k = 2.303K\left(\frac{RT}{\beta}\right)^2 \qquad (4.55)$$

A plot of $\log W$ versus $[\log(P_0/P)]^2$ should give a straight line with an intercept of $\log(\widetilde{V}_0\rho)$, from which $\widetilde{V}_0$, the micropore volume, can be calculated. Linear DR plots over a large relative pressure range can be found for a number of microporous carbons. For many other adsorbents (zeolites are particularly problematic) the linear range is limited over a very narrow relative pressure range. The DR equation often fails in the case where the microporous adsorbent is very heterogeneous with regard to surface chemistry and texture. In such cases the application of a generalized form of the DR equation, i.e., the Dubinin-Asthakov ("DA") equation [16] is of advantage (see chapter 9).

$$\widetilde{V} = \widetilde{V}_0\exp(-A/E)^n) \qquad (4.56)$$

where $A = -RT\ln(P/P_0)$ and n is the Dubinin-Asthakov parameter, which depends not only on the heterogeneity of the adsorbent but of course also on the relative pressure range of the sorption isotherm, where the DA-equation was applied. Further improvements were made by Stoeckli and co-workers, who introduced an alternative to the DA-equation [17]. However, details of the interactions of the adsorptive molecules with the porous material, and their impact on micropore filling (and thus the shape of the adsorption isotherm) were not considered until the Horvath –Kawazoe (HK) theory [18] and related approaches [19] were published (see chapter 9.5 for more details). Although the HK-related methods take into account the effect of pore geometry, and the strength of the attractive adsorptive-adsorbent interaction on the adsorption potential, they still assume incorrectly that the thermophysical properties of the strongly confined liquid-like pore fluid does not differ from the properties of the corresponding bulk liquid. These obstacles can be overcome by applying modern methods of statistical mechanics, which we discuss in the following section, 4.3.3.

4.3.3 Aspects of Modern, Microscopic Theories for Adsorption in Micropores: Density Functional Theory and Molecular Simulation

Density functional theory and computer simulation methods have been developed into powerful methods for the description of the sorption and phase behavior of fluids, confined to porous materials. These methods allow equilibrium density profiles of a fluid adsorbed on surfaces and in pores to be calculated, from which properties such as the adsorption/desorption isotherm, heats of adsorption, neutron scattering patterns and transport properties for model systems can be derived.

Pioneering studies on the application of density functional theory and molecular modeling by computer simulation in order to study the sorption and phase behavior of fluids in pores were performed by Evans and Tarazona [20], Gubbins et al. [21,22], Quirke et al. [23] and Fischer et al. [24] in the time period from 1985 to1989. Seaton et al. [25] were the first to apply Density Functional Theory to calculate the pore size distribution in both the meso-and micropore range. In their approach, the so-called Local Density Functional Theory (LDFT) approach was used, which still represents a significant improvement over the macroscopic, thermodynamic descriptions of pore filling, but is inaccurate for narrow micropores – mainly because LDFT fails to take into account the short-range correlations in these pores. In contrast, the Non-Local Density Functional Theory (NLDFT) and Monte Carlo computer simulation techniques provide a much more accurate description of a fluid confined to narrow pores and both are able to produce the strong oscillation characteristics of a fluid density profile at a solid fluid interface (see Fig. 4.6). The first paper where the non-local density functional theory was used for pore size analysis (of microporous carbons) was published in 1993 by Lastoskie et al. [26]. Since then, the Non-Local Density Theory was quite often employed to calculate the pore size distribution of micro- and mesoporous materials. In order to do so, NLDFT methods dedicated to specific adsorptive/adsorbent pairs had to be developed [27,28]. In particular, Neimark and Ravikovitch [28] confirmed the validity of NLDFT by comparing the calculated pore size distribution curves for mesoporous molecular sieves (e.g., MCM 41, which consist of an array of independent pores) and zeolites (e.g., ZSM 5) with pore size results obtained from other techniques, (e.g., methods based on like XRD, TEM etc.), which are independent of the position of the pore filling step in the adsorption isotherm.

4.3.3.1 Density Functional Theory (DFT). In experimental systems the adsorbed fluid in a pore is in equilibrium with a bulk gas phase. For such a system the grand canonical ensemble provides the appropriate description of the thermodynamics. The local density $\rho(r)$ of the pore fluid is therefore determined by minimizing of the correspondent grand potential $\Sigma[\rho(r)]$. Once $\rho(r)$ is known, other thermodynamic properties, such as the adsorption isotherm, heat of adsorption, free energies, phase transitions, etc. can be calculated. The grand potential function $\Sigma[\rho(r)]$ is given by the following term

$$\Omega[\rho(r)] = F[\rho(r)] - \int dr \rho(r)(\mu - V_{ext}(r)) \qquad (4.57)$$

where $F[\rho(r)]$ is the intrinsic Helmholtz free energy functional in the absence of any external field and $V_{ext}(r)$ is the potential imposed by the walls, i.e., $\Sigma[\rho(r)]$ depends of course on all the interactions.

The parameters of the fluid-fluid interactions are usually determined in a way that they allow to reproduce the bulk properties (e.g., surface tension, gas-liquid coexistence curve etc.). Parameters of the solid-fluid interactions can then be obtained by fitting the calculated adsorption isotherms on a planar surface to the standard (e.g., nitrogen) isotherm. In addition, it is assumed that the fluid is contained in individual pores of simple geometry (e.g., slits or cylinders). For instance, an individual slit pore can be represented as two infinite, parallel graphitic slabs, separated by a width *W*, the distance between the centers of carbon atoms (for a carbon slit-pore).

Fig. 4.6 shows NLDFT density profiles of a fluid confined to a slit-pore of pore width ca. 5σ, where σ is the diameter of one molecule. The fluid shows the characteristic density oscillations (which reflect adsorbed layers) throughout the complete narrow pore space, which demonstrates the strong adsorption potential (in contrast, mesopores reveal a bulk-fluid like region in the core of the pore, see Fig. 4.10). It is also clearly visible that in such a small pore only two adsorbed layers can build up on each pore wall.

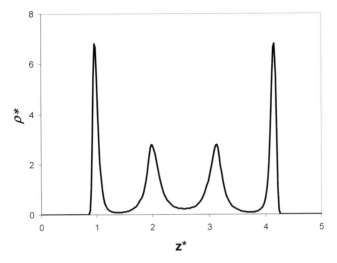

Figure 4.6 Characteristic density profile of a Lennard-Jones fluid in a slit pore of width 5σ, where σ is the diameter of one molecule [29a].

4.3.3.2 Computer Simulation Studies: Monte Carlo Simulation and Molecular Dynamics. The most prominent computer simulation method for the study of adsorption and wetting phenomena of fluids on planar surfaces and in pores is the Grand Canonical Monte Carlo simulation method (GCMC). This technique simulates the situation of an adsorbed fluid (or mixture) in equilibrium with a bulk fluid reservoir, which reflects usually the situation encountered in experimental studies of confined systems. A random number generator is used to move and rotate the molecules in a random fashion, which leads to particular configurations. Such movements and the resulting configurations are then accepted or rejected according to thermodynamic criteria (i.e., based on the temperature and chemical potential). After generating a long sequence of such moves (so-called Markov chain, typically in the order of several millions), they can be averaged (based on equations of statistical mechanics) to obtain the equilibration density profiles and, hence, the adsorption isotherm.

The molecular dynamics (MD) method applies Newton's equations of motion in order to obtain the trajectories and velocities of molecules. This method allows determining the transport as well as equilibrium properties of the system. The method is not as frequently used as GCMC, but some excellent work was performed using this method to study the adsorption and phase behavior of fluids in pores [e.g., 24].

4.3.3.3 NLDFT and Monte Carlo Simulation for Pore Size Analysis.
NLDFT and GCMC are considered to be the most advanced methods with
regard to pore size analysis of micro- and mesoporous materials. For very
narrow micropores the Monte Carlo simulation is considered to be the most
accurate method. In both techniques, a set of isotherms calculated for a set
of pore sizes in a given range for a given adsorptive constitutes the model
database (these isotherms are calculated by integrating the equilibrium
density profiles, $\rho(r)$, of the fluid in the model pore). Such a set of
isotherms, called a kernel, is the basis for pore size analysis by Density
Functional Theory [e.g., 28]. The calculation of the pore size distribution is
based on a solution of the Generalized Adsorption Isotherm equation (GAI),
which correlates the kernel of theoretical adsorption/desorption isotherms
with the experimental sorption isotherm:

$$N(P/P_0) = \int_{W_{MIN}}^{W_{MAX}} N(P/P_0, W) f(W) dW \qquad (4.58)$$

where $N(P/P_0)$ = experimental adsorption isotherm data, W = pore width,
$N(P/P_0,W)$ = isotherm on a single pore of width W and $f(W)$ = pore size
distribution function.

The GAI equation reflects the assumption that the total isotherm
consists of a number of individual "single pore" isotherms multiplied by
their relative distribution, $f(W)$, over a range of pore sizes. The set of
$N(P/P_0,W)$ isotherms (kernel) for a given system (adsorptive/adsorbent) can
be obtained, as indicated above, by either Density Functional Theory or by
Monte Carlo computer simulation. The pore size distribution is then derived
by solving the GAI equation numerically. In general, the solution of the GAI
represents an ill-posed problem, which requires some degree of
regularization. However meaningful and stable solutions of this equation
can be obtained by existing regularization algorithms [e.g., 26- 29]. Because
the equilibrium density profiles are known for each pressure along an
isotherm no assumptions about the pore filling mechanism are required as in
case of the macroscopic, thermodynamic methods. Hence, NLDFT and
GCMC allow describing the adsorption isotherm over the complete range,
and it is possible to obtain with a single method a pore size distribution
which extends over the complete micro-mesopore range.

The application of these advanced methods for micro- and mesopore
size analysis is discussed in chapters 8 and 9.

4.4 ADSORPTION IN MESOPORES

4.4.1 Introduction

The sorption behavior in mesopores (2- 50 nm) depends not only on the fluid-wall attraction, but also on the attractive interactions between fluid molecules. This leads to the occurrence of multilayer adsorption and capillary (pore) condensation. Pore condensation is the phenomenon whereby a gas condenses to a liquid-like phase in a pore at a pressure P less than the saturation pressure P_0 of the bulk liquid. Typically, type IV and V sorption isotherms according to the IUPAC classification (see chapter 3) can be observed. Significant progress was achieved during the last decade with regard to the understanding of sorption phenomena in narrow pores and the subsequent improvement in the pore size analysis of porous materials (which will be discussed in chapter 8). This progress can be primarily attributed to: (i) the discovery of novel ordered mesoporous materials, such as MCM-41, MCM-48, SBA-15 [30], which exhibit a uniform pore structure and morphology and could therefore be used as model adsorbents to test theories of gas adsorption; (ii) carefully performed adsorption experiments and (iii) the development of microscopic methods, such as the Non-Local-Density Functional Theory (NLDFT) or computer simulation methods (e.g. Monte-Carlo – and Molecular-Dynamic simulations), which allow to describe the configuration of adsorbed molecules in pores on a molecular level. In the following chapter we discuss the most important phenomena occurring in mesopores, i.e. multilayer adsorption, phase transition (e.g., pore condensation) and sorption hysteresis in the context of classical approaches and the most recent developments.

4.4.2 Multilayer Adsorption, Pore Condensation and Hysteresis

As described in §4.2, in the complete wetting range a multilayer adsorbed film is produced at the pore walls. For fluids in contact with a planar surface the thickness l of the adsorbed film is expected to increase without limit, i.e., $l \rightarrow \infty$ for $P/P_0 \rightarrow 1$.

$$\Delta\mu_a = \mu_a - \mu_0 = -RT \ln(P/P_0) = -\alpha \, l^{-m} \qquad \text{(cf 4.42)}$$

where α is the fluid-wall interaction parameter, and the l^{-m} law results from the long-range van der Waals' interactions between a fluid molecule and a semi-infinite planar wall. In the case of non-retarded van der Waals' fluid-wall interactions, the exponent m has a theoretical value of 3. However, experimental values for m are often significantly smaller than the theoretical value, even for strongly attractive adsorbents like graphite, i.e., $m = 2.5 - 2.7$ (see chapter 4.2 for a more detailed discussion).

In pores, however, the film thickness cannot grow unlimited. The stability of this film is determined by the attractive fluid-wall interactions, the surface tension and curvature of the liquid-vapor interface. In this case the difference in chemical potential $\Delta\mu = \mu - \mu_0$ between the adsorbed liquid-like film (μ) and the value at gas-liquid coexistence (μ_0) of the bulk fluid is given by

$$\Delta\mu = \Delta\mu_a + \Delta\mu_c \tag{4.59}$$

For small film thickness the first term $\Delta\mu_a$ (equation cf. (4.43)) associated with multilayer adsorption dominates:

$$\Delta\mu_a = -\alpha\, l^{-m} \tag{cf. 4.42}$$

When the adsorbed film becomes thicker, the adsorption potential will become less important, and $\Delta\mu$ will be dominated almost entirely by the curvature contribution $\Delta\mu_c$ (i.e., the Laplace term), which is given for cylindrical pores by

$$\Delta\mu_c = -(\gamma/a\,\Delta\rho) \tag{4.60}$$

where a is the core radius ($a = r - l$; r is the pore radius), γ is the surface tension of the adsorbed liquid-like film (which is assumed to be identical with the liquid), $\Delta\rho = \rho^l - \rho^g$, describes the density difference between the liquid like film and the vapor phase. At a critical thickness, l_C, pore condensation occurs in the core of the pore, controlled by intermolecular forces in the core fluid. Pore condensation represents a first-order phase transition from a gas-like state to a liquid-like state of the pore fluid, occurring at a chemical potential μ less than the value of μ_0 at gas-liquid coexistence of the bulk fluid.

These phenomena are illustrated in Fig.4.7, which depicts a sorption isotherm as it is expected for adsorption/desorption of a pure fluid in a single mesopore of cylindrical shape in combination together with a schematic representation of the appropriate sorption and phase phenomena occurring in the pore. Please note, that the schematic isotherm reveals a vertical pore condensation step; however, a truly vertical step in the adsorption isotherm is not to be expected for any real porous material with a non-vanishing pore-size distribution, i.e. the wider the pore size distribution, the less sharp is the pore condensation step. At lower relative pressures the adsorption mechanism in mesopores is comparable to that on planar surfaces. After completion of the monolayer formation (**A**), multilayer adsorption commences (**B**). After reaching a critical film thickness (**C**), capillary condensation occurs essentially in the core of the pore (transition from configuration **C** to **D**). The plateau region of the isotherm reflects the situation where the pore is completely filled with liquid and separated from the bulk gas phase by a hemispherical meniscus. Pore evaporation therefore occurs by a receding meniscus (**E**) at a pressure, which is less than the pore condensation pressure. The pressure where the hysteresis closes corresponds again to the situation of an adsorbed multilayer film which is in equilibrium with a vapor in the core of the pore and the bulk gas phase. In the relative pressure range between (**F**) and (**A**) adsorption and desorption are reversible.

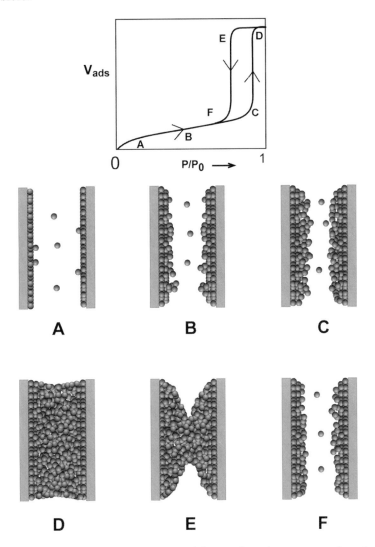

Figure 4.7 Schematic representation of multilayer adsorption, pore condensation and hysteresis in a single cylindrical pore. From [43b].

4.4.3 Pore Condensation: Macroscopic, Thermodynamic Approaches
4.4.3.1 Classical Kelvin Equation. For pores of uniform shape and width pore condensation can be treated on the basis of the Kelvin approach, which relates the shift from bulk coexistence, $\Delta\mu = \mu - \mu_0 = -RT\ln(P/P_0)$, to macroscopic properties such as surface tension, the densities of the bulk gas and liquid and the contact angle θ of the liquid meniscus against the pore wall. The condition for the coexistence of liquid and gas in a cylindrical pore of radius r and temperature T is then given by the Kelvin equation [31]:

$$\Delta\mu = \mu - \mu_0 = -RT\ln\left(P/P_0\right) = -\frac{2\gamma\cos\theta}{r_m\Delta\rho} \qquad (4.61)$$

where R is the universal gas constant, γ is again the surface tension of the liquid, θ the contact angle of the liquid against the pore wall, $\Delta\rho = \rho^l - \rho^g$, where ρ^l represents the orthobaric liquid density at bulk coexistence and ρ^g is the gas density, and r_m is the mean radius of curvature of the meniscus of the pore liquid. In a cylindrical pore the condensed liquid reveals a hemispherical meniscus and the mean radius of curvature corresponds to the pore radius (i.e., the Kelvin radius).

The contact angle θ can be considered as a measure of the relative strength of fluid-wall and fluid-fluid interactions, i.e., the relative strength of fluid-wall and fluid-fluid interactions enters implicitly through the contact angle θ. One would expect the occurrence of pore condensation as long as the contact angle is $< 90°$. For fluids in contact with a single planar wall, one expects complete wetting in the temperature range between the so-called wetting temperature T_w and the critical temperature T_C of the fluid. At temperatures below the wetting temperature incomplete wetting ($\cos\theta < 1$) is observed.

In the case of nitrogen adsorption at 77 K the gas density is small against the liquid density, and ρ^g can be neglected. The liquid density is often given as $1/\rho^l = \overline{V}$, where $\overline{V}$ is the molar volume of the condensed liquid. In addition it is assumed that we have the situation of complete wetting, i.e., the contact angle is assumed to be zero. In this case the Kelvin equation is stated as

$$\ln P/P_0 = \frac{-2\gamma\overline{V}}{rRT} \qquad (4.62)$$

The Kelvin equation relates the equilibrium vapor pressure exerted from the curved meniscus of the pore liquid to the equilibrium pressure of the same liquid on a plane surface. The difference in vapor pressure between the flat and the curved surface is related to the phenomenon of a (mechanical) pressure drop across an interface (having two principal radii of curvature of the surface r_1 and r_2) as described by the Young –Laplace equation, viz $\Delta P = 2\gamma/r_m$, where r_m is the radius of mean curvature, which is given by $1/r_m = \frac{1}{2}[1/r_1 + 1/r_2]$. For a spherical surface $r_m = r_2 = r_1$, and the Laplace equation becomes $\Delta P = 2\gamma/r$, where r is the radius of the spherical surface. The Kelvin equation can be derived by using thermodynamics to assess the effect of a change in mechanical pressure, ΔP, on the molar free energy, which leads to an expression where ΔP is replaced by a function of relative vapor pressures (see for instance derivation in ref. [32]).

An alternative way to derive the Kelvin equation on purely thermodynamic grounds is based on the following: Consider the transfer of ∂n molecules of vapor in equilibrium with the bulk liquid at pressure P_0 into a pore where the equilibrium pressure is P. This process consists of three steps: evaporation from the bulk liquid, expansion of the vapor from P_0 to P, and condensation into the pore. The first and third steps are equilibrium processes and are therefore accompanied by zero free energy change, $\partial G = 0$. The free energy change for the second step is described by

$$\partial G = \left(RT \ln \frac{P}{P_0} \right) \partial n \qquad (4.63)$$

When the adsorbate condenses in the pore, it does so on a previously adsorbed film, thereby decreasing the film-vapor interfacial area. The free energy change associated with the filling of the pore is given by

$$\partial G = -(\gamma \cos\theta)\partial S \qquad (4.64)$$

where γ is again the surface tension of the adsorbed film (assumed to be identical with that of the liquid), ∂S is the change in interfacial area, and θ is again the contact angle, which is taken to be zero, sine the liquid is assumed to wet completely the adsorbed film.

Equations (4.59) and (4.60), when combined using the assumption of a zero wetting angle, yield

$$\frac{\partial n}{\partial S} = \frac{-\gamma}{RT \ln P/P_0} \qquad (4.65)$$

The volume of liquid adsorbate that condenses in a pore of volume V_p is given by

$$\partial V_p = \overline{V} \partial n \qquad\qquad (4.66)$$

where $\overline{V}$ is the molar volume of the liquid adsorbate. Substituting equation (4.61) into equation (4.62) gives

$$\frac{\partial V_p}{\partial S} = \frac{-\gamma \overline{V}}{RT \ln P/P_0} \qquad\qquad (4.67)$$

The ratio of volume to area within a pore depends upon the pore geometry. For example, the volume to area ratios for cylinders, parallel plates, and spheres are, respectively, $r/2$, $r/2$, and $r/3$, where r is the cylinder radius, the sphere radius, or the separation distance between parallel plates. If the pore shapes are highly irregular or consist of a mixture of regular geometries, the volume to area ratio can be too complex to express mathematically. In these cases, or in the absence of specific knowledge of the pore geometry, the assumption of cylindrical pores is usually made. Then equation (4.67) becomes the Kelvin equation

$$\ln P/P_0 = \frac{-2\gamma \overline{V}}{rRT} \qquad\qquad (\text{cf } 4.62)$$

The Kelvin equation provides a correlation between pore diameter and pore condensation pressure, i.e., the smaller the radius, the lower is the P/P_0 value at which pore condensation occurs. In case of real porous materials consisting of pores of different sizes, condensation will occur first in the pores of smaller radii and will progress into the larger pores, at a relative pressure of unity condensation will occur on those surfaces where the radius of curvature is essentially infinite. Conversely, as the relative pressure is decreases evaporation will occur progressively out of pores with decreasing radii.

4.4.3.2 Modified Kelvin Equation. The original Kelvin equation (equation 4.62) does not take into account any fluid-wall interaction parameter, and consequently not the existence of an adsorbed multilayer film prior to pore condensation as illustrated in Fig.4.7. Taking into account that in case of complete wetting the pore walls are covered by a multilayer adsorbed film at the onset of pore condensation, one obtains the modified Kelvin equation [33], which is given for cylindrical pores by:

$$\ln(P/P_0) = \frac{-2\gamma\cos\theta}{RT\Delta\rho(r_p - t_c)} \quad\quad (4.68)$$

where t_c describes the (critical) statistical thickness (see chapter 8) prior to condensation (all other symbols are the same as in case of equation 4.61).

The modified Kelvin equation serves as the basis for many methods applied for pore size analysis of mesoporous materials including the Barrett-Joyner-Halenda method (BJH) [34], which is widely used. In order to account for the preadsorbed multilayer film, the Kelvin equation is combined with a standard isotherm or a so-called t-curve, which usually refers to adsorption measurements on a non-porous solid. Accordingly, the preadsorbed multilayer film is assessed by the statistical (mean) thickness of an adsorbed film on a nonporous solid of a surface similar to that of the sample under consideration (such statistical thickness equations were derived for instance by Halsey, Harkins & Jura and de-Boer (see chapters 8 and 9 for a discussion of these equations). The application of the modified Kelvin equation for mesopore size analysis and its limitations will be discussed in chapter 8.

In contrast to the Kelvin approach, more sophisticated approaches such as the Broeckhoff and de Boer [35] as well as the Cole-Saam theory [36] capture essentially the mechanism of pore condensation and hysteresis as it is described above. These theories take into account the (i) influence of the adsorption potential on the chemical potential where pore condensation occurs in the pores and (ii) the effect of curvature on the thickness of the adsorbed multilayer film. In agreement with experimental observations, these theories predict that an increase in the strength of the attractive fluid-wall interaction, a lowering of the experimental temperature as well as decreasing the pore size will shift the occurrence of pore condensation to lower relative pressures. However all these thermodynamic, macroscopic theories do not take into account the peculiarities of the critical region. In contrast, microscopic theories, such as NLDFT or molecular simulation allow a much more accurate description of the state of the pore fluid, also close to the critical point. This will be discussed in more detail in section 4.47.

4.4.4 Adsorption Hysteresis
4.4.4.1. Classification of Hysteresis Loops. It is widely accepted that there is a correlation between the shape of the hysteresis loop and the texture (e.g., pore size distribution, pore geometry, connectivity) of a mesoporous adsorbent. An empirical classification of hysteresis loops was given by the IUPAC [11], which is based on an earlier classification by de Boer [37]. The IUPAC classification is shown in Fig 4.8. According to the IUPAC classification type H1 is often associated with porous materials consisting of

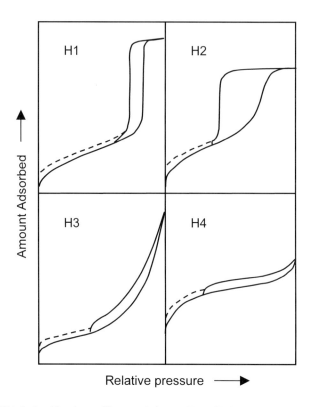

Figure 4.8 IUPAC classifications of hysteresis loops. From [11].

well-defined cylindrical-like pore channels or agglomerates of compacts of approximately uniform spheres. It was found that materials that give rise to H2 hysteresis are often disordered and the distribution of pore size and shape is not well defined. Isotherms revealing type H3 hysteresis do not exhibit any limiting adsorption at high P/P_0, which is observed with non-rigid aggregates of plate-like particles giving rise to slit-shaped pores. The desorption branch for type H3 hysteresis contains also a steep region associated with a (forced) closure of the hysteresis loop, due to the so-called tensile strength effect. This phenomenon occurs for nitrogen at 77K in the relative pressure range from 0.4 – 0.45 (see chapter 8.6.2 for a detailed discussion). Similarly, type H4 loops are also often associated with narrow slit pores, but now including pores in the micropore region (see for instance Fig. 4.3, which depicts the nitrogen sorption isotherm on activated carbon).

The dashed curves in the hysteresis loops shown in Fig. 4.8 reflect low-pressure hysteresis, which may be observable down to very low relative pressure. Low-pressure hysteresis may be associated with the change in volume of the adsorbent, i.e. the swelling of non-rigid pores or with the irreversible uptake of molecules in pores of about the same width as that of the adsorptive molecule. In addition chemisorption will also lead to such "open" hysteresis loops. An interpretation of sorption isotherms showing low-pressure hysteresis is difficult and an accurate pore size analysis is not possible anymore. But also the hysteresis loops usually associated with pore condensation imposes, of course, a difficulty to the pore size analysis of the porous materials and the decision whether the adsorption –or desorption branch should be taken for calculation of the pore size distribution curve depends very much on the reason(s) which caused the hysteresis. Hence, we discuss the origin of pore condensation hysteresis in the following section, 4.4.4.2.

4.4.4.2 Origin of Hysteresis As mentioned before the occurrence of pore condensation/evaporation in mesoporous adsorbents is often accompanied by hysteresis. However, the mechanism and origin of sorption hysteresis is still a matter of discussion. There are essentially three models that contribute to the understanding of sorption hysteresis: (a) *independent (single) pore model* (b) *network model,* and (c) *disordered porous material model.* In the following we will discuss some aspects of these models.

   (a) Independent Pore Model. Sorption hysteresis is considered as an intrinsic property of a phase transition in a single, idealized pore, reflecting the existence of metastable gas states. The hysteresis loop expected for this case is of type H1, according to the IUPAC classification.

   Different approaches, which would explain the occurrence of hysteresis in a single pore, can be found in the literature since *ca.* 1900. Cohan [32] assumed that pore condensation occurs by filling the pore from the wall inward (for a cylindrical pore model). It was suggested that pore condensation would be controlled by a cylindrical meniscus once the pore is filled, whereas evaporation of the liquid would occurs from a hemispherical meniscus, which would lead according to the Kelvin equation to different values of P/P_0 for condensation and evaporation.

   Theories by Foster [38], Cassell [39], Everett [40], Cole and Saam (CS) [35] and Ball and Evans [41] suggested that hysteresis may be caused by the development of metastable states of the pore fluids associated with the capillary condensation transition in a manner analogous to superheating or supercooling of a bulk fluid. These ideas could be essentially confirmed by recent theoretical studies based on Non Local Density Functional Theory (NLDFT) [42]. These studies revealed that the H1 hysteresis can indeed be attributed to the existence of metastable states of the pore fluid, associated with the nucleation of the liquid phase, i.e., pore condensation is delayed. In principle, both pore condensation and pore evaporation can be associated with metastable states of the pore fluid [49]. This is consistent with the classical van der Waals picture, which predicts that the metastable adsorption branch terminates at a vapor-like spinodal, where the limit of stability for the metastable states is achieved and the fluid spontaneously

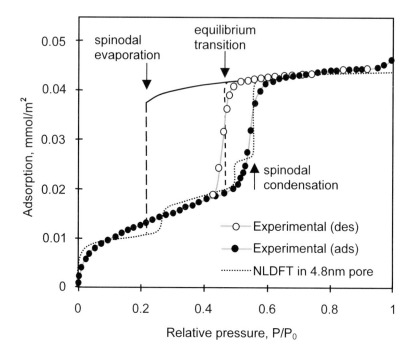

Figure 4.9 NLDFT adsorption isotherm of argon at 87K in a cylindrical pore of diameter 4.8 nm in comparison with the appropriate experimental sorption isotherm on MCM-41. It can be clearly seen that the experimental desorption branch is associated with the equilibrium gas-liquid phase transition, whereas the condensation step corresponds to the spinodal spontaneous transition. From [42].

condenses into a liquid-like state (so-called spinodal condensation). Accordingly, the desorption branch would terminate at a liquid-like spinodal, which corresponds to spontaneous evaporation (spinodal evaporation) In practice however, metastabilities occur only on the adsorption branch. Assuming a pore of finite length (which is always the case in real adsorbents) evaporation can occur via a receding meniscus (see Fig.4.5) and therefore metastability is not expected to occur during desorption. The NLDFT prediction for pore condensation and hysteresis in comparison with the correspondent experimental sorption isotherm of argon at 87 K in MCM-41 silica is shown in Fig. 4.9 (from ref. [42]). The experimental isotherm and the NLDFT isotherm agree quite well and the theoretical prediction for the position of the equilibrium liquid-gas transition (which corresponds to the condition at which the two states have equal grand potential) agrees quite well with the experimentally observed evaporation transition, i.e. the position of the desorption branch of the hysteresis loop. Hence, it was concluded that the desorption branch is associated with the equilibrium gas-liquid phase transition. In such a case the desorption branch should be chosen for pore size analysis if theories/methods are applied which describe the equilibrium transition (e.g., BJH, conventional NLDFT). This will be discussed in more detail in chapter 8.

The small steps in the theoretical isotherm are a consequence of assuming a structureless (i.e., chemically and geometrically smooth) pore wall model, which neglects the heterogeneity of the MCM-41 pore walls.

There is also some evidence that type H1 hysteresis as observed in ordered three dimensional pore systems such as MCM-48 silica [43a] but also in highly ordered porous glasses (such as sol-gel glasses [44] and controlled pore glasses [45]) is predominantly caused by the existence of metastable states associated with pore condensation. The hysteresis loops could be described by applying models based on the independent pore model (e.g., Cole-Saam theory [44,45] NLDFT [43a] etc.). Accordingly, classical networking and pore blocking effects are not necessarily present in an (ordered) interconnected pore system (please see also chapter 8.6).

(b) Network Model. Sorption hysteresis is explained as a consequence of the interconnectivity of a real porous network with a wide distribution of pore sizes. If network and pore blocking effects are present typically a hysteresis loop of type H2 (IUPAC classification) is expected.

Network models take into account that in many materials the pores are connected and form a three-dimensional network. An important feature of the network model is the possibility of pore blocking effects during evaporation, which occurs if a pore has access to the external gas phase only via narrow constrictions (e.g., an ink-bottle pore). The basis for the understanding of sorption hysteresis in inkbottle pores and networks can be found in the work of McBain [46]. The wide inner portion of an inkbottle pore is filled at high relative pressures, but it cannot empty during desorption until the narrow neck of a pore first empties at lower relative pressure. Thus, in a network of inkbottle pores the capillary condensate in the pores is obstructed by liquid in the necks. The relative pressure at which a pore empties now depends on the size of the narrow neck, the connectivity of the network and the state of neighboring pores. Hence, the desorption branch of the hysteresis loops does not (in contrast to the single pore model) occur at thermodynamic equilibrium, but reflects a percolation transition instead. In such a case the desorption branch of the hysteresis loop is much steeper (compared to the adsorption branch) leading to H2 hysteresis according to the IUPAC classification.

Work by Everett [40] and others have led to the development of several specific network models. Advanced network or percolation models were introduced for instance by Mason [47] Wall and Brown [48], Neimark [49], Parlar and Yortsos [50], Ball and Evans [41], Seaton et al. [51] and Rojas et al. [52].

Type H2 hysteresis is observed in many disordered porous materials such as, for instance, porous Vycor® glass, or disordered sol-gel glasses. By combining different experimental techniques such as adsorption measurements (volumetric, gravimetric), ultrasound and light scattering [53] or gas adsorption and *in situ* neutron scattering [54] some evidence for a percolation mechanism associated with pore evaporation could be obtained.

However, the existence of the conventional pore blocking mechanism as described above is under discussion. Sarkisov and Monson [55] concluded that the H2 hysteresis loop (obtained from a molecular dynamics study of adsorption of a simple fluid) typically observed in inkbottle pores is not necessarily caused by the occurrence of conventional pore blocking. The large cavities could be emptied by a diffusional mass transport process from the fluid in the large cavity to the narrow neck and from there into the gas phase, hence the pore body can empty even while the pore neck remains filled. Further experimental and theoretical work by Ravikovitch *et al.* [56] suggests that both conventional pore blocking and so-called cavitation can occur in inkbottle type pores depending on temperature and pore size.

Cavitation corresponds to the situation of spinodal evaporation, i.e., the condensed liquid evaporates when the limit of stability of metastable pore liquid is achieved and the pore fluid spontaneously evaporates into a vapour-like state as shown in Fig. 4.9. In such a case the desorption branch does not reflect the thermodynamic equilibrium liquid-gas transition. The cavitation effect is correlated with the occurrence of a lower limit of hysteresis in the sorption isotherm, which is within the classical picture correlated with the so-called tensile strength effect. This effect is believed to be the cause for the observation that for many disordered porous materials the hysteresis loop for nitrogen adsorption at 77.35 K is forced to close at relative pressure at or above 0.42, apparently independent of the porous material [57, 58]. The existence of a lower closure point affects primarily the position of the desorption branch with regard to its position and steepness. Despite the fact that the reasons for this phenomenon are still not sufficiently understood, it is clear that it leads to complications for pore size calculation, which we will discuss in chapter 8.

(c) *Disordered Porous Material Model.* A more realistic picture takes into account that the thermodynamics of the pore fluid is determined by phenomena spanning the complete pore network. Even with the incorporation of network and percolation effects the adsorption thermodynamics is still modeled at a single pore level, i.e., the behavior of the fluid in the entire pore space is not assessed. In order to achieve this one needs to consider models which attempt to describe the microstructure of porous materials at length scales beyond that of a single pore. According to Gubbins et al. [59] there are two general approaches to construct a model of nanoporous materials by methods of molecular simulation. The first is the so-called mimetic simulation, and involves the development of a simulation strategy, that mimics the development of the pore structure in the materials preparation. In fact, Gelb and Gubbins [60] have reproduced the complex network structure of porous glasses such as Vycor® and controlled-pore glass by applying molecular simulation and have studied the sorption and hysteresis behavior of xenon in such systems. Grand Canonical Monte Carlo simulation results for xenon adsorption in these systems suggest strongly that the shape of the adsorption/desorption hysteresis does not depend on the connectivity of the material model, supporting the hypothesis that in materials of this type (e.g., a porous Vycor® glass with a porosity of 30%) the fluid in different pores behaves quasi-independently, and that no system-spanning phase transitions occur during adsorption or desorption.

The second approach is the reconstruction method. Here one seeks a molecular model, whose structure matches available experimental structure data. Monson and co-workers investigated by Monte Carlo simulation the condensation and hysteresis phenomena of a Lennard-Jones fluid in a reconstructed model of silica xerogel [61]. Their adsorption isotherms exhibited hysteresis loops of type H1 and H2 in agreement with experimental results obtained on the same type of material. The observed hysteresis was attributed with thermodynamic metastability of the low and high density phases of the adsorbed fluid – however these phases span the entire void space of the porous material and are therefore not associated with the individual pores.

However, it was also suggested that in disordered porous glass materials (e.g., porous Vycor® glass) the origin of the hysteresis is associated with long time dynamics, which is so slow that on (experimentally) accessible time scales, the systems appear to be equilibrated, which leads to the observed reproducible results in the observation of the hysteresis loop [62, 63].

Theoretical and experimental work is necessary to (i) clarify what determines the shape of the hysteresis loop in such disordered systems and (ii) to obtain a clearer picture of the nature of phase behavior of fluids in disordered porous systems.

4.4.5 Effects of Temperature and Pore Size: Experiments and Predictions of Modern, Microscopic Theories

As discussed before, the Kelvin approach considers pore condensation as a gas-liquid phase transition in the core of the pore between two homogenous, bulk-like gas and liquid phases. The density difference $\Delta\rho = \rho^l - \rho^g$ is considered to be equal to the difference in orthobaric densities of coexisting bulk phases, i.e., pore condensation and hysteresis are expected to occur up to the bulk critical point, where $\Delta\rho = 0$.

In contrast, microscopic theories such as density functional theory, molecular simulation and lattice model calculations [22-24, 64-68] predict that a fluid confined to a single pore can exist with two possible density profiles corresponding to inhomogeneous gas- and liquid configurations in

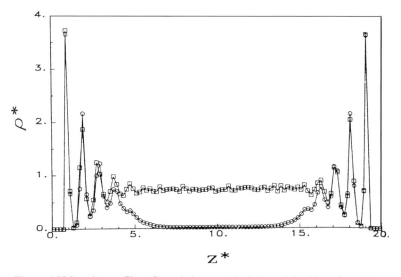

Figure 4.10 Density profiles of coexisting gas (circles)- and liquid configurations (squares) in a slit pore of pore width 20σ, where σ denotes the molecular diameter obtained by Monte-Carlo simulation; $z*$ is the reduced distance from the pore wall and $\rho*$ is the reduced density From [23].

the pore. Corresponding density profiles obtained by Grand Canonical Monte Carlo Simulation (GCMC) for a Lennard-Jones fluid confined to mesoscopic slit-pore are shown in Fig. 4.10. The fluid in the core of pore (in gas and liquid configuration) is – in contrast to the situation in micropores (see Chapter 4.3, Fig. 4.6)- almost structureless, i.e. it does not show the characteristic oscillations observed closer to the pore walls. Hence, in wide mesopores the core fluid can indeed be considered to be bulk-like.

Pore condensation is now understood as first order phase transition between an inhomogeneous gas configuration, which consists of vapor in the core region of the pore in equilibrium with a liquid like adsorbed film (corresponds to configuration **C** in Fig 4.7), and a liquid configuration, where the pore is filled with liquid (corresponds to configuration **D** in Fig. 4.7). At the pore critical point of the confined fluid, these two hitherto distinct fluid configurations will become indistinguishable, i.e., a pore condensation step cannot be observed anymore. The suggested order parameter for this phase transition is the difference in surface excess (or adsorbed amounts at low bulk gas densities, i.e., $\Delta V_{(l,g)} = V_{ads(liquid)} - V_{ads(gas)}$) between the two inhomogeneous gas and liquid phases and not the difference in orthobaric densities $\Delta\rho$ as it is the case for the corresponding bulk phase transition, which occurs between homogeneous gas and liquid phases.

Accordingly, at the pore critical point $\Delta V_{(l,g)} = 0$ and pore condensation cannot be observed anymore. The critical temperature of the confined fluid is shifted to lower temperatures, i.e., in contrast to the predictions of the Kelvin equation pore condensation and hysteresis will vanish already at temperatures below T_c. The shift of the critical temperature can be rationalized by the argument that a fluid in narrow pores is an intermediate between a three-dimensional fluid and a one-dimensional fluid for which no critical point exists at $T > 0$. Hence, the shift of the pore critical temperature is correlated with the pore width, i.e., the more narrow the pore, the lower the pore critical temperature. Consequently, at a given subcritical temperature pore condensation is only possible in pores which are wider than the critical pore size W_c.

Adsorption experiments of pure fluids in porous glasses [67-69], silica gel [70] and MCM-41-type of materials [71-73] revealed that pore condensation and hysteresis indeed disappears below the bulk critical temperature. Furthermore, systematic adsorption studies of SF_6 in controlled-pore glasses indicated that hysteresis already disappears below the capillary critical temperature T_{cc}, i.e., reversible pore condensation could be observed (the criterion applied here to determine pore criticality was the disappearance of the pore condensation step) [69].

An experimental study on nitrogen adsorption in MCM-41 silica in combination with the application of density functional theory clearly revealed that the experimental disappearance of hysteresis at the so-called hysteresis critical temperature T_h is indeed not identical with having achieved the pore critical point [71]. Nitrogen sorption hysteresis (at 77K) disappears when the pore diameter is smaller than 4 nm (see Fig. 8.3, chapter 8), however based on the theoretical results the (pseudo)-pore critical point is achieved at a much smaller pore diameter, i.e. 1.8 nm (the bulk critical temperature of nitrogen is 126.2K). This picture was supported by subsequent experimental sorption studies of pure fluids in ordered mesoporous silica materials [72-74]. For instance, experiments to study the temperature dependence of argon adsorption in MCM-41 materials with pore channels of 2.2 nm diameter revealed a hysteresis critical temperature T_h of ca. 62K. In contrast, the pore critical temperature was located at *ca.* 98K (the bulk critical temperature T_c for argon is 150.7K), i.e., substantial downward shifts in the pore critical and hysteresis critical temperatures are observed for such narrow pores [73].

These systematic studies also revealed that temperature and pore size can be considered as complementary variables with regard to their influence on the occurrence of hysteresis: an increase in temperature has qualitatively a similar effect as a decrease in pore size. Both lead to a decrease in the width of the hysteresis loop, which eventually disappears at a certain critical pore size and temperature (T_h), which is illustrated in Fig 4.11.

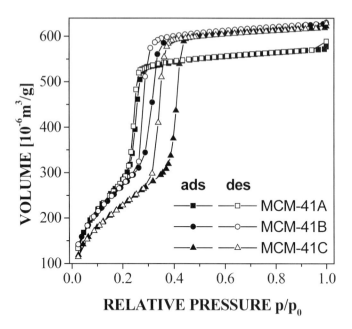

Figure 4.11a Adsorption of argon at 77 K in MCM-41 silica as a function of pore diameter: MCM-41A, d_p = 3.30 nm; MCM-41B, d_p = 3.66 nm; MCM-41C, d_p = 4.25 nm. From [80a].

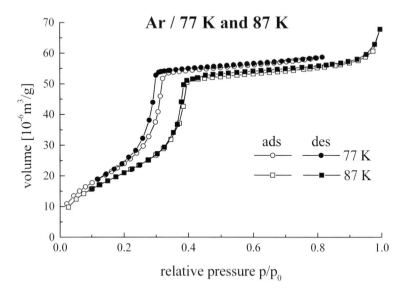

Figure. 4.11b. Effect of temperature on hysteresis. Argon sorption at 77K and 87K on MCM-48 silica (d_p = 4.01 nm). From [80b].

The effect of pore size on hysteresis is shown in Fig. 4.11a, where the hysteresis behavior of argon adsorption isotherms at 77.35K is shown for various MCM-41 silica samples, which exhibit different mode pore diameter (MCM-41A: 3.30 nm, MCM-41 B: 3.66 nm, MCM-41 C: 4.25 nm). It can be clearly seen that the width of the hysteresis loop decreases with decreasing pore size and disappears for the MCM-41 silica A. Fig. 4.11b shows the effect of temperature on hysteresis. Argon adsorption was measured at 77 K and 87 K on an MCM-48 silica sample of pore diameter 4.01 nm. The width of the hysteresis decreases significantly as the temperature is increased from 77 K to 87 K.

In addition to the shift in critical temperature, experiments and theory indicate that as a result of the combined effects of fluid-wall forces and finite-size the freezing temperature and triple point of the pore fluid may also be shifted to lower temperature relative to the bulk triple point if the wall-fluid attraction is not too strong, i.e., the pore wall does not prefer the solid phase [75-79]. This is for instance the case for silica materials. The amount of the shift depends again on the pore size, i.e., the more narrow the pore size the larger the shift of the pore triple point region. Hence, pore condensation can also be observed at temperatures below the bulk triple point temperature [44,88], as it is shown in Figure. 4.11 for argon sorption at 77 K in the narrow pores of MCM-48 and MCM-41 silica. However, systematic sorption experiments of nitrogen and argon adsorption at 77 K and 87 K in mesoporous molecular sieves and controlled pore glasses by Thommes et al. [43a,80] indicate that pore condensation of argon adsorption at 77.35 K cannot be observed anymore if the pore diameter exceeds ca. 15 nm, which limits the range for pores size analysis with argon at 77K. This behavior could be related to confinement effects on the location of the (quasi)-triple point of the pore fluid. The effect of confinement on the phase behavior of a pore fluid is also important for thermoporometry, a technique where the effect of confinement on the suppression of the freezing or melting temperature is used to determine the pore size.

In summary, theoretical and experimental studies have led to the conclusion that the complete coexistence curve of a fluid confined for instance to mesoporous silica materials is shifted to lower temperature and higher mean density [67, 69, 71] (see also the review of Gelb et al [66]). Fig. 4.12 shows a schematic phase diagram of bulk- and pore fluid (confined to different sized, single pores) which illustrates the influence of confinement on the sorption and phase behavior as it can be found for instance in case of mesoporous silica. According to this phase diagram, one can separate the following regimes: (i) continuous pore filling without pore condensation step occurs below a certain critical pore width (w_c) at a given temperature T $< T_c$. For a given pore size (w) continuous pore filling can be observed above the pore critical temperature T_{wc} (and, of course, above the bulk critical temperature); (ii) Reversible pore condensation occurs for pore sizes between the critical pore size w_c and the pore size where hysteresis disappears (w_h), i.e., in the pore size range $w_c < w < w_h$; or in case of fixed pore size in the temperature range between the hysteresis critical temperature T_h and the pore critical temperature T_{Wc}; (iii) Pore condensation with hysteresis occurs for pore sizes larger than w_h at temperatures below T_h.

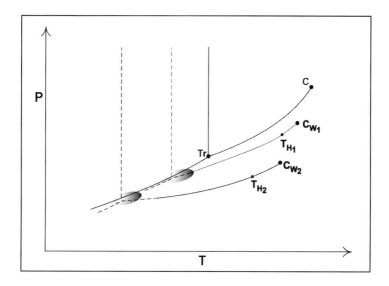

Figure 4.12 Schematic phase diagram of a bulk and pore fluid confined to different sized, single pores of widths $w_1 > w_2$. From [43a]. The pore condensation lines, i.e., the locus of states of the unsaturated vapor at which pore condensation will occur end in the appropriate pore critical points C_{w1} and $C_{w2,}$ with $T_{Cw1} > T_{CW2.}$ For a given experimental temperature, pore condensation will occur first in the pore of width w_2, and a higher (relative) pressure in the larger pore W_1. The temperatures T_{H1} and T_{H2} are the so-called hysteresis critical temperatures, where experimental hysteresis disappears. Details of the sorption and phase behavior below the bulk triple point (Tr) as well as the nature of quasi-triple points is still under investigation [76-81] and these regions of the phase diagram are indicated by dashed lines and the grey areas. Please note, that from a theoretical point of view, real phase transitions and therefore real criticality cannot occur in pseudo-one dimensional cylindrical pores, i.e., pore condensation and the pore critical point should therefore be considered here as pseudo-phase transition and pseudo-critical point, respectively.

These observations clearly reveal that the shape of sorption isotherms does not depend only on the texture of the porous material, but also on the difference of thermodynamic states between the confined fluid and bulk fluid phase. This has to be taken into account for the characterization of porous media by gas adsorption. However, macroscopic, thermodynamic approaches related to the Kelvin equation do not account for these effect (these approaches predict the occurrence of pore condensation up to the bulk critical temperature) and therefore fail to describe correctly the position of the pore condensation step in narrow mesopores, in particular in a temperature and pore size range where hysteresis disappears. In combination with other deficiencies of the Kelvin equation based methods, this leads to significant errors in the pore size analysis (see chapter 8, Fig. 8.3).

4.5 REFERENCES

1. Langmuir I. (1918) *J. Am. Chem. Soc.* **40**, 1368.
2. Brunauer S., Emmett P.H. and Teller E. (1938) *J. Am. Chem. Soc.* **60**, 309.
3. Frenkel J. (1946) *Kinetic Theory of Liquids,* Oxford University Press.
4. Halsey G.D. (1948) *J. Chem. Phys.* **16**, 931.
5. Hill T.L. (1952) *Advances in Catalysis IV*, Academic Press, N.Y, 236.
6. Loering R. and Findenegg G.H. (1984) *J. Chem. Phys.* **81**, *3270.*
7. Carrott P.J., McLeod A.I. and Sing K.S.W. (1982) In *Adsorption at the Gas-Solid and Liquid-Solid Interface* (Rouquerol J. and Sing K.S.W., eds.) Elsevier, Amsterdam.
8. Loering R. and Findenegg G.H. (1981) *J. Colloid Interface Sci.* **84**, 355.
9. Findenegg G.H. and Thommes M. (1997) In *Physical Adsorption: Experiment, Theory, and Application* (Fraissard J. and Conner C., eds.) Kluwer, Dordrecht, Netherlands.
10. Steele W.A. (1974) *The Interaction of Gases with Solid Surfaces,* Chapter 5, Pergamon Press, Oxford.
11. Sing K.S.W., Everett D.H., Haul R.A.W., Moscou L., Pierotti R.A., Rouquerol J. and Siemieniewska T. (1985) *Pure Appl. Chem.* **57**, 603.
12. Marsh H. (1987) *Carbon* **25**, 49.
13. Polanyi M. (1914) *Verh. Dtsch. Phys. Ges.* **16**, 1012.
14. [a]Dubinin M.M. and Timofeev D.P. (1948) *Zh. Fiz. Khim.* **22**, 133; [b]Dubinin M.M. and Radushkevich L.V. (1947) *Dokl. Akad. Nauk. SSSR* **55**, 331.
15. Dubinin M.M. (1960) *Chem. Rev.* **60**, 235.
16. Dubinin M.M. and Astakhov V.A. (1970) *Adv. Chem. Series.* **102**, 69.
17. [a]Stoeckli F. (1977) *J. Colloid Interface Sci.* **59**, 184; [b]Dubinin M.M. and Stoeckli H.F. (1980) *J. Colloid Interface Sci.* **75**, 34; [c]Stoeckli F., Lavanchy A. and Kraehenbuehl F. (1982) In *Adsorption at the Gas-Solid and Liquid-Solid interface* (Rouquerol J. and Sing K.S.W., eds.) Elsevier, Amsterdam.
18. Horvath G. and Kawazoe K. (1983) *J. Chem. Eng. Japan.* **16**, 474.
19. [a]Saito A. and Foley H.C. (1991) *Am. Inst. Chem. Eng. J.* **37**, 429; [b]Saito A. and Foley H.C. (1995) *Microporous Materials* **3**, 531; [c]Cheng L.S. and Yang R.T. (1994) *Chem. Eng. Sci.* **49**, 2599.
20. [a]Evans R., Marconi U.M.B., and Tarazona P. (1986) *J. Chem. Soc. Faraday Trans.* **82**, 1763; [b]Tarazona P., Marconi U.M.B. and Evans R. (1987) *Mol. Phys.* **60**, 573.
21. Peterson B.K., Walton J.P.R.B. and Gubbins K.E. (1986) *J. Chem. Soc. Faraday Trans.* **82**, 1789.
22. [a]Gubbins K.E (1997) In *Physical Adsorption: Experiment, Theory and Applications* (Fraissard J., ed.) Kluwer, Dordrecht, Netherlands; [b]Peterson B.K., Gubbins K.E., Heffelfinger G.S., Marconi U.M.B. and van Swol F. (1988) *J. Chem. Phys.* **88**, 6487.
23. Walton J.P.R.P. and Quirke N. (1989) *Mol. Simul.* **2**, 361.
24. Heinbuch U. and Fischer J. (1987) *Chem. Phys. Lett.* **135**, 587.
25. Seaton N.A., Walton J.R.B. and Quirke N. (1989) *Carbon* **27**, 853.
26. Lastoskie C.M., Gubbins K. and Quirke N. (1993) *J. Phys. Chem.* **97**, 4786.
27. Olivier J.P., Conklin W.B. and v. Szombathley M. (1994) *Stud. Surf. Sci. Catal.* **87**, 81.
28. Neimark A.V. (1995) *Langmuir* **11**, 4183; Neimark A.V., Ravikovitch P.I., Grün M.,

Schüth F. and Unger K.K. (1998) *J. Colloid Interface Sci.* **207,** 159; Ravikovitch P.I, Vishnyakov A. and Neimark A.V. (2001) *Phys. Rev. E,* **64,** :011602.

29. [a]Jagiello J. (2003) *personal communication,* [b]Jagiello J. (1998) In *Fundamentals of Adsorption* **6,** (Meunier F.D., ed.) Elsevier, Amsterdam.

30. Linden M., Schacht S., Schueth F., Steel A. and Unger K. (1998) *J. Porous Mater.* **5,** 177.

31. [a]Thompson W.T. (1871) *Philos. Mag.* **42,** 448; [b]Zsigmondy Z. (1911) *Anorg. Chem.* **71,** 356.

32. Adamson A.W. and Gast A.P. (1997) *Physical Chemistry of Surfaces,* 6th edn , Wiley-Interscience, New York.

33. Cohan L.H. (1938) *J. Am. Chem. Soc.* **60,** 433; Cohan L.H. (1944) *J. Am. Chem. Soc.* **66,** 98.

34. Barrett E.P., Joyner L.G. and Halenda P.P. (1951) *J. Am. Chem. Soc.* **73,** 373.

35. [a]Broekhoff J.C.P. and de Boer J.H. (1967) *J. Catal.* **9,** 8; [b]Broekhoff J.C.P. and de Boer J.H. (1967) *J. Catal.* **9,** 15; [c]Broekhoff J.C.P. and de Boer J.H. (1968) *J. Catal.* **10,** 377; [d]Broekhoff J.C.P. and de Boer J.H. (1968) *J. Catal.* **10** 368.

36. Cole M.W. and Saam W.F. (1974) *Phys. Rev. Lett.* **32,** 985.

37. De Boer J.H. (1958) *The Structure and Properties of Porous Materials,* Butterworths, London.

38. Foster A.G. (1932) *Trans. Faraday Soc.* **28,** 645.

39. Cassell H.M. (1944) *J. Phys. Chem.* **48,** 195.

40. Everett D.H. (1967) In *The Solid-Gas Interface Vol.2* (Flood E.A., ed.) Marcel Decker, New York.

41. Ball P.C. and Evans R. (1989) *Langmuir* **5,** 714.

42. [a]Neimark A.V., Ravikovitch P.I. and Vishnyakov A. (2000) *Phys. Rev. E* **62,** R1493; [b]Neimark A.V. and Ravikovitch P.I. (2001) *Microporous and Mesoporous Materials* **44-56,** 697.

43. [a]Thommes M., Koehn R. and Froeba M. (2002) *Appl. Surf. Sci.* **196,** 239; [b] Thommes M. (in press) In *Nanoporous Materials: Science and Engineering* (Max Lu, ed.) World Scientific, chapter 11.

44. Awschalom D.D., Warnock J. and Shafer M.W. (1986) *Phys. Rev. Lett.* **57,** 1607.

45. Findenegg G.H., Gross S. and Michalski T. (1994) *Stud. Surf. Sci. Catal.* **87,** 71.

46. Mc Bain J.W. (1932) *J. Am. Chem. Soc.* **57,** 699.

47. Mason G. (1982) *J. Colloid Interface Sci.* **88,** 36.

48. Wall G.C. and Brown R.J.C. (1981) *J. Colloid Interface Sci.* **82,** 141.

49. Neimark A.V. (1991) *Stud. Surf. Sci. Catal.* **62,** 67.

50. Parlar M. and Yortsos Y.C. (1988) *J. Colloid Interface Sci.* **124,** 162.

51. [a]Zhu H., Zhang L. and Seaton N.A. (1993) *Langmuir* **9,** 2576; [b]Liu, H. and Seaton N.A. (1994) *Chem. Eng. Sci.* **49,** 1869.

52. Rojas F., Kornhauser I., Felipe C., Esparza J.M., Cordero S., Dominguez A. and Riccardo J.L. (2002) *Phys. Chem. Chem. Phys.* **4,** 2346.

53. Page J.H., Liu L., Abeles B., Herbolzheimer E., Deckmann H.W.and Weitz D.A. (1995) *Phys. Rev. E.* **52,** 2763.

54. Hoinkis E. and Roehl-Kuhn B. (1992) In *Fundamentals of Adsorption* **7**, (Kaneko K., Kanoh H. and Hanzawa Y., eds.) IK International Ltd, Chiba City, Japan.

55. Sarkisov L. and Monson P.A. (2001) *Langmuir* **17,** 7600.

56. Ravikovitch P.I. and Neimark A.V. (2002) *Langmuir* **18,** 9830.

57. Burgess C.G.V. and Everett D.H. (1970) *J. Colloid Interface Sci.* **33,** 611.

58. Gregg S.J. and Sing K.S.W. (1982) *Adsorption, Surface Area and Porosity* 2nd edn, Academic Press, London.

59. Gubbins K.E. (2002) In *Fundamentals of Adsorption* **7** (Kaneko K., Kanoh H. and Hanzawa Y., eds.) IK International Ltd, Chiba City, Japan.

60. [a]Gelb L.D. and Gubbins K.E. (1998) *Langmuir* **14,** 2097; [b]Gelb L.D. (2002) In:

Fundamentals of Adsorption **7** (Kaneko K., Kanoh H. and Hanzawa Y., eds.) IK International Ltd, Chiba City, Japan.

61. Page K.S. and Monson P.A. (1996) *Phys. Rev. E.* **54,** 6557.
62. Kierlik E., Rosinberg M.L., Tarjus G., and Viot P. (2001) *Phys. Chem. Chem. Phys.* **3**, 1201.
63. Woo H.J., Sarkisov L. and Monson P.A. (2001*) Langmuir* **17**, 7472.
64. Evans R.J. (1990) *Phys. Condens. Matter* **2**, 8989.
65. Votyakov E.V., Tovbin Y.U.K., MacElroy J.M.D. and Roche A. (1999) *Langmuir* **15**, 5713.
66. Celstini F. (1997) *Phys. Lett.* **A 228** 84.
67. Gubbins K.E. (1997) In: *Physical Adsorption: Experiment, Theory and Application,* (Fraissard J., ed.) Kluwer, Dordrecht, Netherlands.
68. Gelb L.D., Gubbins K.E., Radhakrishnan R. and Sliwinska-Bartkowiak M. (1999) *Rep. Prog. Phys.* **62**, 1573.
67. Burgess C.G.V., Everett D.H. and Nuttall S. (1989) *Pure Appl. Chem.* **61**, 1845.
68. De Keizer A., Michalski T. and Findenegg G.H. (1991) *Pure Appl. Chem.* **63**, 1495.
69. Thommes M. and Findenegg G.H. (1994) *Langmuir* **10**, 4270.
70. Machin W.D. (1994) *Langmuir* **10,** 1235.
71. Groß S. and Findenegg G.H. (1997) *Ber. Bunsenges.Phys.Chem.* **101**, 1726.
72. Ravikovitch P.I., Domhnaill S.C.O., Neimark A.V., Schueth F. and Unger K.K. (1995) *Langmuir* **11**, 4765.
73. [a]Morishige K. and Shikimi M. (1997) *Langmuir* **13,** 3494; [b]Morishige K. and Shikimi M. (1998) *J. Chem. Phys.* **108**, 7821.
74. [a]Sonwane C.G. and Bhatia S.K. (2000) *J. Phys. Chem B* **104**, 9099; [b]Bhatia S.K. and Sonwane C.G. (1998) *Langmuir* **14**, 1521.
75. Dominguez H., Allen M.P. and Evans R. (1998) *Mol. Phys.* **96**, 209.
76. Fretwell H.M., Duffy J.A., Clarke A.P., Alam M.A. and Evans R. (1995) *J. Phys. Condens. Matter* **7,** L717.
77. Coulomb J.P., Grillet Y., Lewellyn P.L., Martin C. and Andre G. (1998) In: *Fundamentals of Adsorption,* (Meunier F., ed.) Elsevier, Paris.
78. Morishige K., Kawano K. and Hayashigi T. (2000) J. *Phys. Chem B* **104**, 10298.
79. Huber P. and Knorr K. (1999) *Phys. Rev. B* **60,** 12657.
80. [a]Thommes M., Koehn R. and Froeba M. *to be published*; [b]Thommes M., Koehn R. and Froeba M. (2000) *J. Phys. Chem B* **104**, 7932.
81. Myahara M., Sakamoto M., Kandra H. and Higashitani K. (2002) *Stud. Surf. Sci. Catal.* **144,** 411.

5 Surface Area Analysis from the Langmuir and BET Theories

5.1 SPECIFIC SURFACE AREA FROM THE LANGMUIR EQUATION

The Langmuir [1] equation is more applicable to chemisorption (see chapter 12), where a chemisorbed monolayer is formed, but is also often applied to physisorption isotherms of type I. Although this type of isotherm is usually observed with microporous adsorbents, due to the high adsorption potential, a separation between monolayer adsorption and pore filling is not possible for many such adsorbents. A convenient form of the Langmuir equation is

$$\frac{P}{W} = \frac{1}{KW_m} + \frac{P}{W_m} \qquad \text{(cf. 4.12)}$$

where P is the adsorbate equilibrium pressure, and W and W_m are the adsorbed weight and monolayer weights, respectively. The term K is a constant discussed in §4.1.

For type I isotherms, a plot of P/W versus P should give a straight line with $1/W_m$ as the slope. The sample surface area, S_t, is calculated from equation (4.13):

$$S_t = N_m A_x = \frac{W_m \overline{N} A_x}{M} \qquad \text{(cf. 4.13)}$$

where, A_x is the cross-sectional adsorbate area, $\overline{M}$ is the adsorbate molecular weight, and $\overline{N}$ is Avogadro's number. The fact that a Langmuir plot gives a straight line if applied to a type I isotherm is not at all indicative of its success. Without an understanding of the processes occurring within the micropores in terms of adsorption or pore filling, the Langmuir equation may be a correct mathematical description of the isotherm, but the determined monolayer capacity W_m and the corresponding specific surface does not reflect a true surface areas, but rather an equivalent or characteristic surface area.

5.2 SPECIFIC SURFACE AREA FROM THE BET EQUATION

5.2.1 BET-Plot and Calculation of the Specific Surface Area

Although derived over sixty-five years ago, the application of the BET equation is still the most popular approach for the calculation of the specific surface area. The determination of surface areas from the BET theory [2] is a straightforward application of the BET equation, which was derived in chapter 4.

$$\frac{1}{W[P/P_0 - 1]} = \frac{1}{W_m C} + \frac{C-1}{W_m C}\left(\frac{P}{P_0}\right) \tag{cf. 4.38}$$

A plot of $1/W[P_0/P) - 1]$ versus P/P_0, as shown in Fig. 5.1, will yield a straight line usually in the range $0.05 \leq P/P_0 \leq 0.35$.

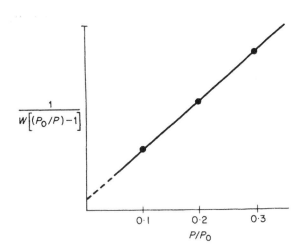

Figure 5.1 Typical BET plot.

The slope s and the intercept i of a BET plot are, respectively,

$$s = \frac{C-1}{W_m C} \tag{5.1}$$

$$i = \frac{1}{W_m C} \tag{5.2}$$

Solving the preceding equations for W_m, the weight adsorbed in a monolayer gives

$$W_m = \frac{1}{s+i}$$ (5.3)

and the solution for C, the BET constant, gives

$$C = \frac{s}{i} + 1$$ (5.4)

The total surface area can be calculated from equation (4.13), *viz.*,

$$S_t = \frac{W_m \overline{N} A_x}{M}$$ (cf 4.13)

where, as before, A_x is the cross-sectional adsorbate area, $\overline{M}$ is the adsorbate molecular weight, and $\overline{N}$ is Avogadro's number. The specific surface area can be determined by dividing S_t by the sample weight.

5.2.2. The Meaning of Monolayer Coverage

Hill [3] has shown that when sufficient adsorption has occurred to cover the surface with exactly one layer of molecules, the fraction of surface, $(\theta_0)_m$, not covered by any molecule is dependent on the BET C value and is given by

$$(\theta_0)_m = \frac{\sqrt{C}-1}{C-1}$$ (5.5)

It is evident from equation (5.5) that when sufficient adsorption has occurred to form a monolayer there is still always some fraction of surface unoccupied. Indeed, only for C values approaching infinity will θ_0 approach zero and in such cases the high adsorbate-surface interaction can only result from chemisorption. For nominal C values, say near 100, the fraction of surface unoccupied, when exactly sufficient adsorption has occurred to form a monolayer, is 0.091. Therefore, on the average each occupied site contains about 1.1 molecules. The implication here is that the BET equation indicates the weight of adsorbate required to form a single molecular layer on the surface, although no such phenomenon as a uniform monolayer exists in the case of physical adsorption.

5.2.3 The BET Constant and Site Occupancy

Equation (5.5) is used to calculate the fraction of surface unoccupied when $W = W_m$, that is, when just a sufficient number of molecules have been adsorbed to give monolayer coverage. Lowell [4] has derived an equation that can be used to calculate the fraction of surface covered by adsorbed molecules of one or more layers in depth. Lowell's equation is

$$(\theta_i)_m = C\left(\frac{\sqrt{C}-1}{C-1}\right)^{i+1}$$
(5.6)

where θ_i represents the fraction of surface covered by layers i molecules deep. The subscript m denotes that equation is valid only when sufficient adsorption has occurred to make $W = W_m$. Table 5.1 shows the fraction of surface covered by layers of various depths, as calculated from equation for $i = 0$ and for $i \neq 0$, as a function of the BET C value.

Equations (5.5) and (5.6) should not be taken to mean that the adsorbate is necessarily arranged in neat stacks of various heights. Rather, it should be understood as an indication of the fraction of surface covered with the equivalent of i molecules regardless of their specific arrangement, lateral mobility, and equilibrium with the vapor phase.

Of further interest is the fact that when the BET equation is solved for the relative pressure corresponding to monomolecular coverage, ($W = W_m$), one obtains

$$\left(\frac{P}{P_0}\right)_m = \frac{\sqrt{C}-1}{C-1}$$
(5.7)

The subscript m above refers to monolayer coverage. Equating (5.6) and (5.7) produces the interesting fact that

$$\theta_0 = \left(\frac{P}{P_0}\right)_m$$
(5.8)

That is, the numerical value of the relative pressure required to make W equal to W_m is also the fraction of surface unoccupied by adsorbate.

Table 5.1 Values for $(\theta)_m$ from equations (5.5) and (5.6).

	C = 1000	C = 100	C = 10	C = 1
0	0.0307	0.0909	0.2403	0.5000
1	0.9396	0.8264	0.5772	0.2500
2	0.0288	0.0751	0.1387	0.1250
3	0.0009	0.0068	0.0333	0.0625
4		0.0006	0.0080	0.0313
5		0.0001	0.0019	0.0156
6			0.0005	0.0078
7			0.0001	0.0039
8				0.0019
9				0.0009
10				0.0005
11				0.0002
				0.0001

5.2.4 The Single Point BET Method

The BET theory requires that a plot of $1/W[(P_0/P)-1]$ versus P/P_0 be linear with a finite intercept (see equation (4.38) and Fig. 5.1). By reducing the experimental requirement to only one data point, the single point method offers the advantages of simplicity and speed often with little loss in accuracy. The slope s and the intercept i of a BET plot are

$$s = \frac{C-1}{W_m C} \qquad \text{(cf.5.1)}$$

$$i = \frac{1}{W_m C} \qquad \text{(cf. 5.2)}$$

Then

$$\frac{s}{i} = C-1 \qquad (5.9)$$

For reasonably high values of C the intercept is small compared to the slope and in many instances may be taken as zero. With this approximation, equation (4.38), the BET equation, becomes

$$\frac{1}{W\left[(P/P_0)-1\right]} = \frac{C-1}{W_m C}\left(\frac{P}{P_0}\right) \tag{5.10}$$

Since $1/W_m C$, the intercept, is assumed to vanish, equation (5.11) reduces to

$$W_m = W\left(1 - P/P_0\right) \tag{5.11}$$

The total surface area as measured by the single point method, is then calculated as:

$$S_t = W\left(1 - \frac{P}{P_0}\right)\frac{\overline{N}}{\overline{M}} A_x \tag{5.12}$$

5.2.5 Comparison of the Single Point and Multipoint Methods

The error introduced by the single point method can be evaluated by examining the difference between W_m as determined by equations (5.11) and (4.38), the BET equation. Solving equation (4.38) for W_m gives

$$W_m = W\left(\frac{P_0}{P} - 1\right)\left[\frac{1}{C} + \frac{C-1}{C}\left(\frac{P}{P_0}\right)\right] \tag{5.13}$$

Subtracting equation (5.11) from (5.13) and dividing by equation gives the relative error associated with the single point method, that is,

$$\frac{\left(W_m\right)_{mp} - \left(W_m\right)_{sp}}{\left(W_m\right)_{mp}} = \frac{1 - P/P_0}{1 + (C-1)\, P/P_0} \tag{5.14}$$

The subscripts mp and sp refer to the multi- and single point methods, respectively. Table 5.2 shows the relative error of the single point method compared to the multipoint method as a function of P/P_0 as calculated from equation (5.14). The last column of Table 5.2 is established by substituting equation (4.46) into equation (5.12) for the special case when $P/P_0 = (P/P_0)_m$, thus

$$\frac{\left(W_m\right)_{mp} - \left(W_m\right)_{sp}}{\left(W_m\right)_{mp}} = \frac{\sqrt{C}-1}{C-1} = \left(\frac{P}{P_0}\right)_m = \theta_0 \tag{5.15}$$

Table 5.2 Relative errors using the single point method at various relative pressures.

C	$P/P_0 = 0.1$	$P/P_0 = 0.2$	$P/P_0 = 0.3$	$(P/P_0)_m$*
1	0.90	0.80	0.70	0.50
10	0.47	0.29	0.17	0.24
50	0.17	0.07	0.04	0.12
100	0.08	0.04	0.02	0.09
1000	0.009	0.004	0.002	0.003

*$(P/P_0)_m$ is the relative pressure that gives monolayer coverage according to a multipoint determination.

The surprising relationship above shows that when a single point analysis is made using the relative pressure that would give monolayer coverage according to the multipoint theory, the relative error will be equal to the relative pressure employed. The error will also, according to equation (4.44), be equal to the fraction of surface unoccupied. A more explicit insight into the mathematical differences of the multi- and single point methods is obtained by considering a single point analysis using a relative pressure of 0.3 with a corresponding multipoint C value of 100. From equation (5.11), the single point BET equation, one obtains

$$\left(W_m\right)_{sp} = 0.7W_{0.3} \qquad (5.16)$$

The term $\left(W_m\right)_{sp}$ refers to the monolayer weight as determined by the single point method, and $W_{0.3}$ is the experimental weight adsorbed at a relative pressure of 0.3. From equation (4.46) the relative pressure required for monolayer coverage is

$$\left(\frac{P}{P_0}\right)_m = \frac{\sqrt{100}-1}{100-1} = 0.0909 \qquad (5.17)$$

Using equation (5.13), the multipoint equation, to find W_m gives

$$W_m = W_{0.3}(3.33-1)\left[\frac{1}{100} + \frac{100-1}{100}(0.3)\right] = 0.715W_{0.3} \qquad (5.18)$$

Comparison of equations (5.18) and (5.16) shows that the difference between the single and multipoint methods is identical to that shown in Table 5.2 for $P/P_0 = 0.3$ and $C = 100$; *viz.*,

$$\frac{0.715-0.700}{0.715} = 0.02 \tag{5.19}$$

The above analysis discloses that, when the BET C value is 100, the single point method using a relative pressure more than three times that required for monolayer coverage causes an error of only 2%. To further understand the BET equation and the relationship between the C value and the single point error it is useful to rewrite the equation (4.38), the BET equation, as

$$\frac{W}{W_m} = \frac{C(P/P_0)}{\left[1+(C-1)P/P_0\right](1-P/P_0)} \tag{5.20}$$

Using the method of partial fractions, the right side of equation (5.20) can be written as

$$\frac{C(P/P_0)}{\left[1+(C-1)P/P_0\right](1-P/P_0)} = \frac{X}{1-P/P_0} - \frac{Z}{1+(C-1)P/P_0} \tag{5.21}$$

Recognizing that $X = Z = 1$ is a solution, gives

$$\frac{W}{W_m} = \frac{1}{1-P/P_0} - \frac{1}{1+(C-1)P/P_0} \tag{5.22}$$

Equation (5.22) is the BET equation expressed as the difference between two rectangular hyperbolas. If the value of C is taken as infinity, equation (5.22) immediately reduces to equation (5.11), the single point BET equation. The hyperbolae referred to above are shown below in Fig. 5.2.

As indicated in Fig. 5.2, curve Y, the BET curve for an arbitrary C value, approaches curve X, the single point curve, as the value of C increases. In the limiting case of $C \to \infty$, the BET curve is coincident with the single point curve. For all other C values, the single point curve lies above the BET curve and their difference vanishes as the relative pressure approaches unity. Thus, as the value of C increases, the knee of the isotherm becomes sharper and moves toward lower relative pressures (see also Fig. 5.3). For lower C values, curves X and Y diverge and higher relative pressures must be used to make single point surface areas conform to those obtained by the multipoint method.

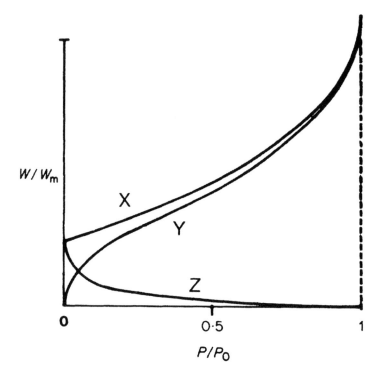

Figure 5.2 Plot of the hyperbola from equation (5.14) using an arbitrary C-value. Curve X = $1/(1 - P/P_0)$; curve Z = $1/(1 + (C-1)P/P_0)$; curve Y = X − Z.

The extent of divergence of curves X and Y is controlled, in a mathematical sense, by the second term in equation (5.21). This term contains the C value.

Table 5.2 indicates that, regardless of the C value, using higher relative pressures *within the linear range* reduces the relative error. Similarly, Fig. 5.2 shows that at sufficiently high relative pressure the BET curve and the single point curve merge regardless of the C value. It would appear that all single point analyses should be performed at the highest possible relative pressures. Although theoretically sound, the use of relative pressures above 0.3 can lead to serious errors on the large number of samples that contain pores. In a later chapter, the influence of pores is discussed, but here it is sufficient to note that once condensation in pores commences, the BET equation, which deals only with adsorption, fails adequately to describe the state of the system. Ample evidence is available to indicate that many adsorbents possess pores which causes condensation at relative pressures as small as 0.3 and in some cases at even lower values. Therefore, relative pressures of 0.3 may be considered sufficiently high to give good agreement with multipoint measurements on most surfaces while avoiding condensation in all but microporous samples.

When used for quality control, the error associated with the single point method can be eliminated or greatly reduced if an initial multipoint analysis is performed to obtain the correct C value. Then equation (5.14) can be used to correct the results. Even an approximate value of C can be used to estimate the single point error. However, on the great majority of surfaces, the C value is sufficiently high to reduce the single point error to less than 5%.

5.2.6 Applicability of the BET Theory

As already stated before, the BET theory continues to be almost universally used because of its simplicity, its definitiveness, and its ability to accommodate each of the five isotherm types. The mathematical nature of the BET equation in its most general form, equation (4.39) gives the Langmuir or type I isotherm when $n = 1$. Plots of W/W_m versus P/P_0 using equation (4.38) conforms to type II or type III isotherms for C values greater than and less than 2, respectively. Fig. 5.3 shows the shape of several isotherms for various values of C. The data for Fig. 5.3 are shown in Table 5.3 with values of W/W_m calculated from equation after rearrangement to

$$\frac{W}{W_m} = \left(1 - \frac{P}{P_0}\right) + \frac{1}{C}\left(\frac{P}{P_0} + \frac{P_0}{P} - 2\right) \qquad (5.23)$$

The remaining two isotherms, types IV and V, are modifications of the type II and type III isotherms due to the presence of pores.

Rarely, if ever, does the BET theory exactly match an experimental isotherm over its entire range of relative pressures. In a qualitative sense, however, it does provide theoretical foundation for the various isotherm shapes.

Of equal significance is the fact that in the region of relative pressures near completed monolayers ($0.05 \leq P/P_0 \leq 0.3$) the BET theory and experimental isotherms do agree very well, leading to a powerful and extremely useful method of surface area determination. The fact that most monolayers are completed in the range $0.05 \leq P/P_0 \leq 0.3$ reflects the value of most C constants. As shown in Table 5.3, the value of W/W_m equals unity in the previous range of relative pressures for C values between 3 and 1000, which covers the great majority of all isotherms.

The sparsity of data regarding type III isotherms, with C values of 2 or less, leaves open the question of the usefulness of the BET method for determining surface areas when type III isotherms are encountered. Often

Table 5.3 Values of W/W_m and relative pressures for various values of C.

P/P_0	$C = 0.05$	$C = 0.5$	$C = 1$	$C = 2$	$C = 3$	$C = 10$	$C = 100$	$C = 1000$
0.02	0.001	0.010	0.020	0.040	0.059	0.173	0.685	0.973
0.05	0.003	0.027	0.052	0.100	0.143	0.362	0.884	1.030
0.10	0.006	0.058	0.111	0.202	0.278	0.585	1.020	1.100
0.20	0.015	0.139	0.250	0.417	0.536	0.893	1.200	1.250
0.30	0.030	0.253	0.429	0.660	0.804	1.160	1.400	1.430
0.40	0.054	0.417	0.667	0.952	1.110	1.450	1.640	1.660
0.50	0.095	0.667	1.000	1.330	1.500	1.820	1.980	2.000
0.60	0.172	1.060	1.490	1.870	2.040	2.340	2.480	2.500
0.70	0.345	1.790	2.330	2.740	2.910	3.190	3.320	3.330
0.80	0.833	3.330	4.000	4.440	4.620	4.880	4.990	5.000
0.90	3.330	8.330	9.090	9.520	9.680	9.900	9.990	10.000
0.94	7.350	14.700	15.700	16.200	16.300	16.600	16.700	16.700

in this case it is possible to change the adsorbate to one with a higher C value, thereby changing the isotherm shape. Brunauer *et al* [5], however, point to considerable success in calculating the surface area from type III isotherms as well as predicting the temperature coefficient of the same isotherms.

Despite of the success of the BET theory, some of the assumptions upon which it is founded are not above criticism. One questionable assumption is that of an energetically homogeneous surface, that is, all the adsorption sites are energetically identical. Further, the BET model ignores the influence of lateral adsorbate interactions.

Brunauer [6] answers these criticisms by pointing out that lateral interaction between adsorbate molecules necessarily increases as the surface becomes more completely covered. The interaction with the surface, however, decreases with increasing adsorption up to monolayer coverage since on an energetically heterogeneous surface the high energy sites will be occupied at lower relative pressures. In this situation, occupancy of the lower energy sites occurs nearer to completion of the monolayer.

Fig.5.4 illustrates how the lateral interactions and the surface interactions can sum to a nearly constant overall adsorption energy up to completion of the monolayer, an implicit assumption of the BET theory. This results in a constant C value from equation (4.22)

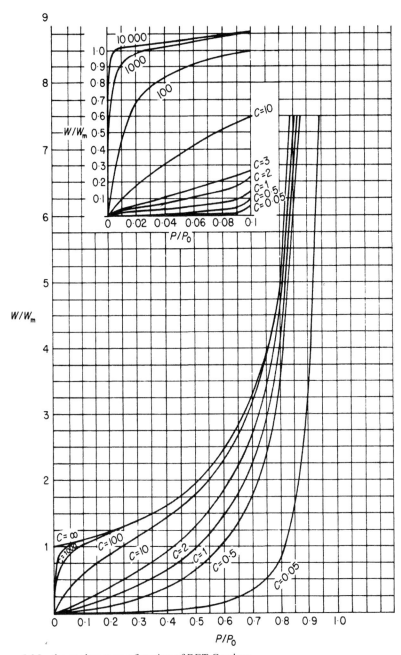

Figure 5.3 Isotherm shapes as a function of BET C values.

$$C = \frac{A_1 V_2}{A_2 V_1} e^{(E-L)/RT}$$ (cf. 4.22)

The dotted lines in Fig. 5.4 indicate the influence of very high adsorption potentials which can account, at least in part, for the usual nonlinearity of BET plots at very low relative pressures ($P/P_0 \leq 0.05$).

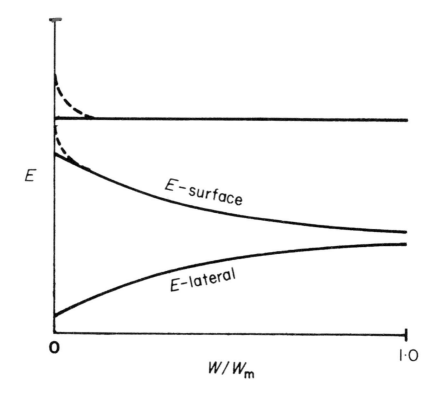

Figure 5.4 Variation in adsorption and lateral interaction potentials.

A further criticism of the BET theory is the assumption that the heat of adsorption of the second and higher layers is equal to the heat of liquefaction. It seems reasonable to expect that polarization forces would induce a higher heat of adsorption in the second layer than in the third, and so forth. Only after several layers are adsorbed should the heat of adsorption equal the heat of liquefaction. It is, therefore, difficult to resolve a model of molecules adsorbed in stacks while postulating that all layers above the first are thermodynamically a true liquid structure. The apparent validity of these criticisms contributes to the failure of the BET equation at high relative pressures (P/P_0>0.3). However, in the range of relative pressure leading to coverage near $W/W_m = 1$, the BET C values usually give heats of adsorption that are reasonable. Thus, for the great majority of isotherms the range of relative pressures between 0.05 and 0.3, the linear BET range, apparently represents a condition in which the very high energy sites have been occupied and extensive multilayer adsorption has not yet commenced. It is within these limits that the BET theory is generally valid. Instances have been found in which BET plots were noted to be linear to relative pressures as high as 0.5, and in other cases the linear range is found only below relative pressures of 0.1 [e.g., 7].

In addition to effects arising from the chemical and geometrical heterogeneity of the surface, the porosity (i.e. existence of micro- or mesopores) plays here an important role. The BET equation is applicable for surface area analysis of nonporous- and mesoporous materials consisting of pores of wide pore diameter, but is in a strict sense not applicable to microporous adsorbents. Micropores exhibit unusually high adsorption potentials due to the overlapping potential from the walls of the micropore, and it is therefore difficult to separate the processes of mono-multilayer adsorption from micropore filling. Pore filling is usually completed at relative pressures below than 0.1, and linear BET plots are found at even lower relative pressures. In this case the obtained surface area does not reflect the true internal surface area but should be considered as a kind of "characteristic or equivalent BET area", and it is of course obligatory to report the range of linearity for the BET plot!

In the case of mesoporous materials consisting of pores of wide pore diameter, pore filling (i.e., here pore condensation) occurs in the multilayer region of the sorption isotherms (see chapter 4) and usually no special difficulty arises in application of the BET-theory. However, the application of the BET approach is problematic for estimating the surface area of adsorbents exhibiting mesopores in the pore size range between 20 Å and 40Å (e.g., mesoporous molecular sieves of type MCM-41, MCM-48 (e.g., [8]). In this case pore filling is observed at pressures very close to the pressure range where monolayer-multilayer formation on the pore walls occurs, which may lead to a significant overestimation of the monolayer capacity In the case of an BET analysis.

In addition, the obtained result depends very much on the cross-sectional area A_x assumed for the calculation, which will be discussed in the next section.

5.2.7 Importance of the Cross-Sectional Area

Using the BET equation to determine W_m, the monolayer weight, and with reasonable estimates of the adsorbate cross-sectional area, A_x, the total sample surface area, S_t, in square meters, can be calculated from equation (4.13).

$$S_t = \frac{W_m \overline{N} A_x}{\overline{M}} \times 10^{-20}\, m^2 \qquad\qquad (\text{cf. } 4.13)$$

with W_m in grams, $\overline{M}$ is the adsorbate molecular weight, $\overline{N}$ is Avogadro's number $(6.022 \times 10^{23}$ molecules per mole$)$ and A_x in square Ångströms per molecule. Division by the sample weight converts S_t to S, the specific surface area.

A reasonable approximation of the cross-sectional area of adsorbate molecules was proposed by Emmett and Brunauer [9]. They assumed the adsorbate molecules to be spherical and using the bulk liquid properties (at the temperature of the adsorption experiment) they calculated the cross-sectional area from

$$A_x = \left(\frac{\overline{V}}{\overline{N}}\right)^{\frac{2}{3}} \times 10^{16}\, \text{Å}^2 \qquad\qquad (5.24)$$

where $\overline{V}$ is the liquid molar volume. Equation (5.24) must be amended to reflect the molecular packing on the surface. Assuming that the liquid is structured as spheres with 12 nearest neighbors, 6 in a plane, in the usual close packed hexagonal arrangement shown in Fig. 6.1, and that the adsorbate has the same structure on the adsorbent surface, equation (5.24) becomes

$$A_x = 1.091 \left(\frac{\overline{V}}{\overline{N}}\right)^{\frac{2}{3}} \times 10^{16}\, \text{Å}^2 \qquad\qquad (5.25)$$

The factor 1.091 in equation (5.25) arises from the characteristics of close packed hexagonal structures. If D is the distance between centers of adjacent spheres, the spacing between the centers of adjacent rows in a plane is $\sqrt{3}\,D/2$. The spacing between centers of adjacent planes is $\sqrt{2/3}D$ [10]. Allowing N_x and N_y to represent the number of spheres along the X and Y axes of a plane of spheres, the planar area, A_p is given by

$$A_p = \frac{\sqrt{3}}{2} D^2 N_x N_y \qquad\qquad (5.26)$$

If N_z is the number of planes or layers, then the volume, V, containing $N_xN_yN_z$ spheres is given by

$$V = \frac{\sqrt{3}}{2}\sqrt{\frac{2}{3}}D^3N_xN_yN_z \qquad (5.27)$$

Since $N_xN_yN_z$ represents the total number of spheres, N, in the volume, V, equation (5.27) can be expressed as

$$V = \frac{\sqrt{3}}{2}\sqrt{\frac{2}{3}}D^3N \qquad (5.28)$$

Then

$$\left(\frac{V}{N}\right)^{\frac{2}{3}} = \frac{3^{\frac{1}{3}}}{2^{\frac{2}{3}}}\left(\frac{2}{3}\right)^{\frac{1}{3}}D^2 \qquad (5.29)$$

Substituting for D^2 into equation (5.26) gives

$$A_p = 1.091N_xN_y\left(\frac{V}{N}\right)^{\frac{2}{3}} \qquad (5.30)$$

The molecular cross-sectional area A_x then can be obtained by dividing the planar area, A_p, by N_xN_y, the number of molecules in a plane. Thus, dividing both numerator and denominator of the fraction, V/N, by the number of moles yields

$$A_x = 1.091\left(\frac{\overline{V}}{\overline{N}}\right)^{\frac{2}{3}} \times 10^{16} \qquad (\text{cf.5.25})$$

That the adsorbate resides on the adsorbent surface with a structure similar to a plane of molecules within the bulk liquid, as depicted in Fig. 5.5 is a simplified view of the real situation on surfaces. Factors that make this model and therefore equation (5.25) of limited value include the following:

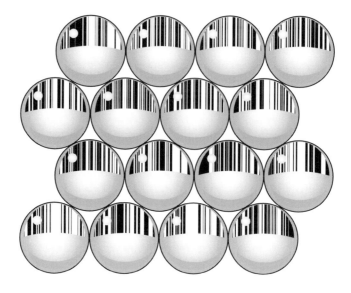

Figure 5.5 Sectional view of a close-packed hexagonal arrangement of spheres.

1. Weak interactions with the surface lead to lateral mobility of the adsorbate on the surface, which will tend to disrupt any tendency for the development of an organized structure, i.e., will prevent a definite arrangement of adsorbate on the surface.

2. Complex molecules, which rotate about several bond axes, can undergo conformational changes on various surfaces and thereby exhibit different cross-sectional areas.

3. Orientation of polar molecules produces different surface arrangements depending on the polarity of the adsorbent.

4. Strong interactions with the surface lead to localized adsorption, which constrains the adsorbate to a specific site. This type of 'epitaxial' adsorption will lead to decreasing measured surface areas relative to the true BET value as the surface sites become more widely spaced The effective adsorbate cross-sectional area will then reflect the spacing between sites rather than the actual adsorbate dimensions.

5. Fine pores may not be accessible to the adsorbate, sot that a substantial portion of the surface is inaccessible to measurement. This would be particularly true for large adsorbate molecules.

Based on the points discussed before it is obvious to assume a relationship between the BET C constant and the cross-sectional area. Indeed, Kiselev and Eltekov [11] established that the BET C value influences the adsorbate cross-sectional area. They measured the surface area of a number of adsorbents using nitrogen. When the surface areas of the same adsorbents were measured using *n*-pentane as the adsorbate the cross-sectional areas of *n*-pentane had to be revised in order to match the surface areas measured using nitrogen. It was found that the revised areas increased hyperbolically as the *n*-pentane C value decreased, as show in Fig. 5.6.

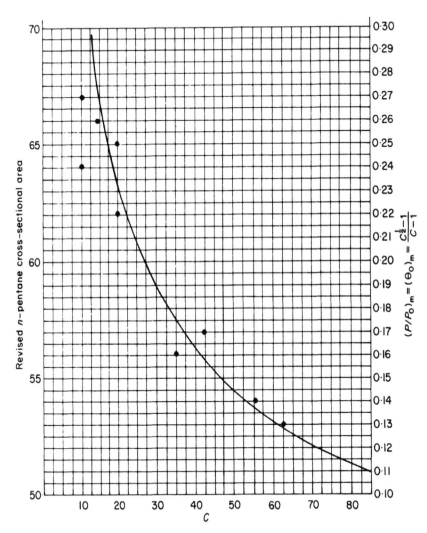

Figure 5.6 Variation of *n*-pentane cross-sectional area with the BET C constant (points) and $\left(\theta_0\right)_m$ (solid line).

A similar relationship between n-butane cross-sectional areas and the BET C constant has been reported [12]. A plot of the revised cross-sectional areas of n-butane versus the BET constant is shown in Fig. 5.7. Figs. 5.6 and 5.7 show that plots of $\left(\sqrt{C}-1\right)\big/(C-1)$ versus C give hyperbolae which also match the cross-sectional area data. A plausible explanation for the observation that the fraction of surface not covered by adsorbate, (θ_0), increases at low C values, leading to high apparent cross-sectional area, is that the two hydrocarbons used as adsorbates interact weakly with the adsorbent. Thus, they behave as two-dimensional gases on the surface. Therefore, their cross-sectional areas may reflect the area swept out by the adsorbate molecules during their residence time on the surface rather than their actual cross-sectional areas.

In those instances of very high C values, the fraction of surface uncovered by adsorbate again increases, as a result of epitaxial deposition on specific surface sites, which when widely spaced, would lead to high apparent cross-sectional areas. A complete plot of cross-sectional area versus the BET C value would then be parabolic in shape, with the most suitable values of cross- sectional areas lying near the minimum of the parabola. For the great majority of adsorbents, the C constant for nitrogen lies in the rage from about 50 to 300. Interactions leading to C values as low as 10 or 20 are not found with nitrogen nor is nitrogen chemisorbed, which would lead to adsorption on specific sites. Thus, nitrogen is uniquely suited as a desirable adsorbate, since its C value is not found at the extremes at each end of the parabola.

Since $\left(\sqrt{C}-1\right)\big/(C-1)=\left(\theta_0\right)_m$ can be calculated from a BET plot, there exists a potential means of predicting the cross-sectional area variation relative to nitrogen. On surfaces that contain extensive porosity, which exclude large adsorbate molecules from some pores while admitting smaller ones, it becomes even more difficult to predict any variation in the adsorbate cross-sectional area by comparison to a standard [13,14].

Summarizing, it can be said that the effective cross-sectional area depends on the temperature, the nature of the adsorbate-adsorbate and adsorbent-adsorbate interaction and the texture of the adsorbent surface. The surface areas calculated from equation (4.13) usually give different results depending upon the adsorbate used. If the cross-sectional areas are arbitrarily revised to give surface area conformity on one sample, the revised values generally will not give surface area agreement when the adsorbent is changed. With regard to cross-sectional areas, it must be kept in mind that the area occupied by a molecule or atom can often be many times its true area and the terms *effective area* or *occupied area* are more appropriate and less misleading.

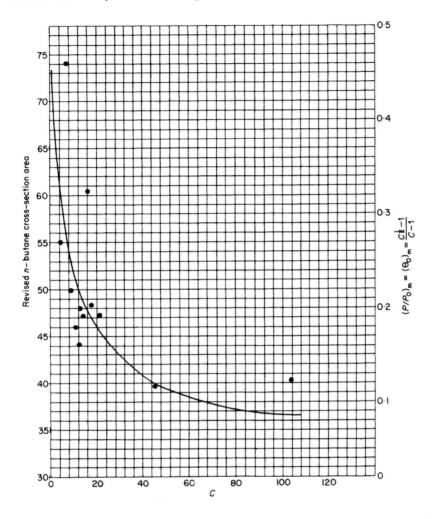

Figure 5.7 Variation of n-butane-cross- sectional area with the BET C constant (points) and $\left(\theta_0\right)_m$ (solid line).

Table 5.4 lists cross-sectional areas for some frequently used adsorptives. The data in the first column are taken from the work of McCellan and Harnsberger [15] who compared and discussed the cross-sectional areas reported for a wide range of adsorption systems. Very often the cross-sectional area was obtained by assigning a value to each adsorptive, as required, in order to make measured surface areas agree with the nitrogen value [16].

Table 5.4 Cross-sectional areas of some frequently used adsorptives.

Adsorptive Temperature		Cross-sectional area $(Å^2)[15]$	Customary Value $(Å^2)$
Nitrogen	77.35 K	13.0 - 20.0	16.2
Argon	77.35 K	10.0 - 19.0	13.8
Argon	87.27 K	9.7 - 18.5	14.2
Krypton	77.35 K	17.6 - 22.8	20.2
Xenon	77.35 K	6.5 - 29.9	16.8
Carbon Dioxide		14 - 22.0	
	195 K		19.5
	273 K		21.0
Oxygen	77.35 K	13 - 20	14.1
Water	298.15 K	6 - 19	12.5
n-Butane	273.15 K	36 - 54	44.4
Benzene	293.15 K	73 - 49	43.0

5.2.8 Nitrogen as the Standard Adsorptive for Surface Area Measurements

Due to the uncertainty in calculating absolute cross-sectional areas, the variation in cross-sectional areas with the BET C value, and the fact that on porous surfaces less area is available for larger adsorbate molecules, there is a need for a universal, although possibly arbitrary, standard adsorptive. The unique properties of nitrogen have led to its acceptance in this role with an assigned cross-sectional area of 16.2 $Å^2$ at its boiling point of 77.35 K. In addition the availability of liquid nitrogen, has also led to the situation that nitrogen is now internationally accepted as the standard BET adsorptive. This is demonstrated in the IUPAC recommendations [17], but also in numerous standards from international, and national standardization institutions (e.g., ISO, ASTM International).

The cross-sectional area of 16.2 $Å^2$ is based on the assumption that at 77 K the nitrogen monolayer is in a close-packed "liquid state", which appears to be quite accurate in the case for hydrocarbon surfaces. The fact, that nitrogen has a permanent quadrupole moment is important because it is responsible for the formation of a well-defined monolayer on most surfaces. However, in the case of surfaces of high polarity the nitrogen adsorption and the orientation of the adsorbate molecules on the surface is affected by specific interactions between the polar groups on the adsorbent surface and the quadrupole moment of the nitrogen molecule. The possibility of such a problem was already mentioned in the paper by McClellan and Harnsberger [15]. Indeed, recent experimental sorption studies on highly ordered mesoporous silica materials such as MCM-41 (which consists of independent cylindrical-like pores), suggest strongly that the cross-sectional area of nitrogen on a hydroxylated surface might differ from the commonly adopted value of 0.162 nm^2 [18]. Similar observations were already made in the past and it was assumed that the quadrupole moment of the nitrogen molecule leads to specific interactions with the hydroxyl groups on the surface causing an orientating effect on the adsorbed nitrogen molecule [19]. But it was only in the last ten years that an accurate cross-sectional area, (i.e., 0.135 nm^2), valid for nitrogen adsorption on a hydroxylated silica surface, could be proposed [20]. This value was obtained by measuring the volume of N_2 adsorbed on silica spheres of known diameter. If one uses the standard cross-sectional area (0.162 nm^2) the BET surface area of hydroxylated silica surfaces can be overestimated by *ca.* 20 % [18].

In contrast to nitrogen, argon has no quadrupole moment and the above-mentioned problems do not occur when argon is used as the adsorptive. In contrast to argon adsorption at liquid argon temperature (i.e., 87.27 K), the use of argon adsorption at the liquid nitrogen temperature is more problematic. Firstly, argon is here *ca.* 6.5 K below the bulk triple point temperature (T_r = 83.81 K), hence the bulk reference state is in doubt. However, for surface area analysis the saturation pressure of supercooled liquid argon (P_0 = 230 torr) is used. In addition, argon sorption at 77 K is much more sensitive to the details of the surface structure, and type VI sorption isotherms (see chapter 3) have been observed on homogeneous surfaces [21].

5.2.9 Low Surface Area Analysis

Using highly accurate volumetric adsorption equipment, it is possible to measure absolute surface areas as low as approximately 0.5 – 1 m^2 with nitrogen as the adsorptive. In order to measure even lower surface areas the number of molecules trapped in the void volume of the sample cell needs to be reduced (see Chapter 14 for details). This can be achieved by applying krypton adsorption at liquid nitrogen temperature for the surface area analysis. Krypton at ~77 K is *ca.* 38.5 K below its triple point temperature (T_r = 115.35 K), and it sublimates (i.e., $P_{0,solid}$) at *ca.* 1.6 torr. However, it has become customary to adopt the saturation pressure of supercooled liquid krypton for the application of the BET equation, i.e., one assumes that despite the fact that the sorption measurement is performed that far below the bulk triple point temperature, the adsorbed krypton layer is liquid-like. The saturation pressure of the supercooled liquid krypton is 2.63 torr, i.e., the number of molecules in the free space of the sample cell is significantly reduced (to 1/300th) compared to the conditions of nitrogen adsorption at liquid nitrogen temperature. Hence, krypton adsorption at ~77 K is much more sensitive, and can be applied to assess surface areas down to at least 0.05 m^2.

Problems are of course associated with the fact that the nature and the thermodynamic state (solid or liquid?) of the adsorbed layer(s) is not well defined, and hence the reference state to calculate P/P_0. Connected with this is some uncertainty with regard to the wetting behavior of the adsorbed krypton phase that far below the bulk triple point temperature (i.e., in the BET approach a complete wetting of the adsorbate phase is assumed). Whereas in the case of nitrogen adsorption (at its boiling temperature) for almost all materials a complete wetting behavior can be assumed, this situation may be different for of adsorption below the triple point temperature [21, 22]. This might also contribute to the fact that the effective cross-sectional area of krypton depends very much on the adsorbent surface and is therefore not well established.

The cross-sectional area calculated from the density of the supercooled liquid krypton is 0.152 nm^2 (15.2 Å^2), but the higher cross-sectional area of 0.202 nm^2 (20.2 Å^2) is commonly used [15, 23].

However, despite these deficiencies, it must be clearly stated that krypton adsorption at ~ 77 K is considered to be a very useful tool for routine surface area measurements of materials with low-surface area.

5.3 REFERENCES

1.   Langmuir I. (1918) *J. Am. Chem.Soc.* **40**, 1368.
2.   Brunauer S., Emmett P.H. and Teller E. (1938) *J. Amer. Chem. Soc.* **60**, 309.
3.   Hill T.L. (1946) *J. Chem. Phys.* **14**, 268.
4.   Lowell S. (1975) *Powder Technol.* **12**, 291.
5.   Brunauer S., Copeland L.E. and Kantro D.L. (1967) *The Gas Solid Interface,* Vol. 1, Dekker, New York, chapter 3.
6.   Brunauer S. (1961) *Solid Surfaces and the Gas Solid Interface,* Advances in Chemistry Series, No. 33, American Chemical Society, Washington, D.C.
7.   MacIver D.S. and Emmett P.H. (1956) *J. Am. Chem. Soc.* **60**, 824.
8.   Kruk M., Jaroniec M. and Sayari A. (1997) *J. Phys. Chem. B* **101**, 583.
9.   Emmett P.H. and Brunauer S. (1937) *J. Am. Chem. Soc.* **59**, 1553.
10.   Moelwyn-Hughes E.A. (1961) *Physical Chemistry*, 2nd edn, Pergamon Press, New York, p545.
11.   Kiselev A.V. and Eltekov Y.A. (1957) *International Congress on Surface Activity* II, Butterworths, London, p228.
12.   Lowell S., Shields J.E., Charalambous G. and Manzione J. (1982) *J. Colloid Interface Sci.* **86**, 191.
13.   Harris B.L. and Emmett P.H. (1949) *J. Phys. Chem.* **53**, 811.
14.   Davis R.T., DeWitt T.W. and Emmett P.H. (1947) *J. Phys. Chem.* **51**, 1232.
15.   McClellan A.L. and Harnsberger H.F. (1967) *J. Colloid Interface Sci.* **23**, 577.
16.   Livingston H.K. (1949) *J. Colloid Sci.* **4**, 450.
17.   Rouquerol F., Rouquerol J. and Sing K.S.W. (1999) *Adsorption by Powders & Porous Solids*, Academic Press, London.
18.   Galarneau A., Desplantier D., Dutartre R. and Di Renzo F. (1999) *Microporous-*

Mesoporous Mater. **27**, 297.
19. Rouquerol F., Rouquerol J., Peres C., Grillet Y. and Boudellal M. (1979) In *Characterization of Porous Solids* (Gregg S.J., Sing K.S.W. and Stoeckli H.F., eds.) The Society of Chemical Industry, Luton, UK, p107.
20. Jelinek L. and Kovats E.S. (1994) *Langmuir* **10**, 4225.
21. Dash J.G. (1975) *Films on Solid Surfaces*, Academic Press, New York.
22. Dominguez H., Allen M.P. and Evans R. (1998) *Mol. Phys.* **96**, 209.
23. Gregg S.J. and Sing K.S.W. (1982) *Adsorption, Surface Area, and Porosity*, 2nd edn, Academic Press, New York.

6 Other Surface Area Methods

6.1 INTRODUCTION

Because of its simplicity and straightforward applicability, the BET theory is almost universally employed for surface area measurements. However, other methods [e.g., 1, 2] including methods based on small angle x-ray scattering and small neutron scattering [3] have been developed. Whereas the scattering methods cannot be used in routine operations (to date at least), immersion calorimetry and in particular permeability measurements are more frequently used in various applications. We will discuss below some aspects of the latter two methods together with the so-called Harkins Jura relative method, which is based on gas adsorption and is applied in a relative pressure range P/P_0, which is similar as in case of the BET theory (i.e., 0.05 – 0.3). However, no attempt is made to derive and discuss these alternate methods completely, but rather to present their essential features and to indicate how they may be used to calculate surface areas.

6.2 GAS ADSORPTION: HARKINS AND JURA RELATIVE METHOD [4]

When a thin film of fatty acid is spread on the surface of water, the surface tension of the water is reduced from γ_0 to γ. A barrier placed between pure water and water with a surface film will experience a pressure difference resulting from the tendency of the film to spread. This 'surface pressure', π, is given by

$$\pi = \gamma_0 - \gamma \qquad (6.1)$$

Langmuir [5], in 1917, constructed the 'film balance' for the measurement of the 'surface' or 'spreading' pressure. Thus, it became possible to experimentally observe that adsorbed films pass through several states of molecular arrangement [6]. The various states resemble that of a two-dimensional gas, a low-density liquid, and finally a higher density or condensed liquid state. In the latter case, the spreading pressure can be described by the linear relationship,

$$\pi = \alpha - \beta A_x \qquad (6.2)$$

where A_x is the effective cross-sectional area of the adsorbed molecules and

α and β are constants related to the film's compressibility.

A fundamental equation derived by Gibbs [7] is used to calculate the spreading pressure of films on solids where, unlike films on liquids, it cannot be determined experimentally. Guggenheim and Adam [8] reduced Gibbs' general adsorption equation to equation (6.3) for the special case of gas adsorption.

$$d\pi = \frac{RTW}{MS_t} d\ln P \tag{6.3}$$

The terms in equation (6.3) have previously been defined as W = weight of adsorbate, $\overline{M}$ = adsorbate molecular weight, S_t = solid surface area and P = equilibrium pressure.

Differentiating equation (6.2) and substituting for $d\pi$ in equation (6.3) yields

$$-\beta dA_x = \frac{RTW}{\overline{MS_t}} d\ln P \tag{6.4}$$

The implicit assumption made in deriving equation (6.4) is that the behavior of an adsorbed gas on a solid surface is similar to that of a thin film of fatty acid on the surface of water. Rewriting equation (4.13) as

$$A_x = \frac{S_t \overline{M}}{W\overline{N}} \tag{cf.4.13}$$

and differentiating A_x with respect to W yields

$$dA_x = -\left(\frac{S_t \overline{M}}{\overline{N}}\right)\frac{dW}{W^2} \tag{6.5}$$

Replacing dA_x in equation (6.4) by equation (6.5) and rearranging terms gives

$$d\ln P = \left(\frac{\beta \overline{M}^2 S_t^2}{RT\overline{N}}\right)\frac{dW}{W^3} \tag{6.6}$$

Integrating equation (6.6) and subtracting $\ln P_0$ from each side yields

$$\ln\left(\frac{P}{P_0}\right) = -\left(\frac{\beta \overline{M}^2 S_t^{\,2}}{2RT\overline{N}}\right)\frac{1}{W^2} + \text{const} - \ln P_0 \tag{6.7}$$

or

$$\ln\left(\frac{P}{P_0}\right) = A - \frac{B}{W^2} \tag{6.7a}$$

According to equation (6.7a), the Harkins-Jura equation, a plot of $\ln(P/P_0)$ versus $1/W^2$ should give a straight line with a slope equal to $-B$ and an intercept equal to A. The surface area is then calculated as

$$S_t = \frac{1}{\overline{M}}\sqrt{\frac{2\,\overline{N}RT}{\beta}}\sqrt{B} \tag{6.8}$$

or

$$S_t = K\sqrt{B} \tag{6.8a}$$

where

$$K = \frac{1}{\overline{M}}\sqrt{\frac{2\,\overline{N}RT}{\beta}} \tag{6.9}$$

The term K in equation (6.9) is the Harkins-Jura (HJ) constant and is assumed to be independent of the adsorbent and dependent only on the adsorbate. In some instances, Harkins and Jura found two or more linear regions of different slopes when $\ln(P/P_0)$ is plotted versus $1/W^2$. This indicates the existence of two or more liquid condensed states in which different molecular packing occurs. When this situation appears, the slope that gives best agreement with an alternate method, for example, must be chosen. Alternatively, the temperature or the adsorbate can be changed to eliminate the ambiguity.

    Emmett [9] has shown that linear HJ plots are obtained for relative pressures between 0.05 and about 0.30, the usual BET range, when the BET C value lies between 50 and 250. For C values below 10, the linear HJ region lies above relative pressures of 0.4; for $C = 1000$, the HJ plot is linear between 0.01 and 0.18. Emmett also found that as C varies between 50 and

250, the nitrogen cross-sectional area must be adjusted from 13.6Å^2 to 18.6Å^2 in order to obtain the same surface area as the HJ method using 4.06 as the value for K.

6.3. IMMERSION CALORIMETRY: HARKINS AND JURA ABSOLUTE METHOD [10]

In addition to the relative method, Harkins and Jura have also developed an absolute method for surface area measurement that is independent of the adsorption isotherm and is based entirely upon calorimetric data. Consider a system in which one gram of solid and α moles of vapor are transformed into a state in which the solid is immersed in a liquid made by condensing the vapor. As shown in Fig. 6.1, this transformation can be accomplished along two possible paths.

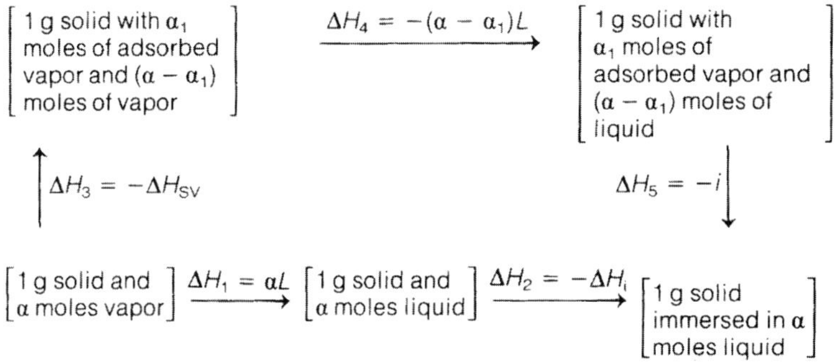

Figure 6.1 Enthalpy relationships between solid, liquid and vapor.

The terms ΔH_i, L, ΔH_{SV} and i used in Fig. 6.1 are all enthalpy changes defined as follows: ΔH_i is the heat of immersion of the solid into the liquid, L is the latent heat of condensation, ΔH_{SV} is the heat of adsorption when the solid is equilibrated with saturated vapor, and i is the heat liberated when solid in equilibrium with saturated vapor is immersed into liquid. Using Hess's law of heat summation

$$\Delta H_1 + \Delta H_2 = \Delta H_3 + \Delta H_4 + \Delta H_5 \qquad (6.10)$$

then

$$\Delta H_i = \left(\Delta H_{SV} - \alpha_1 L\right) + i \qquad (6.11)$$

The quantity $(\Delta H_{SV} - \alpha_1 L)$ is the integral heat of adsorption. This value, as well as the value of i, can be measured using calorimetry. The value of i is actually zero if the isotherm approaches the ordinate asymptotically. If the isotherm cuts the ordinate at a finite angle, i will be finite but small.

For powder samples that differ only in their surface area, the heat of immersion will be proportional to the surface area. Thus,

$$H_i = h_i S \qquad\qquad (6.12)$$

where h_i is the heat of immersion per unit area of solid. Gregg and Sing [11] for instance have tabulated values of h_i for various solids and liquids.

Please note that serious errors can be encountered if close attention is not paid to the liquid purity since the measured heat may reflect a quantity of strongly adsorbed impurities. With regard to the solid itself, even identical chemical analyses will not guarantee that two samples will have the same h_i value. The same material, prepared differently or with different histories, can posses varying amounts of lattice strain, which will produce meaningful differences in h_i values. Samples with differing pore sizes but otherwise identical will also exhibit varying h_i values due to the variation in the potential fields within the pores. Ideally therefore, it is desirable to work with carefully annealed nonporous materials when using the heat of immersion method. These types of samples are rarely encountered and their low surface areas will yield small heat values, making the experimental determination of their surface areas a tedious and difficult procedure.

These difficulties indicate clearly that the immersion method is of questionable value as a rapid or routine method for surface area measurements. However it should be noted, that in general immersion calorimetry appears in principle to be useful to assess the surface area of solids [e.g., 12, 14]. Denoyel *et al.* could show that there exist a direct relation between the energy of immersion and the total area of the microporous material [12]. It could be shown that even for microporous materials the enthalpy of immersion is proportional to the extent of the surface area including the walls of the micropores. The method could be successfully applied to activated carbon materials [12,15].

6.4. PERMEAMETRY

In a permeametry apparatus a known amount of air is drawn through a compacted bed of powder. The resistance to the flow of the air is a function of the surface area. It should be noted, that different types of gas permeability apparatus were developed, which differ only in some detail (see [2] chapter 3 and references therein).

One of these methods, the so-called 'Blaine test', and other flow test methods are used to characterize, in particular, building materials and are

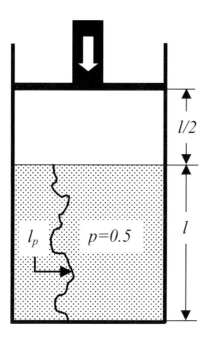

Figure 6.2 Piston displacement forcing gas to flow along a tortuous path through a packed powder bed.

standardized in many countries [e.g., 2]. In the following section we discuss some principles of using permeametry to measure the surface area of solids.

Consider a gas, near ambient pressure and temperature, forced by a small pressure gradient to flow through the channels of a packed bed of powder. At room temperature, a gas molecule can be adsorbed on a solid surface for an extremely short time but not less than the time required for one vibrational cycle or about 10^{-13} seconds. When the adsorbed molecule leaves the surface it will, on the average, have a zero velocity component in the direction of flow. After undergoing one or several gas phase collisions, it will soon acquire a drift velocity equal to the linear flow velocity. These collisions and corresponding momentum exchanges will occur within one or, at most a few mean free path lengths from the surface with the net effect of decreasing the linear velocity of those molecules flowing near the surface. As the area to volume ratio of the channels in the powder bed increases, the viscous drag effect will also increase and the rate of flow, for a given pressure gradient, will decrease. Clearly then, the surface area of the particles that constitute the channel walls is related to the gas flow rate and the pressure gradient. This type of flow is called viscous flow and is described by Poiseuille's law [16].

When the mean free path is approximately the same as the channel diameter, as with coarse powders at reduced gas pressures or fine powders with the gas at atmospheric pressures, the gas will behave as though there were slippage at the channel walls. This occurs because collisions between molecules rebounding from a wall and the flowing molecules occur uniformly across the diameter of the channel. Therefore, there appears to be no preferential retardation of flow near the channel wall when compared to the center of the channel.

A third type of flow occurs at significantly reduced pressures where the mean free path of the gas molecules is greater than the channel diameter. Viscosity plays no part in this type of flow since the molecular collisions with the channel walls far outnumber the gas phase collisions. This type of molecular flow is diffusion process.

Depending on which of the preceding three types of flow is employed, somewhat different measurements of the surface area of a sample are obtained. Viscous flow measurements tend to ignore blind pores and produce only the 'envelope' area of the particles. At the other extreme, diffusional flow senses the blind pores and often gives good agreement with surface areas measured by the BET method. At intermediate pressures, with so-called slip flow, the very small blind pores are ignored while the larger ones can contribute to the measured surface area.

Darcy's law [17] asserts that the average flow velocity, v, of a fluid through a packed bed is proportional to the pressure gradient, ΔP, across the bed and is inversely proportional to the length, ℓ, of the bed. Thus,

$$v = K\frac{\Delta P}{\ell} \tag{6.13}$$

Poiseuille [16] has shown that the viscosity, η, of a fluid can be expressed in terms of the volume V (cm^3) flowing through a tube of radius r (cm) and length ℓ (cm) in time t (s) under a pressure gradient of ΔP dyne/cm^2, that is,

$$\eta = \frac{\pi r^4 \Delta P}{8V\ell} \tag{6.14}$$

Equation (6.14) can be rearranged, by substituting $\pi r^2 \ell$ for V and v for ℓ/t

$$v = \frac{D^2 \Delta P}{32\eta\ell} \tag{6.15}$$

where v is the linear flow velocity and D is the tube diameter. The similarity in form of equations (6.13) and (6.15) suggests that flow through a packed powder bed is equivalent to fluid through many small capillaries or channels. Kozeny [18] recognized this equivalency when he derived an equation for flow through a packed bed. Kozeny used the ratio of volume V to the area A of a tube, *viz.*,

$$\frac{V}{A} = \frac{D}{4} \qquad (6.16)$$

If many parallel capillaries of equal length and diameters are used to carry the fluid, the ratio of V to A remains constant and equation (6.16) becomes

$$D_e = \frac{4V}{A} \qquad (6.17)$$

where D_e is the diameter equivalent to one large tube.

In a powder bed, the void volume V_v is defined as the volume not occupied by solids. Further, the porosity p is defined by

$$p = \frac{V_v}{V_v + V_s} \qquad (6.18)$$

where $V_v + V_s$ is the total volume of the powder bed and V_v is the volume occupied by the solid. Rearranging equation (6.18) gives

$$V_v = \left(\frac{p}{1-p} \right) V_s \qquad (6.19)$$

Substituting V_v from equation (6.19) into equation (6.17) gives

$$D_e = 4 \left(\frac{p}{1-p} \right) \frac{V_s}{A} \qquad (6.20)$$

and replacing D in equation (6.15) with D_e yields

$$V_p = \left(\frac{V_s^2 \Delta P}{2 A^2 \eta \ell_p} \right) \left(\frac{p}{1-p} \right)^2 \qquad (6.21)$$

The term V_p is the average flow velocity through the channels in the bed and ℓ_p is the average channel length.

Because the approach velocity v, that is, the gas velocity prior to entering the powder bed is experimentally simple to measure, it is desirable to state equation (6.21) in terms of v rather than v_p. To make this conversion, consider Fig. 6.2. The unshaded area represents gas contained between a piston and the surface of the powder bed. If the porosity p of the bed is 0.5, then in order to contain an equal volume of gas, the powder bed must be twice as long as the distance the piston can move. Clearly then, if the piston is displaced downward at a uniform rate the linear velocity through the bed must be twice the approach velocity, or

$$v_p = \frac{v}{p} \tag{6.22}$$

However, equation (6.22) assumes that the channels in the powder bed are straight through. In fact they are tortuous in shape, as indicated in the diagram. Actually, in the time required to travel through the powder bed an element of gas volume traverses a distance ℓ_p/ℓ greater than the length of the powder bed. This increases v_p and equation (6.22) is corrected to

$$v_p = \frac{v}{p}\left(\frac{\ell_p}{\ell}\right) \tag{6.23}$$

Replacing v_p in equation (6.21) with v from equation (6.23) gives

$$v = \left(\frac{\ell}{\ell_p}\right)\left(\frac{V_s^2 \Delta P}{2A^2 \eta \ell_p}\right)\left[\frac{p^3}{(1-p)^2}\right] \tag{6.24}$$

and

$$A^2 = \frac{1}{2}\left(\frac{\ell}{\ell_p}\right)^2\left(\frac{V_s^2 \Delta P}{\eta v \ell}\right)\left[\frac{p^3}{(1-p)^2}\right] \tag{6.25}$$

Further, if V_s^2 is replaced by the ratio of mass m to density ρ, equation (6.25) becomes

$$\left(\frac{A}{m}\right)^2 = \frac{1}{2}\left(\frac{\ell}{\ell_p}\right)^2\left(\frac{\Delta P}{\rho^2 \eta v \ell}\right)\left[\frac{p^3}{(1-p)^2}\right] \tag{6.26}$$

The area A represents the total surface area of all channels in the powder bed that offer resistance to flow. This excludes the surface area generated by close-ended or blind pores and crevices since they make no contribution to the frictional area. It follows then that equation (6.26) will give surface area values always less than BET or other adsorption techniques.

Recognizing that the ratio A/m in equation (6.26) is the specific surface area and calling $2(\ell_p/\ell)^2$ the 'aspect factor' f, which is determined by the particle's geometry, equation (6.26) can be written as

$$S^2 = \left(\frac{\Delta P}{\ell}\right)\left[\frac{p^3}{(1-p)^2}\right]\left(\frac{1}{f\rho^2\eta v}\right)$$

(6.27)

Equation (7.27) will hold for incompressible fluids and for compressible fluids with small values of ΔP. If the pressure gradient across the bed is large and the fluid is compressible, equation (6.27) takes the form

$$S^2 = \left(\frac{P_2^2 - P_1^2}{2P_1}\right)\left[\frac{p^3}{(1-p)^2}\right]\left(\frac{1}{f\rho^2\ell\eta v}\right)$$

(6.28)

where P_1 and P_2 are the exit and inlet pressures, respectively.

Experimentally the terms in equations (6.27) or (6.28) can be determined as follows:

1. ΔP can be measured by placing a sensitive pressure gauge or manometer at the inlet to the powder bed while venting to atmospheric pressure.
2. ρ, the true powder density, can be determined by a variety of methods, one of which in particular (gas pycnometry) is discussed in Chapter 19.
3. v can be determined by dividing the volumetric flow rate by the cross-sectional area of the powder bed.
4. p is calculated by subtracting the volume of powder from the total bed volume and dividing by the total bed volume.

The aspect factor f, equal to $2(\ell_p/\ell)^2$, arises from the size and shape of the cross-sectional areas that make up the channels. Therefore, f is highly dependent upon the particle size and shape. Carman [19] studied numerous materials and found a value of 5 was suitable in most cases. Various theoretical and experimental values for f usually are found to be in the range of 3 to 6.

Permeation measurements will be prone to serious errors if the average channel diameter is not within a factor of two of the largest diameter.

Large diameter channels tend to give area to volume ratios excessively low (compare equation (6.17)), while their contribution to the average flow rate is in excess of their number. Also, agglomerated particles behave as one particle, the flow measuring only the envelope of the aggregate.

Several modifications of Poiseuille's equation have been attempted by various authors [20-22] to describe permeability in the transitional region between viscous and diffusional flow. The assumptions underlying these modifications are often questionable and the results obtained offer little or no theoretical or experimental advantage over the BET theory for surface area measurements. Allen [23] discusses these modifications as well as diffusional flow at low pressures.

Viscous flow permeametry measured near atmospheric pressure offers the advantages of experimental simplicity and a means of measuring the external or envelope area of a powder sample, which is otherwise not readily available by any adsorption method. The usefulness of measuring the external surface area rather than the BET or total surface area becomes evident if the data is to be correlated with fluid flow through a powder bed or with the average particle size.

6.4 REFERENCES

1. Rouquerol J., Avnir D., Fairbridge C.W., Everett D.H., Haynes J.H., Pernicone N., Ramsay J.D., Sing K.S.W. and Unger K.K. (1994) *Pure Appl. Chem.* **66**, 1739.
2. Mikhail R.Sh. and Robens E.Z. (1983) *Microstructure and Thermal Analysis of Solid Surfaces*, Wiley, Chichester.
3. Smarsly B., Goeltner C., Antonietti M., Ruland W. and Hoinkis E. (2001) *J. Phys. Chem B.* **105**, 831.
4. Harkins W.D. and Jura G. (1944) *J. Am. Chem. Soc.* **66**, 919.
5. Langmuir I. (1917) *J. Am. Chem. Soc.* **39**, 1848.
6. Adamson A.W. (1982) *Physical Chemistry of Surfaces*, 4th edn, Wiley Interscience, New York, chapter 3.
7. Gibbs J.W. (1931) *The Collected Works of J. W. Gibbs*, Vol. 1, Longmans Green, New York.
8. Guggenheim E.A. and Adam N.K. (1936) *Proc. Roy. Soc. London* **154A**, 608.
9. Emmett P.H. (1946) *J. Am. Chem. Soc.* **68**, 1784.
10. Harkins W.D. and Jura G. (1944) *J. Am. Chem. Soc.* **66**, 1362.
11. Gregg S.J. and Sing K.S.W. (1967) *Adsorption, Surface Area and Porosity*, Academic Press, New York, p301.
12. Denoyel R., Fernandez-Colinas J., Grillet Y. and Rouquerol J. (1993) *Langmuir* **9**, 515.
13. [a]Zettlemoyer A.C. and Chessick J. (1959) *Adv. Catal.* **11**, 263; [b]Zettlemoyer A.C., Young G.J., Chessick J.J. and Healey F.H. (1953) *J. Phys. Chem.* **57**, 649.
14. Rouquerol F., Rouquerol J. and Sing K.S.W. (1999) *Adsorption by Powders & Porous Solids*, Academic Press, London.
15. Gonzales M.T., Sepulveda-Escribano A., Gonzales M.J. and Rodriguez-Reinoso F.

(1995) *Langmuir* **11**, 2354.

16. Poiseuille J.L.M. (1846) *Inst. France Acad. Sci.* **9**, 433.
17. Darcy H.P.G. (1856) *Les Fontaines Publiques de la Ville de Dijon*, Victor Dalmont.
18. Kozeny J. (1927) *Ber. Wien Akad.* **136A**, 271.
19. Carman P.C. (1938) *J. Soc. Chem. Ind. London, Trans. Commun.* **57**, 225.
20. Ridgen P.J. (1954) *Road Res. Pap.* No. 28 (NMSO).
21. Lea F.M. and Nurse R.W. (1947) *Trans. Inst. Chem. Eng.* **25**, 47.
22. Carman P.C. (1956) *Flow of Gases Through Porous Media*, Butterworths, London.
23. Allen T. (1990) *Particle Size Measurement*, 4th edn, Chapman and Hall, London, p615.

7 Evaluation of Fractal Dimension by Gas Adsorption

7.1 INTRODUCTION

The concepts of fractal geometry elaborated by Mandelbrot [1] can be applied successfully to the study of solid surfaces. Fractal objects are self-similar, i.e., they look similar at all levels of magnification. The geometric topography (roughness) of the surface structure of many solids can be characterized by the fractal dimension, D. In the case of a Euclidean surface D is 2, however for an irregular (real) surface D may vary between 2 and 3. The magnitude of D may depend on the degree of roughness of the surface and/or the porosity. There exist several experimental methods to determine the fractal dimension, e.g., small-angle X-ray (SAXS) and small-angle neutron scattering measurements (SANS), adsorption techniques and mercury porosimetry. All these techniques search for a simple scaling power law of the type: *Amount of surface property* $\propto$ *resolution of analysis* ^{D} [2], where D is the fractal dimension of the surface for which the property is relevant. *Amount of surface property* can for instance be related to the intensity of scattered radiation, pore volume or monolayer capacity. The change in resolution is here achieved by changing the scattering angle, pore radius or the size of the adsorbate.

Different techniques and calculating procedures have meanwhile been developed to obtain the fractal dimension from gas sorption and mercury porosimetry data. The determination of the surface roughness in the scale range of molecular sizes (usually less than 1 nm) can be obtained by means of the method of molecular tiling, which was introduced by Pfeiffer and Avnir [3, 4]. The scale range from 1 nm to 100 nm can be investigated by means of the modified Frenkel-Halsey Hill method [5] and a thermodynamic method (the so-called Neimark-Kiselev method) [6].

7.2 METHOD OF MOLECULAR TILING

The basis of this method involves a comparison of the monolayer capacities of different adsorbents. Pfeifer and Avnir [4] have shown that if a set of adsorbents with different sizes is considered, the monolayer capacity N_m, which is a function of the cross-sectional area σ of the adsorptive, satisfies the equation:

$$N_m = k\sigma^{-D/2} \tag{7.1}$$

where k is a constant and D is the fractal dimension of the surface accessible for adsorption.

Equation (7.1) is based on the idea that the number of molecules required to cover the surface with a single layer depends on the size of the adsorptive. Whereas small molecules can be adsorbed on small irregularities of the surface, large molecules may be too big to detect small surface defects. Therefore, the amount of adsorptive necessary to form a monolayer depends on the size of the molecule, which acts as a kind of *molecular yardstick*, and on the fractal dimension of the surface. Hence, by changing the *yardstick*, one can determine the fractal dimension of the surface. Consequently one needs to perform adsorption experiments involving different adsorptives, i.e., gas molecules of different size.

In order to solve equation (7.1) the monolayer capacity has to be determined. This can be achieved by applying the BET equation to the adsorption data. Because the method of molecular tiling is based on the determination of the monolayer capacity, the range of detectable scales available with this method is that of molecular level usually between 0.4 nm and 1 nm. In order to look at larger scales, the FHH-method or the thermodynamic method have to be employed.

7.3 THE FRENKEL-HALSEY-HILL METHOD

The Frenkel-Halsey-Hill (FHH) theory describes the range of multilayer adsorption on a homogeneous planar surface. The classical FHH equation [7] is given by:

$$N \propto \left[\ln\left(P_0/P\right)\right]^{-1/s} \tag{7.2}$$

where N is the adsorbed amount at the relative pressure P/P_0, and s is related to the adsorbate/substrate and adsorbate/adsorbate potential. For non-retarded van der Waals' interactions theory predicts that $s = 3$. However, experimentally determined s values are usually smaller than 3 (see discussion of FHH in §4.3) and it was assumed that the observed deviations from the theoretical value are caused by the roughness of a "real" adsorbent surface.

Accordingly, Pfeifer *et al.* [5] generalized the classical FHH equation to fractal surfaces and postulated that the FHH exponent s is related to the fractal dimension D:

$$N \propto \left[ln\left(P_0/P\right)\right]^{-1/m} \tag{7.3}$$

with m = $s/3$ – D. This fractal FHH equation can be applied only at early stages of multilayer formation, where the substrate potential is dominating. In the pore condensation regime, liquid-gas surface tension forces predominate and the relationship between D and s is given by

$$N \propto \left[\ln\left(P_0/P\right)\right]^{D-3} \tag{7.4}$$

A plot of $\ln N$ vs. $\ln\ln(P_0/P)$ should yield a straight line with negative slope s within the multilayer region of the isotherm in both cases.

7.4 THE THERMODYNAMIC METHOD

A thermodynamic method (the Neimark-Kiselev (NK) method) has been proposed by Neimark [6] for calculating the surface fractal dimension from the adsorption or desorption isotherm in the region of capillary condensation. Basic assumptions are that the role of scale or gauge is here not played by the molecular size of a molecule, but by the mean radius of curvature a_c of the meniscus at the interface between condensed adsorbate and gas as given by the Kelvin equation,

$$a_c = \frac{2\gamma V_m}{RT \ln\left(P_0/P\right)} \tag{7.5}$$

where V_m is the adsorbate molar volume, γ is the surface tension, R is the universal gas constant and T is the adsorption temperature. The adsorbate-vapour interfacial area, S_{lg}, can be calculated using an integral thermodynamic relationship known as the Kiselev equation, which balances the work of adsorption and the work involved in the formation of the interface [6]:

$$S_{lg} = \frac{RT}{\gamma} \int_n^{n_{max}} \ln\left(P_0/P\right) dn \tag{7.6}$$

where n and n_{max} are the amounts of gas adsorbed at a given P/P_0. Based on equation (7.6) one can interpret the surface area of the adsorbed film as that of the adsorbent that would be measured by spheres with radius a_c. For a fractal surface, S_{lg} is related to the radius of curvature according to the following equation:

$$S_{lg} = Ka_c^{2-D} \tag{7.7}$$

where D is again the fractal dimension and K is a constant.

According to equation (7.7) a plot of $\ln S_{lg}$ vs. $\ln a_c$ should yield a straight line from which the fractal dimension D can be readily calculated. Please note, that the thermodynamic method can be in principle also applied to the intrusion of nonwetting fluids, i.e., mercury porosimetry.

7.5 COMMENTS ABOUT FRACTAL DIMENSIONS OBTAINED FROM GAS ADSORPTION

The method of molecular tiling (*molecular yardstick method*) was applied for a number of materials to assess their surface fractal dimension [4]. As indicated in section 7.2 the successful application of this method depends very much on the accuracy of the determination of the monolayer capacity. Xu et al [8] used the method of molecular tiling successfully to determine the surface fractal dimension of various graphitized carbon blacks. Gases of different molecular sizes including nitrogen, argon, ethane, propane and butane were used as probe molecules. Studies performed with methods such as Atomic Force Microscopy (AFM) revealed that these materials consist of a very homogeneous, smooth surface and in agreement with this expectation the molecular yardstick method gave a fractal dimension of 2.0. The fractal FHH approach was also applied, but failed to determine the fractal dimension correctly.

One problem of the FHH-fractal approach is, that deviations of the exponent s from the theoretical value of 3 cannot be entirely attributed to the roughness or the fractal nature of the surface. The fractal dimension obtained depends very much on the FHH-exponent s for a given non-porous system and factors associated with the details of adsorptive-adsorbent interactions. The relative pressure range over which the FHH method was applied [7,9] also affects the value of the FHH exponent s (see also chapter 4.3 where the FHH method is extensively discussed). Of course, also the occurrence of capillary condensation in mesoporous materials (or observed as interparticle condensation in very fine powder) restricts the range over which the FHH theory can be applied. An advantage of the fractal FHH method is, that it allows one to determine the fractal dimension from a single adsorption isotherm, whereas the method of molecular tiling requires adsorption data for more than one adsorbate. Among others [10-12], Krim et al. [13] tested the applicability of the fractal FHH equation and could successfully determine the surface fractal dimension of smooth and rough silver substrates from nitrogen and oxygen adsorption isotherms obtained at 77 K. However, their experiments as well as the work performed by Sahouli et al. [14] revealed again the problems associated with the use of the fractal

FHH equation, i.e., the fractal dimension depends very much on the FHH-exponent *s* for a given non-porous system.

The fractal FHH method can be considered as a particular case of the more general thermodynamic method. Neimark's thermodynamic method has been applied for different materials including silica gels, porous glasses, apatite, and coal [15]. It was found that the roughness of the internal surface of mesoporous materials could be characterized by using the fractal dimension up to the characteristic size of pore channels. Hence, in this case the fractality reflects the roughness of the pore walls, but not the geometry of pore network. In this way a surface fractal dimension 2.20 – 2.22 was found for a mesoporous silica gel (LiChrospher Si 300) of mean pore size 30 nm. Similar results were obtained for a controlled-pore glass (CPG 240) of mean pore diameter 24 nm (the surface fractal dimension was determined to lie around 2.10 - 2.14).

Rudzinki *et al.* [16] developed a novel theoretical approach, which apparently overcomes some of the problems associated with the application of the fractal FHH equation. This approach correlates the geometric heterogeneity with energetic surface heterogeneities, i.e., they found that the differential pore size distribution reduces to a fractal pore size distribution in the limit of very small pores or when the fractal dimension approaches 3. In addition, they put forward a generalized fractal BET equation based on preliminary work of Fripiat *et al* [17], which can be applied even in the multilayer region. This new approach was applied successfully on silica and activated carbon, but more tests have to be performed before more general conclusions about its applicability can be drawn.

Lazlo *et al.* [18] determined fractal dimensions of carbonaceous compo- site materials by applying small angle x-ray scattering and sorption measurements. Differences in the fractal dimensions obtained from these different methods were attributed to the fact that X-ray diffraction reveals both the closed and open pores, which may not be available for a specific adsorbate as well. Weidler *et al.* [19] used fractal analysis by gas sorption and SAXS in order to study the surface roughness created by acidic dissolution of synthetic goethite. Reasonable agreement between the fractal dimensions obtained by these two methods (the FHH method was used to analyze the gas adsorption data) was observed.

Wong and co-workers [20] compared in systematic experiments the fractal dimension D for three shale samples obtained by small angle neutron scattering (SANS) with the values obtained from the analysis of gas adsorption isotherms using the fractal FHH equation. They found that the D-values obtained from the FHH approach were always significantly lower (ca. 15 –20 %) as compared to the SANS data. Wong *et al.* concluded that these differences were due to the above-mentioned problems of the fractal FHH approach and that the occurrence of a crossover between multilayer adsorption and capillary condensation precludes the use of gas adsorption isotherms to determine the fractal dimension. However, it should be noted that the shale samples used in their study are not well defined with regard to pore geometry, pore size, and pore size distribution. In contrast, mesoporous molecular sieves such as MCM-41 silica materials are well defined with regard to their surface and pore structural properties.

MCM-41 samples were used by Bhatia and co-workers [21] in order to perform systematic studies of the roughness of well-defined pores in MCM-41 silica by using methods of fractal analysis. Their paper indicated clearly that different techniques like gas sorption, scattering methods, and mercury porosimetry might probe different length scales of roughness. This has to be taken into account when the results obtained with different techniques are compared. MCM-41 comprises several levels of structure, i.e., that of mesopores, crystallites, grains and particles – spanning four decades of resolution, each having independent surface properties at characteristic length scales. These different length scales were tested each with an appropriate experimental technique. Length scales in the range of molecular resolutions (0.3 - 0.7 nm) were tested by gas sorption using the method of molecular tiling. A fractal dimension of 2 was found indicating that MCM-41 silica is smooth at molecular resolution. The length scale range of 2 – 5 nm (corresponding to the mesopore channels) was probed again by gas adsorption. The gas adsorption data were analyzed using all three methods (molecular tiling, FHH-method thermodynamic method) and compared to the results obtained with a new method developed by Bhatia *et al.*, which incorporates the effects of attractive van der Waals' forces on the adsorbate film and meniscus curvature in an improved way. It was found that the results of this new method agree best with Neimark's thermodynamic method and it was concluded that the surface has a fractal dimension of 2.4- 2.6. Small angle scattering methods (SANS and SAXS) were used to study length scales associated with the crystallite size (8 – 25 nm) and the corresponding fractal dimension was found to lie between 2 and 3. Mercury porosimetry was then used to study roughness at very low resolution, i.e., corresponding to the grain size (0.1 – 0.4 μm) and a fractal dimension of about 3 was found.

Summarizing, it can be stated that the fractal dimension obtained by gas adsorption should be considered as a useful and characteristic parameter, which complements the surface and pore size characterization. However, the interpretation of these empirical fractal dimension parameters in the context of fractal self-similarity (and/or self-affinity) is not always justified. If one compares results for fractal dimensions obtained with different experimental techniques, one has to make re that the techniques under comparison have probed the same range of length scales.

7.6 REFERENCES

1. Mandelbrot B. (1982) *Fractal Geometry of Nature*, Freeman, San Francisco.
2. Rouquerol J., Avnir D., Fairbridge C.W., Everett D.H., Haynes J.H., Pernicone N., Ramsay J.D.F., Sing K.S.W. and Unger K.K. (1994) *Pure Appl. Chem.* **66**, 1740.
3. Farin D. and Avnir D. (1989) In *The Fractal Approach to Heterogeneous Chemistry* (D. Avnir, ed) Wiley, Chichester.
4. [a]Pfeifer P. and Avnir D. (1983) *J. Chem. Phys.* **79**, 3558; [b]Avnir D., Farin D. and Pfeifer P. (1984) *Nature* **33**, 261.
5. Pfeifer P., Kenntner J. and Cole M.W. (1991) In: *Fundamentals of Adsorption* (A.B. Mersmann and S.E. Sholl, eds.) Engineering Foundation, New York.
6. [a]Neimark A.V. (1991) *Adv. Sci. Tech.* **7**, 210; [b]Neimark A.V. (1992) *Physica A* **191**, 258; [c]Neimark A.V. (1993) In *Multifunctional Mesoporous Inorganic Solids*, (C.A.C. Sequeria and M.J. Hudson, eds) Kluwer, Dordrecht.
7. Gregg S.J. and Sing K.S.W. (1982) *Adsorption, Surface Area and Porosity*, Academic Press, London.
8. Xu W., Zerda T.W., Yang H. and Gerspacher M. (1996) *Carbon* **34**, 165.
9. Rudzinski W. and Everett D.H. (1992) *Adsorption of Gases on Heterogeneous Surfaces*, Academic Press, London.
10. Avnir D. and Jaroniec M. (1989) *Langmuir* **5**, 1431.
11. Yin Y. (1991) *Langmuir* **7**, 216.
12. Pfeifer P., Wu Y.J., Cole M.W. and Krim J. (1989) *Phys. Rev. Lett.* **62**, 1997.
13. Krim J. and Panella V. (1991) In: *Characterization of Porous Solids II* (Rodriguez-Reinoso F., Rouquerol J., Sing K.S.W. and Unger K.K., eds) Elsevier, Amsterdam.
14. [a]Sahouli B., Blacher S. and Brouers F. (1996) *Langmuir* **12**, 2872; [b]Sahouli B., Blacher S. and Brouers F. (1997) *Langmuir* **13**, 94.
15. Neimark A.V. and Unger K.K. (1993) *J. Colloid Interface Sci.* **158**, 412.
16. [a]Rudzinski W., Lee S.-L., Panczyk T. and Yan C.-C. S. (2001) *J.Phys. Chem B* **105**, 10847; [b]Rudzinski W., Lee S.-L., Panczyk T. and Yan C.-C. S. (2001) *J. Phys. Chem B* **105**, 10857.
17. Fripiat J.J., Gatineau L. and Van Damme H. (1986) *Langmuir* **2**, 562.
18. Laszlo K., Bota A., Nagy L.G., Subklew G. and Schwuger M. (1998) *J. Colloids Surf. A: Physicochemical and Engineering Aspects* **138**, 29.
19. Weidler P., Degovics G. and Laggner P. (1998) *J. Colloid Interface Sci.* **197**, 1.
20. [a]Qi H., Ma. J. and Wong P. (2001) *Phys. Rev. E* **63**, 41601; [b]Qi H., Ma J. and Wong P. (2002) *Colloids Surf. A: Physicochemical and Engineering Aspects* **206**, 401.
21. Sonwane C.G., Bhatia S.K. and Calos N.J. (1999) *Langmuir* **15**, 4603.

8 Mesopore Analysis

8.1 INTRODUCTION

As discussed in chapter 4, the state and thermodynamic stability of pure fluids in mesopores depends on the interplay between the strength of fluid-wall and fluid-fluid interactions on the one hand, and the effects of confined pore space on the other hand. The most prominent phenomenon observed in mesopores is pore condensation, which represents a first-order phase transition from a gas-like state to a liquid-like state of the pore fluid occurring at a pressure P less than the corresponding saturation pressure P_0 of the bulk fluid, i.e., pore condensation occurs at a chemical potential μ less than the value μ_0 at gas-liquid coexistence of the bulk fluid. The relative pressure where this condensation occurs depends on the pore diameter. The relationship between the pore size and the relative pressure where capillary condensation occurs can be described by the classical Kelvin equation. However, in the classical Kelvin equation the shift from bulk coexistence $(\mu_0 - \mu)$, is expressed in terms of macroscopic quantities, whereas a more comprehensive understanding of the underlying physics was achieved only recently by applying microscopic approaches based on the Density Functional Theory (DFT), and computer simulation studies (Monte Carlo and Molecular Dynamics). We have discussed these different approaches from a more theoretical point of view in chapter 4. Here, we will discuss their significance for the pore size analysis of mesoporous materials.

8.2 METHODS BASED ON THE KELVIN EQUATION

Adsorption studies leading to measurements of pore size and pore size distributions generally make use of the Kelvin equation [1], which was discussed in chapter 4. The Kelvin equation relates the equilibrium vapor pressure of a curved surface, such as that of a liquid in a capillary or pore (P), to the equilibrium pressure of the same liquid on a planar surface (P_0). For a cylindrical pore the Kelvin equation is given by

$$\ln P/P_0 = \frac{-2\gamma\overline{V}}{rRT} \tag{cf 4.58}$$

where γ is the surface tension of the liquid, $\overline{V}$ is the molar volume of the condensed liquid contained in a narrow pore of radius r, R is the gas constant and T is the temperature.

For nitrogen as adsorptive at its boiling temperature (~77 K), the Kelvin equation can be written as

$$r_k = \frac{4.15}{\log(P_0/P)} \quad \text{Å} \tag{8.1}$$

The term r_k indicates the radius into which condensation occurs at the required relative pressure. This radius, called the Kelvin radius or the critical radius, is not the actual pore radius since (see chapter 4.4) some adsorption has already occurred on the pore wall prior to condensation, leaving a center core of radius r_k. Conversely, an adsorbed film remains on the wall during desorption, when evaporation of the center core takes place. The incorporation of the preadsorbed film leads to the so-called modified Kelvin equation [2] (see chapter 4, §4.4, equation (4.62)).

If in a pore of given size the thickness of the adsorbed film is l_c when condensation or evaporation occurs, then the actual pore radius, r_p, is given by

$$r_p = r_k + l_c \tag{8.2}$$

for a cylindrical pore model. In the case of a slit-pore the pore width w is given as follows:

$$w = r_k + 2l_c \tag{8.3}$$

Using the assumption that the adsorbed film thickness in a pore is the same as that on a plane surface (which is true in the limit of $r_p \to \infty$) for large pores) for any value of relative pressure, usually the depth of the adsorbed multilayer film is expressed in form of a statistical thickness t:

$$t = \left(\frac{W_a}{W_m}\right)\tau \tag{8.4}$$

where W_a and W_m are, respectively, the weight adsorbed at a particular relative pressure and the weight corresponding to the BET monolayer and τ is thickness of one layer. Essentially, equation (8.4) asserts that the thickness of the adsorbed film is simply the number of layers times the thickness of one layer, τ, regardless of whether the film is in a pore or on a plane surface. The value of τ can be calculated by considering the area S and volume

$\overline{V}$ occupied by one mole of liquid nitrogen if it were spread over a surface to the depth of one molecular layer:

$$S = (16.2)(6.02 \times 10^{23}) = 97.5 \times 10^{23} \text{ Å}^2 \tag{8.5}$$

$$\overline{V} = (34.6 \times 10^{24}) \text{Å}^3 \tag{8.6}$$

Then

$$\tau = \frac{V}{S} = 3.54 \text{ Å} \tag{8.7}$$

This value of 3.54 Å is somewhat less than the diameter of a nitrogen molecule based on the cross-sectional area of 16.2 Å². This is as it should be, since the liquid structure is considered close-packed hexagonal, with each nitrogen molecule sitting in the depression between three molecules in the layers above and below. Equation (8.7) can now be written as

$$t = 3.54 \frac{W_a}{W_m} \text{ Å} \tag{8.8}$$

On nonporous surfaces, it has been shown that when W_a/W_m is plotted versus P/P_0, the data approximate a common type II curve above a relative pressure of 0.3 [2-7]. This implies that when $W_a/W_m=3$, for example, the statistical layer thickness, t, will be 10.62 Å regardless of the adsorbent. The common curve can be described for instance by the Halsey equation [8], which for nitrogen can be written as

$$t = 3.54 \left(\frac{5}{\ln(P_0/P)} \right)^{1/3} \text{ Å} \tag{8.9}$$

Other thickness equations were obtained by de Boer et al. [9], Harkins and Jura [10], Broeckhoff de Boer [11] and others. The latter equations assume an oxidic surface (e.g., silica, aluminas etc.). In the case of carbon materials thickness equations dedicated to adsorption on carbons should be used (e.g., the equation used by Kaneko et al. [12]). We will discuss some of these equations in more detail in chapter 9 (§9.21).

To calculate the pore size distribution, consider the work sheet, shown in Table 8.1, and the corresponding explanation of each column. The adsorbed volumes are from a hypothetical isotherm. The procedure used is the numerical method of Pierce [5] as modified by Orr and DallaValle [13] with regard to calculating the thickness of the adsorbed film. This method, as well as the Barrett, Joyner, and Halenda (BJH) numerical integration method [14] and the related Dollimore–Heal approach [15], takes advantage of Wheeler's theory [16] that condensation occurs in pores when a critical relative pressure is reached corresponding to the Kelvin radius, r_k. This model also assumes that a multilayer of adsorbed film, with the same depth as the adsorbed film on a nonporous surface, exists on the pore wall when evaporation or condensation occurs. Among these different approaches, the Barett-Joyner-Halenda method can be considered as the most popular method for mesopores size analysis.

The typical procedure shown in Table 8.1 uses data from either the adsorption or desorption isotherm. However, the desorption curve is usually employed except in those cases where pore blocking or percolation phenomena affect the position of the desorption branch (see chapter 4, §4.4 and chapter 8, §8.6). In either case, for ease of presentation, the data is evaluated downward from high to low pressures.

Columns 1 and 2 of Table 8.1 contain data obtained directly from the isotherm. The desorbed volumes are normalized for one gram of adsorbent. Relative pressures are chosen using small decrements at high values, where r_k is very sensitive to small changes in relative pressure, and where the slope of the isotherm is large, such that small changes in relative pressure produce large changes in volume.

Column 3, the Kelvin radius, is calculated from the Kelvin equation assuming a zero wetting angle. If nitrogen is the adsorbate, equation (8.1) can be used.

Column 4, the film thickness t, is calculated using equation (8.9), the Halsey equation.

Column 5 gives the pore radius, r_p, obtained from equation (8.2).

Columns 6 and 7, $\bar{r}_k$ and $\bar{r}_p$, are prepared by calculating the mean value in each decrement from successive entries.

Column 8, the change in film thickness, is calculated by taking the difference between successive values of t from column 4.

Column 9, ΔV_{gas}, is the change in adsorbed volume between successive P/P_0 values and is determined by subtracting successive values from column 2.

Column 10, ΔV_{liq}, is the volume of liquid corresponding to ΔV_{gas}. The most direct way to convert ΔV_{gas} to ΔV_{liq} is to calculate the moles of gas and multiply by the liquid molar volume. For nitrogen at standard temperature and pressure, this is given by

$$\Delta V_{liq} = \frac{\Delta V_{gas}}{22.4 \times 10^3} \times 34.6 = \Delta V_{gas}\left(1.54 \times 10^{-3}\right) \text{ cm}^3 \qquad (8.10)$$

Column 11 represents the volume change of the adsorbed film remaining on the walls of the pores from which the center core has previously evaporated. This volume is the product of the film area ΣS and the decrease in the film depth Δt. By assuming no pores are present larger than 950 Å ($P/P_0 = 0.99$), the first entry in column 11 is zero since there exists no film area from previously emptied pores. The error introduced by this assumption is negligible because the area produced by pores larger than 950 Å will be small compared to their volume. Subsequent entries in column 11 are calculated as the product of Δt for a decrement ΣS from the row above corresponding to the adsorbed film area exposed by evaporation of the center cores during all the previous decrements.

Column 12, the actual pore volume, is evaluated by recalling that the volume of liquid, column 10, is composed of the volume evaporated out of the center cores plus the volume desorbed from the film left on the pore walls. For a pore of length ℓ

$$\Delta V_{liq} = \pi \bar{r}_k^2 \ell + \Delta t \Sigma S \tag{8.11}$$

and, since

$$V_p = \pi \bar{r}_p^2 \ell \tag{8.12}$$

by combining the above two equations,

$$V_p = \left(\frac{\bar{r}_p}{\bar{r}_k}\right)^2 \left[\Delta V_{liq} - \left(\Delta t \Sigma S \times 10^{-4}\right)\right] \text{cm}^3 \tag{8.13}$$

Column 13 is the surface area of the pore walls calculated from the pore volume by

$$S = \frac{2V_p}{\bar{r}_p} \times 10^4 \text{ m}^2 \tag{8.14}$$

with V_p in cubic centimeters and $\bar{r}_p$ in angstroms. It is this value of S that is summed in column 14. The summation is multiplied by Δt from the following decrement to calculate the film volume decrease in column 11. If the utmost rigor were used, it would be correct to modify the area contributed by previously emptied pores, since their statistical thickness t diminishes with each successive decrement. However, this procedure would

Table 8.1 Pore size distribution worktable. In this example, $\Sigma V_p = 0.28$ cm³/g and $\Sigma S = 212.1$ m²/g.

1	2	3	4	5	6	7	8	9	10	11	12	13	14
$\dfrac{P}{P_0}$	V_{gas} STP (cm³/g)	r_k (Å)	T (Å)	r_P (Å)	$\bar r_k$ (Å)	$\bar r_p$ (Å)	Δt (Å)	ΔV_{gas} STP (cm³/g)	ΔV_{liq} ×10³ (cm³/g)	$\Delta t\Sigma S$ ×10³ (cm³/g)	V_p ×10³ (cm³/g)	S (m²/g)	ΣS (m²/g)
0.99	161.7	950	28.0	978									
					711	737	5.8	0.2	0.31	0.00	0.33	0.01	0.01
0.98	161.5	473	22.2	495									
					394	414	2.8	0.5	0.77	0.00	0.85	0.04	0.05
0.97	161.0	314	19.4	333									
					250	268	3.1	0.8	1.23	0.02	1.40	0.10	0.15
0.95	160.2	186	16.3	202									
					138	153	3.5	1.4	2.16	0.05	2.59	0.34	0.49
0.90	158.8	90.7	12.8	104									
					74.8	87.0	1.7	1.6	2.46	0.08	3.22	0.74	1.23
0.85	157.2	58.8	11.1	69.9									
					50.8	61.4	1.1	2.0	3.08	0.14	4.30	1.40	2.63
0.80	155.2	42.8	10.0	52.8									
					39.7	49.5	0.5	2.3	3.54	0.13	5.30	2.14	4.77
0.77	152.9	36.6	9.5	46.1									
					34.9	44.3	0.3	4.0	6.16	0.14	9.70	4.38	9.15
0.75	148.9	33.2	9.2	42.4									
					31.8	40.9	0.3	3.8	5.85	0.27	9.22	4.51	13.66
0.73	145.1	30.4	8.9	39.3									
					29.2	38.0	0.2	4.2	6.47	0.27	10.49	5.52	19.18
0.71	140.9	27.9	8.7	36.6									
					26.9	35.4	0.3	5.0	7.70	0.58	12.34	6.97	26.15

0.69	135.9	25.8	8.4	34.2
0.67	130.0	23.9	8.2	3 2.1
0.65	123.9	22.2	8.0	30.2
0.63	117.3	20.7	7.8	28.5
0.61	110.1	19.3	7.7	27.0
0.59	102.6	18.1	7.5	25.6
0.57	95.0	17.0	7.3	24.3
0.55	86.9	16.0	7.2	23.2
0.53	78.8	15.1	7.0	22.1
0.51	71.5	14.2	6.9	21.1
0.49	65.4	13.4	6.8	20.2
0.45	57.3	12.0	6.5	18.5
0.40	51.7	10.4	6.2	16.6
0.35	47.4	9.1	6.0	15.1

24.9	33.2	0.2	5.9	9.09	0.52	15.23	9.17	35.32
23.1	31.2	0.2	6.1	9.39	0.71	15.84	10.15	45.47
21.5	29.4	0.2	6.6	10.16	0.91	17.30	11.77	57.24
20.0	27.8	0.1	7.2	11.09	0.57	20.32	14.62	71.86
18.7	26.3	0.2	7.5	11.55	1.44	20.00	15.21	87.07
17.6	25.0	0.2	7.6	11.70	1.74	20.09	16.07	103.1
16.5	23.8	0.1	8.1	12.47	1.03	23.80	20.00	123.1
15.6	22.7	0.2	8.1	12.47	2.46	21.19	18.67	141.8
14.7	21.6	0.1	7.3	11.24	1.42	21.21	19.64	161.4
13.8	20.7	0.1	6.1	9.39	1.61	17.50	16.90	178.3
12.7	19.4	0.3	8.1	12.47	5.35	16.62	17.13	195.4
11.2	17.6	0.3	5.6	8.62	5.86	6.81	7.74	203.1
9.8	15.9	0.2	4.3	6.62	4.06	6.73	8.47	212.1

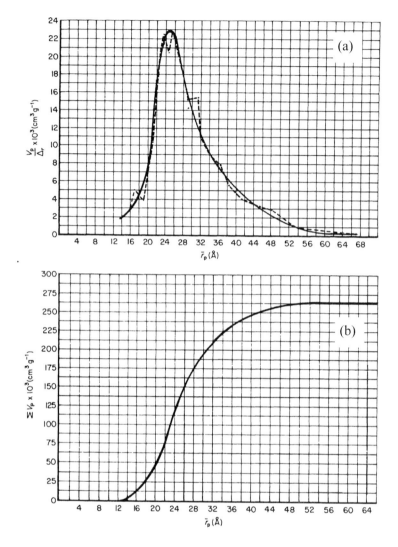

Figure 8.1 (a) Pore size distribution curve from table 8.1. Raw data: --------; smoothed data ──────; (b) Cumulative pore volume plot from table 8.1.

be cumbersome and of questionable value in view of the many other assumptions that have been made. Nevertheless, the BJH method [12] attempts to make this modification by introducing an average inner core based on its variation with each decrement of relative pressure.

Table 8.1 discloses that the volume of all pores greater than 15.1Å is 0.28 cm^3/g. This does not mean that micropores with smaller radii are absent. It means that the validity of the Kelvin equation becomes questionable because of the uncertainty regarding molar volumes and surface tension, etc. [17,18] when only a few molecular diameters are involved. In some analyses, it is found that the volume desorbed from the film ($\Delta t \Sigma S$) becomes equal to the total amount desorbed, indicating that only desorption is occurring in the narrower pores.

Please note that in the absence of evaporation from center cores, the Kelvin equation is not applicable. The volume of micropores, if present, can be evaluated by the difference between the total pore volume (see equation (8.20)) and the sum of column 12, ΣV_p. The total area of pores to 15.1 Å radius (column 14) is 212.1 m^2/g. This area is usually less than the BET area, since it does not include the surface contributed by micropores. An area larger than the BET area would be exhibited by inkbottle pores, in which a larger volume of gas is condensed in pores having a relatively small area. This is the case in the example shown in Table 8.1.

An implicit assumption, hidden in the method, is that no surface exists other than that within pores. The volume desorbed in any decrement is assumed to originate from either the center core or the adsorbed film on the pore wall.

8.3 MODELLESS PORE SIZE ANALYSIS

Wheeler's ideas [16] that condensation and evaporation occur within a center pore during adsorption and desorption and that an adsorbed film is present on the pore wall have led to proposals of various methods for pore size analysis. In addition to Pierce [17] and BJH [13] techniques, other schemes have been suggested, including those by Shull [3], Oulton [19], Innes [20] and Cranston and Inkley [21]. These ideas are all based upon some assumption regarding pore shape. Brunauer et al. [22] have developed a means of determining the pore volume distribution wherein the pore shape has a negligible influence. Kiselev [23] used the relationship between the moles of adsorptive condensed into pores which is given by

$$S = \frac{1}{\gamma} \int RT \ln\left(P/P_0\right) dn \qquad (8.15)$$

Using this equation by graphical integration of the isotherm between the limits of saturation and hysteresis loop closure, Kiselev was able to calculate surface areas for wide pore samples in good agreement with BET measured areas.

Brunauer's modelless method uses pore volume and pore area not as functions of the Kelvin radius but rather as functions of hydraulic radii that he defines as

$$r_h = \frac{V}{S} \qquad (8.16)$$

where V and S are the pore volume and surface area, respectively, regardless of their shape. For cylinders and parallel plates, r_h is one-half the radius or distance between plates. As in the method of Pierce [17], Brunauer assumes that in the first decrement, say between $P/P_0 = 1$ and 0.95, desorption occurs only from the center core of the largest pores regardless of their shape. The hydraulic radius of the core is calculated by dividing the liquid volume which has evaporated out of the pore by the core area, as determined by graphically integrating equation (8.16), using $P/P_0 = 1$ and 0.95 as the limits of integration. In the second decrement, the liquid volume desorbed, $(V_{liq})_2$, must be corrected for the decrease in the adsorbed film depth remaining on the walls of previously emptied pores. By assuming the pores are cylindrical, the core volume, $(V_k)_2$, can be calculated from the decrease in statistical thickness, t, as

$$(V_k)_2 = \left(V_{liq}\right)_2 - S_1(t_1 - t_2) - \frac{S_1}{4(r_h)_1}(t_1 - t_2)^2 \qquad (8.17)$$

where S_1 and $(r_h)_1$ are the core surface and hydraulic radius calculated from the first decrement. The terms t_1 and t_2 refer to the adsorbed film depths at the beginning and end of the second decrement. These values can be obtained from equation (8.9), the Halsey equation. The volume $(V_k)_2$ is divided by S_2, calculated from equation (8.15) using the relative pressures at the beginning and end of the second decrement as limits for the graphical integration. This ratio gives the second hydraulic radius, $(r_h)_2$.

To calculate the hydraulic radius on the third decrement, the desorbed volume must be corrected for two factors. The first accounts for the contribution made by the film on the walls of pores emptied during the first decrement and decreased in depth during the second decrement. The second is for the contribution made by the change made in film depth within pores emptied in the second decrement. The total correction factor is

$$(V_k)_3 = (V_{liq})_3 - S_1(t_1 - t_3) - \frac{S_1}{4(r_h)_1}(t_1 - t_3)^2 - S_1(t_1 - t_2) -$$
$$\frac{S_1}{4(r_h)_1}(t_1 - t_2)^2 - S_2(t_2 - t_3) - \frac{S_2}{4(r_h)_2}(t_2 - t_3)^2 \qquad (8.18)$$

It becomes painfully and quickly obvious that for an analysis requiring many data points, the number of correction terms becomes cumbersome.

However, Brunauer was able to show that the squared terms do not contribute significantly, and thus the above equation can be rewritten as

$$(V_k)_3 = (V_{liq})_3 - S_1(t_1 - t_3) - S_1(t_1 - t_2) - S_2(t_2 - t_3) \tag{8.19}$$

Equation (8.19) is identical to the correction factor for parallel plate pores. Using the above corrections, the Brunauer method is not modelless. However, it does offer a means of employing the same correction factor for pores as diverse as parallel plates and cylinders. In those instances where the pore geometry does differ considerably from parallel plates or cylinders, the error introduced by assuming either of these shapes is small. Brunauer confirmed this by plotting pore distributions, (V_k/r_h) versus r_h, using corrected and uncorrected core volumes. These plots differed only slightly, and the hydraulic radius remained essentially unchanged. Accordingly, if uncorrected hydraulic core volumes are used, the method is entirely modeless and little accuracy is sacrificed.

8.4 TOTAL PORE VOLUME AND AVERAGE PORE SIZE

If the sorption isotherm exhibits a distinct plateau as in case of type IV and V isotherms, the total specific pore volume is defined as the liquid volume at a certain predetermined P/P_0 (usually at $P/P_0 = 0.95$; the relative pressure chosen for the calculation of total pore volume should of course always be located after the pore condensation step). In this case, the adsorbed amount reflects the adsorption capacity and the total specific pore volume can be calculated by converting the amount adsorbed into liquid volume assuming that the density of the adsorbate is equal to the bulk liquid density at saturation, i.e., if in such case W_s is the adsorbate weight, then $W_s/\rho_\ell = V_\ell$, where V_ℓ and ρ_ℓ are the volume of liquid adsorbate at saturation and the liquid density, respectively. Various studies [24-26] have shown that at saturation, the liquid volume of different adsorbates, when measured on porous adsorbents, is essentially constant and is independent of the adsorptive. The constancy of adsorbed liquid volume at saturation is known as the Gurvich rule [27].

The pore volume, V_p, is given by

$$V_p = \frac{W_a}{\rho_\ell} \tag{8.20}$$

where W_a is the adsorbed amount (in grams).

Assuming, that no surface exists, other than the inner walls of the pores and that the pore is of cylindrical geometry, the average pore radius, $\bar{r}_p$, can be calculated from the ratio of the total pore volume and the BET surface area from the following equation:

$$2V_p/(S_{BET}) = \bar{r}_p \tag{8.21}$$

If one includes the external surface area, equation (8.21) becomes

$$2V_p(S_{total} - S_{ext}) = \bar{r}_p \tag{8.22}$$

The total surface area S_{total} can be calculated either from the BET method or from t-plot or α_s comparison plot methods (see chapter 9). The external surface area is determined from comparison plot in the region of relative pressure above the pore condensation step. Please note, that the results obtained with the Gurvich method do not only depend on the assumed pore geometry, they depend also on the validity of the determined surface area, which depends of course on the accuracy of the cross-sectional area.

The determined pore volumes and the related pore diameter have to be interpreted in connection with the shape of the sorption isotherm. If the isotherm does not reveal a plateau (e.g., type H3 isotherm), a total pore volume cannot be determined, and the calculated pore volume depends on the measured upper limit of the pore size distribution. This can be calculated using the Kelvin equation for a given adsorptive at a given temperature. For instance, the largest pore radius r, which could be assessed in a nitrogen sorption isotherm at P/P_0 of 0.99 corresponds to a pore radius of

$$r = \frac{-2(8.85)(34.6)}{(8.314 \times 10^7)(77)\ln(0.99)} = 950 \times 10^{-8} \text{ cm} \tag{8.23}$$

where 8.85 erg/cm^2 is the surface tension and 34.6 cm^3/mol is the molar volume of liquid nitrogen at 77K. Equations (8.20) and (8.23) state that the total volume of all pores up to 950 Å is V_p cm^3.

8.5 CLASSICAL, MACROSCOPIC THERMODYNAMIC METHODS VERSUS MODERN, MICROSCOPIC MODELS FOR PORE SIZE ANALYSIS

All methods related to the original BJH approach are based on the modified Kelvin equation and the accuracy of the calculated PSD depends on the applicability and the deficiencies of the Kelvin equation, which were already discussed to some extent in chapter 4. In narrow pores, attractive fluid-wall interactions are dominant and the macroscopic, thermodynamic concept of a smooth liquid-vapor interface and bulk-like core fluid cannot realistically be applied. In addition, methods based on the modified Kelvin equation do not take into account the influence of the adsorption potential on the position of the pore condensation transition. It is further assumed that the pore fluid has essentially the same thermophysical properties as the correspondent bulk fluid. For instance, the surface tension of the pore liquid is thought to be equal to the properties of the corresponding bulk liquid, but the surface tension of the pore liquid depends on the radius of curvature. Therefore, significant deviations from the bulk surface tension are to be expected in narrow mesopores [28,29].

Another problem is that the thickness of the preadsorbed multilayer film is assessed by the statistical thickness of an adsorbed film on a nonporous solid of a surface similar to that of the sample under consideration. However, in particular for narrow pores of widths < 10 nm this mean thickness does not reflect the real thickness of the preadsorbed multilayer film, because curvature effects are not taken into account.

A direct experimental test of the validity of the Kelvin equation and its modifications was possible after ordered mesoporous molecular sieves (e.g., MCM-41, SBA-15 etc.) became available, which exhibit a uniform pore structure and morphology and can therefore be used as model adsorbents to test theories of gas adsorption. For these ordered materials the pore diameter can be derived by independent methods (based on x-ray-diffraction, high resolution transmission electronic microscopy etc.). It was found that the BJH- and related approaches based on the modified Kelvin equation significantly underestimate the pore size. Necessary corrections to the modified Kelvin equation, in the spirit of the Broeckhoff - de Boer [32] and Cole-Saam approach [33], were developed by many researchers (e.g., 34 - 36). These improved classical methods were tested using ordered mesoporous molecular sieves such as MCM-41and/or SBA-15 materials (an overview of these unique ordered mesoporous materials is given in ref. [37]) in combination with pore size data derived from the aforementioned independent experimental methods. In general, good agreement was found, but only over a limited pore size range. Kruk *et al* [38] proposed a method (KJS) based on a corrected (modified) BJH equation, which was properly calibrated using MCM-41 mesoporous silica materials. A series of MCM-41 silica samples of different pore sizes (2 and ca. 7 nm) were originally used to establish a relation between capillary condensation/evaporation pressures and pore size. A pore size assessment independent from the pore condensation pressure could be obtained from XRD interplanar spacing and the primary mesopore volume. However, the KJS approach is strictly valid only for MCM-41 type silica materials over a pore size range which reflects the range over which an accurate calibration curve exists.

However, an accurate pore size analysis over the complete micro- and mesopore size range is possible by applying microscopic methods based on statistical mechanics, such as Density Functional Theory [e.g., 39] and the Grand Canonical Monte Carlo simulation (GCMC) [40]. These methods correctly describe the local fluid structure near curved solid walls on a microscopic level. These methods also correctly capture (at least qualitatively) that the thermodynamics of the confined fluid is altered as compared to the bulk fluid (e.g., critical point shifts), which affects the pore condensation and hysteresis behavior in narrow mesopores (see chapter 4, §4.5).

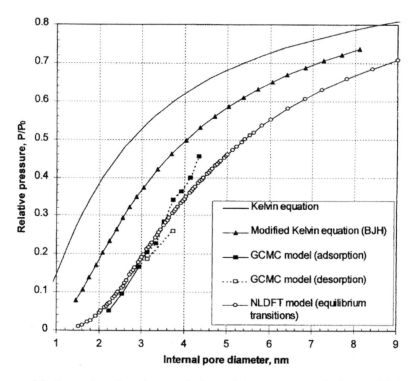

Figure 8.2 Pore size dependence of the relative pressure of the equilibrium condensation/desorption transition for N_2 in cylindrical pores at 77 K is displayed for different theoretical models. From [41].

Fig. 8.2 clearly shows, that the BJH method significantly underestimates the pore diameter (for a given pore condensation/evaporation pressure) compared to the predictions of the NLDFT and GCMC methods. A comparison of the pore size distribution curves obtained by applying the BJH and the NLDFT methods on nitrogen (77K) sorption data in MCM-41 silica is shown in Fig. 8.3.

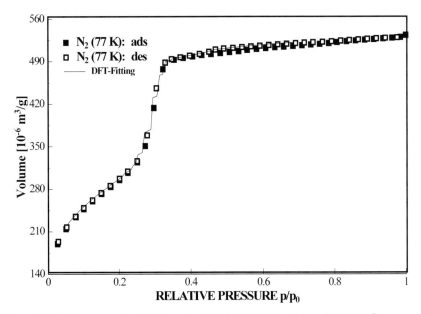

Figure 8.3a Nitrogen sorption isotherm (at 77K) in MCM-41 silica and NLDFT fit.

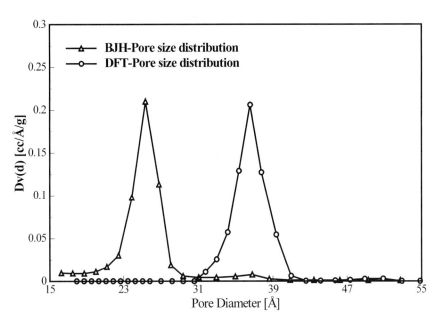

Figure 8.3b BJH and NLDFT pore size distribution curves obtained from the sorption isotherm shown in Fig. 8.3a.

Fig.8.3a shows the nitrogen sorption isotherm together with the correspondent theoretical NLDFT isotherm. The sorption isotherm does not show hysteresis as to be expected for a pore size below 4 nm. The pore size distribution curves as calculated by the BJH and the NLDFT method are shown in Fig. 8.3b. The shapes of the pore size distribution curves are very similar, but it can be clearly seen that the BJH pore size distribution curve is significantly shifted to a smaller pore diameter as compared to the NLDFT results. That is, the mode diameter (maximum of the pore size distribution curve) predicted by the BJH method is ca. 10 Å (i.e. 25 %) smaller as the pore diameter obtained by NLDFT. An extensive comparison of the results for pore size calculations for MCM-41, and SBA-15 materials as obtained by the Kelvin-approach, NLDFT and methods independent from the position of the pore condensation/evaporation step (e.g., XRD, geometrical method etc.) was conducted by di Renzo *et al* [42] and by Ravikovitch *et al* [43]. Good agreement was found between the experimental pore sizes and the ones determined by the NLDFT method. It was also again confirmed that the BJH method significantly underestimates the pore sizes over a wide pore diameter range. Better agreement is obtained with methods based on the Broeckhoff-de Boer approach. However, deviations between the Broeckhoff-de Boer method and NLDFT occur for pore diameters < 7 nm [43], mainly due to the fact that the Broeckhoff-de Boer method cannot predict the existence of pore criticality and the associated disappearance of sorption hysteresis below a certain critical pore diameter at a given temperature.

Summarizing, it can be stated that the microscopic methods allow one to obtain an accurate pore size analysis over a wide pore size range (i.e., combined micro/mesopore analysis). In contrast, classical methods based on macroscopic thermodynamic assumptions (e.g., BJH) underestimate the pore size up to 25 % (for pore sizes < 10 nm), if not properly corrected or calibrated. Microscopic methods for pore size analysis need to take into account details of the fluid-fluid interactions and the adsorption potential (which depends on the strength of the fluid-wall interactions and the pore geometry). However, appropriate methods for pore size analysis based on NLDFT and GCMC are meanwhile (commercially) available for many important fluid/substrate systems. It should be emphasized that the application of these advanced methods is only useful if the given experimental adsorptive/adsorbent system is compatible with the NLDFT or GCMC kernel available!

Because the equilibrium density profiles are known for each pressure along an isotherm (i.e., these isotherms are calculated by integrating of the equilibrium density profiles, $\rho(r)$, of the fluid in the model pores) no assumptions about the pore filling mechanism are required as in case of the macroscopic methods. As a consequence, these microscopic methods can be applied for pore size analysis over a large range of pore widths, i.e., from micropores up to well into the meso-macropore range (see chapter 9.10). A drawback of the currently available NLDFT and GCMC methods is that they do not take sufficiently into account the chemical and geometrical heterogeneity of the pore walls, i.e., usually a structureless (i.e., chemically and geometrically smooth) pore wall model is assumed, which leads to small steps in the calculated isotherms due to layering transition, which cannot be observed in experimental sorption isotherms obtained on adsorbents with chemical and geometrical heterogeneous surfaces. In addition, these methods are like all other available method for pore size analysis still based on the single pore model, which might not be accurate for the description of heterogeneous micro- and mesoporous solids consisting of a disordered network of pores. More work is clearly needed in the future to overcome these deficiencies, but it should be clearly stated that the pore size analysis method based on NLDFT and GCMC is meanwhile widely used and are considered as to be the most accurate methods for micro- and mesopore size analysis.

8.6 MESOPORE ANALYSIS AND HYSTERESIS

8.6.1 Use of the Adsorption or Desorption Branch for Pore Size Calculation?

The presence of the hysteresis loop introduces, as already discussed before, a considerable complication, and the question arises whether the adsorption branch or the desorption branch of a hysteretic sorption isotherm should be used for the pore size analysis. In the case of ordered materials such as MCM-41 (which essentially consists of independent, cylindrical-like pores) hysteresis is quite well understood. The typically observed type H1 hysteresis loop can be completely described within the so-called independent pore model (see chapter 4). Sorption hysteresis is considered to be an intrinsic property of a phase transition in a single, idealized pore, reflecting the existence of metastable gas states.

Theoretical studies applying Non Local Density Functional Theory (NLDFT) revealed that for an ideal single pore of given geometry (cylinder or slit) of finite pore length, pore condensation is associated with metastable states of the pore fluid [e.g., 44- 46]. As discussed in chapter 4, this is consistent with the classical van der Waals picture, which predicts that the metastable adsorption branch terminates at a vapor-like spinodal, where the limit of stability for the metastable states is achieved and the fluid spontaneously condenses into a liquid-like state. However, assuming a pore of finite length (which is always the case in real adsorbents) vaporization can occur via a receding meniscus and therefore metastability is not expected to occur during desorption (evaporation).

Hence, within this picture the desorption branch of the hysteresis loop reflects the equilibrium phase transition, i.e., theories and methods which describe the equilibrium phase transition (e.g., BJH, NLDFT etc.) have to be applied to the desorption branch in order to calculate the pore size correctly. The adsorption branch can also be taken for pore size analysis if a theory (method) is applied which takes metastability into account and provides a correlation between the relative pressure where (spinodal) condensation occurs and the pore size. Such a method based on NLDFT (the NLDFT-spinodal condensation method) was suggested by Ravikovitch and Neimark [46]. Another possibility is to apply an empirical method such as the KJS approach [38], which, however, can only be applied for pore size analysis over a limited pore size range.

The situation is more complex for materials consisting of a three-dimensional network of pores. However, it appears that pore condensation hysteresis of type H1 has been observed in ordered three dimensional pore systems (e.g., MCM-48 [47(a)], some ordered sol-gel glasses [48]). It could be shown that also here the observed hysteresis is still predominantly caused by metastabilities associated with the occurrence of pore condensation, i.e., the hysteresis loops could be described by applying models, which are based on the independent pore model. Consequently, in such cases the desorption branch of the hysteresis loop can be associated with the equilibrium gas-liquid phase transition (see §4.4.4.2), and should be chosen for pore size analysis if theories/methods are applied which are based on equilibrium thermodynamics (e.g., methods based on the Kelvin equation).

In case of highly disordered materials (e.g., porous Vycor® glass, some silica gels), the occurrence of hysteresis is, as discussed before in chapter 4, associated with a variety of effects including metastable states of the pore fluid, potential pore blocking and percolation phenomena, effects from possible system-spanning transitions, long time dynamics etc. The significance of each mechanism for the shape of hysteresis depends on details of the texture of the porous material. Very often sorption hysteresis of type H2 is observed in such cases. In contrast to the situation in ordered materials, the desorption branch of the hysteresis loops is then not necessarily correlated with the pore size. This is also true if the evaporation of the pore liquid occurs close to the lower limit of hysteresis (e.g., tensile strength effect). This is discussed in §8.6.2.

Having NLDFT methods available that describe (based on the independent pore model) both the position of the pore condensation (NLDFT-spinodal condensation method, i.e. adsorption branch kernel) and the evaporation (NLDFT-equilibrium transition method, i.e., desorption branch kernel), the consistency of the pore size analysis can be verified [46]. As indicated before, the NLDFT adsorption branch kernel correctly takes into account the effect of pore size on the pressure range over which metastable pore fluid extends in association with the pore condensation transition.

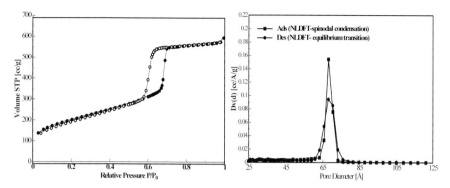

Figure. 8.4 Nitrogen adsorption/desorption at 77.35 K in ordered SBA-15 and NLDFT pore size distributions (NLDFT for N_2/silica [46]) from adsorption- (spinodal condensation) and desorption (equilibrium transition) branch. From [49].

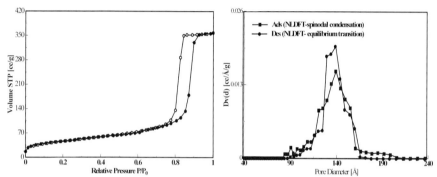

Figure 8.5 Nitrogen adsorption/desorption at 77.35 K in controlled pore glass and NLDFT pore size distribution (NLDFT for N_2/silica [46]) from adsorption- (spinodal condensation) and desorption (equilibrium transition) branch. From [49].

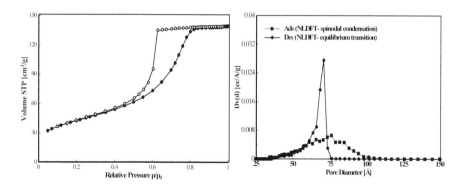

Figure 8.6 Nitrogen adsorption/desorption at 77.35 K in porous Vycor® glass and NLDFT pore size distribution (NLDFT for N_2/silica [46]) from adsorption- (spinodal condensation) and desorption (equilibrium transition) branch. From [49].

This allows comparison of theoretical and experimental "widths" of the hysteresis loops within the framework of the independent pore model. Hence, hysteresis loops, which are wider than those predicted theoretically (i.e., the pore size distribution curves calculated from adsorption and desorption branch do not agree), indicate that hysteresis cannot be solely explained within the framework of the independent pore model. This approach has been applied on nitrogen sorption data obtained at 77.35 K on various ordered and disordered silica materials, i.e. SBA-15, controlled-pore glass (CPG) and porous Vycor® glass.

The observed hysteresis loops for SBA-15 as well as for controlled pore glass, is of type H1 hysteresis, whereas the porous Vycor® glass clearly exhibits type H2 hysteresis. The pore size distribution curves obtained from the adsorption branch (by applying the NLDFT-spinodal condensation method) and desorption branch (by applying the NLDFT equilibrium method) are in perfect agreement for SBA-15. The results are shown in Figs. 8.4 – 8.6. This confirms that sorption hysteresis in this SBA-15 silica sample is more or less entirely caused by delayed condensation (i.e., by metastable states of the pore fluid occurring during adsorption/ condensation).

Good agreement between the PSD's calculated from the adsorption and desorption branches can also be found for the controlled-pore glass sample. Although controlled-pore glasses consist of a network of cylindrical-like pores, the hysteresis appears to be to more or less entirely due to metastability effects, i.e., the observed hysteresis can also here be described within the independent pore model.

However, the situation is different for the Vycor® glass sample. The pore size distribution obtained from the desorption branch is artificially sharp leading to the observed H2 type hysteresis. The disagreement between the pore size distribution curves obtained from adsorption and desorption branch indicates that hysteresis cannot be described within the single pore model. As discussed, H2 hysteresis is often attributed to the occurrence of pore blocking and percolation phenomena, and it is believed that such phenomena are only associated with the evaporation (desorption) process, but not with the pore condensation. Hence, the pore size distribution curve calculated from the adsorption branch can here be considered to be more realistic.

The examples discussed here confirm that the IUPAC classification of hysteresis loops (see chapter 4, Fig. 4.8) can be in principle quite helpful to determine the "correct" branch of the hysteresis loop for pore size analysis. It seems that in ordered systems very often H1 hysteresis occurs, whereas H2 and H3 hysteresis is typically observed for disordered mesoporous systems. There is some evidence that in many cases where type H1 hysteresis occurs – even if observed in materials consisting of three dimensional pore networks - the desorption branch is correlated with the equilibrium phase transition, and therefore with the pore size if methods based on the determination of the gas-liquid phase transition (e.g., BJH, equilibrium-NLDFT) are applied for pore size analysis. This is not the case for H2 and H3 hysteresis, and the analysis of the adsorption branch by applying a proper method (e.g., NLDFT-spinodal condensation method or Kelvin equation based approach calibrated for adsorption branch) may lead here to a more accurate pore size analysis.

However, it should be clearly stated, that the suggestions given above are certainly not justified for highly disordered materials, where a clear decision with regard to the type of hysteresis loop is not always possible. In such cases it may be helpful to include in the sorption experiment the measurement of so-called scanning curves (of the hysteresis loop), which reveal details of the mechanisms of pore condensation and evaporation (e.g., effects of connectivity, existence of cooperative processes etc. [50, 51]). In addition, the factors which determine the shape of the hysteresis loop are still not completely known for disordered, connected pore systems [52, 53] More work is needed to correlate such theoretical predictions with sorption experiments on disordered porous materials, where the texture can be explored by independent methods (e.g., SANS, SAXS) [54].

8.6.2 Lower Limit of the Hysteresis Loop - Tensile Strength Hypothesis

It was observed a long time ago, that the hysteresis loop for nitrogen adsorption at 77.35 K closed at relative pressure at or above 0.42, apparently independent of the porous material [55, 56]. Experimental results with other adsorptives also supported the view that for a given temperature this lower closure point of hysteresis is never located below a certain (critical) relative pressure. Based on the experimental results it was also concluded that the lower closure point depends mainly on the nature of the adsorptive and temperature. Further, it was suggested that the lower closure point of hysteresis is determined by the tensile strength of the capillary condensed liquid, i.e., there exists a mechanical stability limit, below which a macroscopic meniscus cannot exist anymore and where a spontaneous evaporation of the pore liquid occurs. This is in contrast to the situation of pore blocking in case of inkbottle pores, where the evaporation of the liquid is controlled by the diameter of the necks. More recent work also indicates that the lower limit of the hysteresis loop also depends on the pore geometry, which questions the conventional assumption that the lower limit of the hysteresis loop is a unique function of the adsorptive and the temperature [57]. A correlation between the lower closure point of hysteresis and the hysteresis critical temperature was also suggested [58,59].

Despite the fact that the occurrence of the lower closure point of hysteresis is still not sufficiently understood, it is clear that it leads to an important implication for pore size calculations. The existence of a lower closure point affects primarily the position of the desorption branch with regard to its position and steepness, i.e., the desorption isotherm exhibits a characteristic step.

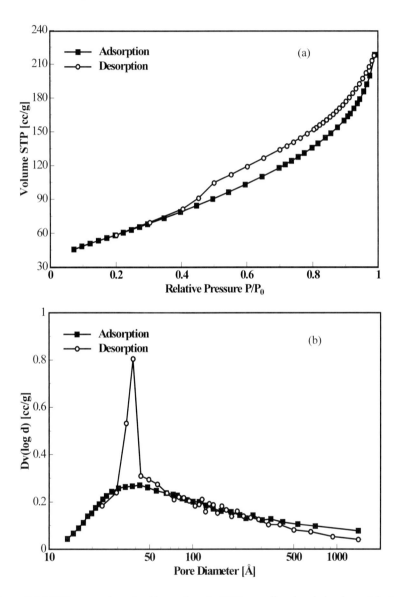

Figure 8.7 (a) Nitrogen adsorption/desorption at ~77 K on a disordered alumina catalyst; (b) BJH pore size distribution curves from adsorption and desorption branches.

Hence, a pore size calculation based on an analysis of the desorption branch is here not straightforward. A typical example is given in Fig. 8.7, which shows nitrogen sorption data on a highly disordered alumina catalyst sample together with BJH pore size distribution curves from both adsorption and desorption branches. As it can be clearly seen, the hysteresis loop closes at a relative pressure of ca. 0.4 - 0.45 and exhibits the mentioned characteristic step down. This step is not associated with the evaporation of pore liquid from a specific group of pores, i.e., the spike in the desorption PSD reflects an artifact, believed to be caused by the spontaneous evaporation of metastable pore liquid (i.e., is due to the tensile strength effect). In contrast, the PSD derived from the adsorption branch does not reveal this artificial peak and reveals the wide pore size distribution, characteristic for such a disordered sample.Hence, in this case it is believed that a more realistic pore size can be obtained from an analysis of the adsorption branch.

8.7 ADSORPTIVES OTHER THAN NITROGEN FOR MESOPORE ANALYSIS

Generally, nitrogen adsorption at liquid nitrogen temperature (~77 K) is used for surface and pore size characterization. Krypton adsorption at ~77 K is more or less exclusively used for low surface area analysis [52-54]. Argon adsorption at ~77 K and liquid argon temperature (~87 K) is also often used for micro- and mesopore size analysis. The use of argon adsorption is of advantage for the pore size analysis of zeolites and other microporous materials because the filling of pores of dimension 0.5 - 1 nm occurs at much higher relative pressure as compared to nitrogen adsorption (see chapter 9). However, a combined and complete micro- and mesopore size analysis with argon is not possible at liquid nitrogen temperature because it is ca. 6.5 K below the triple point temperature of bulk argon.

Systematic sorption experiments [47,49] indicate that the pore size analysis of mesoporous silica by argon adsorption at ~77 K is limited to pore diameters smaller than ca. 15 nm, i.e., pore condensation cannot be observed above this pore size limit [47]. This behavior is related to confinement effects on the location of the (quasi)-triple point of the pore fluid (see chapter 4.4). Of course, such a limitation does not exist for argon sorption at 87.27 K; pore filling and pore condensation can be observed here over the complete micro- and mesopore size range.

Argon isotherms obtained at 87 K and 77 K in various mesoporous silica materials (MCM-41, SBA-15 and controlled pore glasses) are shown in Figs. 8.8 and 8.9. For reference, pore size distribution curves for these materials (Fig. 8.10b) were obtained from nitrogen isotherms (Fig. 8.10a). The argon data reveal that, as expected, pore condensation shifts to higher relative pressures with increasing pore diameter. Hysteresis occurs in all materials, with the exception of MCM-41A (pore size: 3.3.nm), which exhibits reversible pore condensation. With increasing pore diameter, hysteresis begins to develop and is present for MCM-41C (4.2 nm), and a wide hysteresis loop is observed for SBA-15 (6.7 nm).

Importantly, the data also clearly reveal that the width of the hysteresis loop increases with increasing pore size (see chapter 4, §4.4.5). Lowering the measurement temperature to ~77 K (Fig. 8.9) leads to a widening of the hysteresis loops for the MCM-41 and SBA-15 and hysteresis is now evident for MCM-41A. However, pore condensation and hysteresis can no longer be observed in the controlled pore glass of (BJH) mode pore diameter 16 nm. A detailed analysis of the data has led to the conclusion that pore condensation no longer occurs for pore sizes larger than ca. 15 nm under said conditions, which limits (as stated above) the pore diameter range over which mesopore size analysis can be performed using argon as adsorptive at ~77 K.

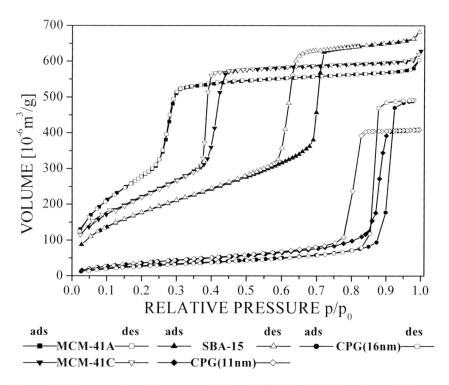

Figure 8.8 Argon sorption isotherms at ~87 K on MCM-41, SBA-15 and controlled-pore glasses (CPG). From [47a].

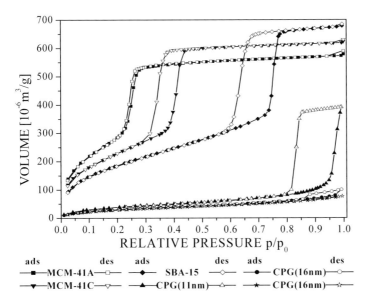

Figure 8.9 Argon sorption at ~77 K on MCM-41, SBA-15 and CPG. (P_0 refers to solid argon). From [47a].

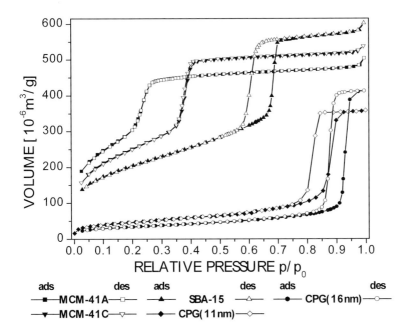

Fig. 8.10a Nitrogen sorption at ~77 K on MCM-41, SBA-15 and CPG. From [47a].

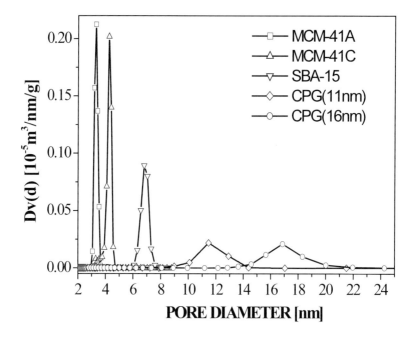

Figure 8.10b Pore size distribution curves for mesoporous molecular sieves (MCM-41 SBA 15, PSD by NLDFT method), and controlled-pore glasses (PSD by BJH method) obtained from N_2 (~77K) sorption isotherms. From [47a].

Hence, nitrogen adsorption at ~77 K and argon adsorption at ~87 K are recommended for mesopore size analysis.

8.8. REFERENCES

1.   [a]Thompson W.T. (1871) *Philos. Mag.* **42**, 448; [b]Zsigmondy Z. (1911) *Anorg. Chem.* **71**, 356.
2.   [a]Cohan L.H. (1938) *J. Am. Chem. Soc.* **60**, 433; [b]Cohan L.H. (1944) *J. Am. Chem. Soc.* **66**, 98.
3.   Shull C.G. (1958) *J. Am. Chem. Soc.* **70**, 1405.
4.   Lippens B.C. and de Boer J.H. (1965) *J. Catal.* **4**, 319.
5.   Pierce C. (1953) *J. Phys. Chem.* **57**, 149.
6.   Harris M.R. and Sing K.S.W. (1959) *Chem Ind. (London)* 11.
7.   Cranston R.W. and Inkley F.A. (1957) *Adv. Catal.* **9**, 143.
8.   Halsey G.D. (1948) *J. Chem. Phys.* **16**, 931.
9.   Lippens B.C., Linsen B.G. and de Boer J.H. (1964) *J. Catal.* **3**, 32.
10. Harkins W.D. and Jura G. (1944) *J. Am. Chem. Soc.* **66**, 1366.
11. [a]Lippens B.C. and de Boer J.H. (1965) *J. Catal.* **4**, 319; [b]de Boer J.H., Lippens B.C., Linsen B.G., Broekhoff J.C.P., van den Heuvel A. and Onsinga T.V. (1966) *J. Colloid Interface Sci.* **21**, 405.
12. Kaneko K. (1994) *J. Membrane Sci.* **96**, 59.
13. Orr Jr C. and DallaValle J.M. (1959) *Fine Particle Measurement*, Macmillan, New York, chapter 10.

14. Barrett E.P., Joyner L.G. and Halenda P.P. (1951) *J. Am. Chem. Soc.* **73**, 373.
15 Dollimore D. and Heal G.R. (1964) *J. Appl. Chem.* **14**, 109.
16. Wheeler A. (1945, 1946) Catalyst Symposia, Gibson Island A.A.A.S. Conference.
17. Pierce C. and Smith R.N. (1953) *J. Phys. Chem.* **57**, 64.
18 Carman P.C. (1953) *J. Phys. Chem.* **57**, 56.
19. Oulton T.D. (1948) *J. Phys. Colloid Chem.* **52**, 1296.
20. Innes W.B. (1957) *Anal. Chem.* **29**, 1069.
21. Cranston R.W and Inkley F.A. (1957) *Adv. Catal.* **9**, 143.
22. Brunauer S., Mikhail R.S. and Bodor E.E. (1967) *J. Colloid Interface Sci.* **24**, 451.
23. Kiselev A.V. (1945) *Usp. Khim.* **14**, 367.
24. [a]Broad D.W. and Foster A.G. (1946) *J. Chem. Soc.* 447; [b]Foster A.G. (1934) *Proc. Roy. Soc. London* **147A**, 128.
25. McKee D.W. (1959) *J. Phys. Chem.* **63**, 1256.
26. Brown M J. and Foster A.G. (1951) *J. Chem. Soc.* 1139.
27. Gurvich L. (1915) *J. Phys. Chem. Soc. Russ.* **47**, 805.
28. Melrose J.C. (1966) *Am. Inst. Chem. Eng.* **12**, 986.
29. AhnW.S., Jhou M.S., Pak H. and Chang S. (1972) *J. Colloid Interface Sci.* **38**, 605.
30. Zhu H.Y., Ni L.A. and Lu G.Q. (1999) *Langmuir* **15**, 3632.
31. Galarneau A., Desplantier D., Dutartre R. and Di Renzo F. (1999) *Microporous Mesoporous Mater.* **27**, 297.
32. [a]Broekhoff J.C.P. and de Boer J.H. (1968) *J. Catal.* **10**, 377; [b]Broekhoff J.C.P. and de Boer J.H. (1968) *J. Catal.* **10**, 391; [c]Broekhoff J.C.P. and de Boer J.H. (1967) *J. Catal.* **9**, 15.
33. Cole M.W. and Saam W.F. (1974) *Phys.Rev.Lett.* **32**, 985.
34. Zhu H.Y., Ni L.A. and Lu G.Q. (1999) *Langmuir* **15**, 3632.
35. Sonwane C.G. and Bhatia S.K. (2002) In *Fundamentals of Adsorption* **7** (Kaneko K., Kanoh H. and Hanzawa Y., eds.) IK International Ltd, Chiba City, Japan, p999.
36. Groen J.C., Doorn M.C. and Peffer L.A.A. (2000) In *Adsorption Science and Technology* (Do D.D., ed.) World Scientific, Singapore, p229.
37. Linden M., Schacht S., Schueth F., Steele A. and Unger K. (1998) *J. Porous Mater.* **5**, 177.
38. [a]Kruk M., Jaroniec M. and Sayari A. (1997) *J. Phys. Chem.* B **101**, 583; [b]Kruk M. and Jaroniec M. (2000) *Chem. Mater.* **12**, 222.
39. Ravikovitch P.I., Vishnyakov A. and Neimark A.V. (2001) *Phys. Rev. E*, **64**, :011602
40. Gubbins K.E (1997) In *Physical Adsorption: Experiment, Theory and Application* (Fraissard J., ed) Kluwer, Dordrecht, p65.
41. Neimark A.V., Ravikovitch P.I., Grün M., Schüth F. and Unger K.K. (1998) *J. Coll. Interface Sci.* **207**, 159.
42. Di Renzo F., Galarneau A., Trens P., Tanchoux N. and Fajula F. (2002) *Stud. Surf. Sci. Catal.* **142**, 1057.
43. Neimark A.V., Ravikovitch P.I. and Vishnyakov A. (2003) *J. Phys. Condens. Matter* **1**, 347.
44. Evans R. (1990) *J. Phys. Condens. Matter* **2**, 8989.
45. Ravikovitch P.I. and Neimark A.V. (2001) *Colloids and Surfaces A*: *Physicochemical and Engineering Aspects* **187**, 11.
46. Neimark A.V. and Ravikovitch P.I. (2001) *Microporous Mesoporous Mater.* **44-56**, 697.
47. [a]Thommes M., Koehn R. and Froeba M. (2002) *Appl. Surf. Sci.* **196**, 239; [b]Thommes M., Koehn R. and Froeba M. (2000) *J. Phys. Chem B* **104**, 7932.
48. Awschalom D.D., Warnock J. and Shafer M.W. (1986) *Phys. Rev. Lett.* **57**, 1607.
49. Thommes M. (2004, in print) In *Nanoporous Materials – Science and Engineering* (Lu G.Q., ed.) World Scientific, Sydney, Chapter 11.
50. Rojas F., Kornhauser I., Felipe C., Esparza J.M., Cordero S., Dominguez A. and Riccardo J.L. (2002) *Phys. Chem. Chem. Phys.* **4**, 2346.

51. Mason G. (1982) *J. Colloid Interface Sci.* **88**, 36.
52. Kierlik E., Rosinberg M.L., Tarjus G. and Viot P. (2001) *Phys. Chem.Chem. Phys* **3**, 1201.
53. Woo H.J., Sarkisov L. and Monson P.A. (2001) *Langmuir* **17**, 7472.
54. Smarsly B., Goeltner C., Antonietti M., Ruland W. and Hoinkis E. (2001) *J. Phys. Chem.B* **105**, 831.
55. Gregg S.J. and Sing K.S.W. (1982) *Adsorption, Surface Area and Porosity,* Academic Press, London.
56. Burgess C.G.V. and Everett D.H. (1970) *J. Colloid Interface Sci.* **33**, 611.
57. Ravikovitch P.I. and Neimark A.V. (2002) *Langmuir* **18**, 9830.
58. Schreiber A., Reinhardt S. and Findenegg G.H. (2002) *Stud. Surf. Sci. Catal.* **144**, 177.
59. Sonwane C.G. and Bhatia S.K. (1999) *Langmuir* **15**, 5347.

9 Micropore Analysis

9.1 INTRODUCTION

As discussed in chapters 3 and 4, IUPAC [1] classifies pores as macropores for pore widths greater than 50 nm, mesopores for the pore range 2 to 50 nm and micropores for the pores in the range less than 2 nm.

Further, micropores are classified into ultramicropores (pore widths < 0.7 nm) and supermicropores (pore widths 0.7 to 2 nm). Whereas mesopores show type IV and type V sorption isotherms, microporous materials exhibit in ideal cases type I isotherms. The characteristic feature of a type I isotherm is a long horizontal plateau, which extends up to relatively high P/P_0. Such sorption isotherms can be described by the Langmuir equation (4.12), which was developed on the assumption that adsorption was limited to at most one monolayer. Hence, very often the Langmuir equation is used for the determination of specific surface area of microporous materials. However, despite the often-observed good fit of the Langmuir equation to the experimental data, the obtained surface area results do not reflect a true surface area (see §5.1). *Any* factor that can limit the quantity adsorbed to a few monolayers will also produce a type I isotherm, as is the case for micropores, where the small pore width prevents multilayer adsorption and therefore limits the amount adsorbed. We have discussed the theoretical background of adsorption in micropores in chapter 4, §4.3. Here we discuss the application of these theoretical approaches for the pore size/volume analysis of microporous materials.

9.2 MICROPORE ANALYSIS BY ISOTHERM COMPARISON

9.2.1 Concept of V-t Curves

An adsorbent is never covered with an adsorbed film of uniform thickness, but rather with a characteristic density profile, which depends very much on temperature. Despite this, it is often assumed that the film thickness on pore walls is uniform, which enables one to obtain the so-called *statistical thickness, t,* from the gas adsorption isotherms. This was already discussed in chapter 8 in connection with the application of the Kelvin equation for pore size analysis (see §8.2).

Shull [2] showed that on a number of nonporous solids, the ratio of the weight adsorbed, W_a, to the weight corresponding to the formation of a monolayer, W_m, could be closely represented by a single curve, regardless of the solid, when plotted versus the relative pressure. The curve produced by Shull is a typical type II isotherm. Similar plots were made by others [3-6] on a variety of nonporous materials and in each case, the points fit a common curve reasonably well, particularly at relative pressures above 0.3. The common curve is also fitted by the Halsey equation (already discussed in chapters 4 and 8; see equation (8.9) and ref. [7]) at higher relative pressures. If the monolayer is envisioned as being uniformly one molecule in depth, then a plot of W_a/W_m versus P/P_0 discloses the relative pressures corresponding to surface coverage by any number of monolayers. Therefore, if the adsorbate diameter is known, the statistical depth t can be calculated by multiplying the number of monolayers by the adsorbate diameter. Shull [2] assumed that the adsorbate molecules packed one on top of the other in the film and deduced the monolayer depth to be 4.3 Å for nitrogen. A more realistic assumption is that the film structure is close packed hexagonal, leading to a monolayer depth of 3.54 Å as shown by equation (8.7). Therefore, the statistical thickness, t, of the adsorbed film is

$$t = 3.54 \frac{W_a}{W_m} \ \text{Å} \qquad\qquad (\text{cf. } 8.8)$$

If the volume adsorbed is expressed as the corresponding liquid volume, then

$$t = \frac{V_{liq}}{S} \times 10^4 \ \text{Å} \qquad\qquad (9.1)$$

where S is the total surface area and V_{liq} is the adsorbed liquid volume; $V_{liq} = V_{ads} (\text{STP}) \times 15.47$ in case of nitrogen adsorption at 77 K.

Lippens and de Boer [3] have shown that (in case of a type II isotherm) a plot of the volume adsorbed, V_{liq}, versus t calculated from equation will yield a straight line through the origin. Plots of this nature are termed V-t curves and the surface area calculated from the slope (equation (9.1) generally give surface areas comparable to BET values.

9.2.2 The t-Method

The utilization of the technique of comparing an isotherm of a microporous material with a standard type II isotherm is the t-plot method as it was proposed by Lippens and de Boer [3, 8]. This method permits the determination of micropore volume and surface area and, in principle, information about the average pore size. The t-method employs a composite t-standard (reference) curve, obtained from data on a number of nonporous adsorbents with BET C constants similar to that of the microporous sample being tested. The experimental test isotherm is then redrawn as a t-curve, i.e., a plot of the volume of gas adsorbed as a function of t, i.e., the standard multilayer thickness on the reference non-porous material at the corresponding P/P_0. These t values are in practice calculated with the help of a thickness equation that describes the particular standard (reference) curve.

A popular thickness equation was obtained by de Boer (as mentioned before) [3, 8], which represents nitrogen sorption at 77 K on non-porous adsorbents with oxidic surfaces like, for example, siliceous materials:

$$t = \left[\frac{13.99}{\log(P_0/P)+0.034} \right]^{1/2} \text{Å} \qquad (9.2)$$

A similar thickness equation, which is also very often used for the analysis of zeolitic materials, is the Harkins-Jura equation. This thickness equation is based on adsorption data obtained on nonporous Al_2O_3 [9].

As previously mentioned in chapter 8 the Halsey equation is another commonly employed alternative. The Halsey equation for nitrogen adsorption at 77 K can be expressed as

$$t = 3.54 \left[\frac{5}{\ln(P_0/P)} \right]^{1/3} \text{Å} \qquad (\text{cf. 8.9})$$

or, in a generalized form (for other adsorbates, adsorbents and temperatures) as

$$t = a \left[\frac{1}{\ln(P_0/P)} \right]^{1/b} \text{Å} \qquad (9.3)$$

where, for nitrogen adsorption on oxidic surfaces at 77 K, the pre-exponential term, a, and the exponential term, b, are 6.0533 and 3.0, respectively. The pre-factor a and the exponent b can be obtained by applying the FHH approach on adsorption data obtained on a non-porous adsorbent (see chapter 4) which should have the same surface chemistry as the porous sample.

Lecloux's "n-method" t-plot takes into account the nature of the adsorbent surface by using the value of the BET C constant in the calculations of the statistical film thickness curve [10]. A critical evaluation of the t-plot method, in particular with regard to the characterization of zeolites, was recently published by Hudec *et al* [11].

In the case of carbon-like adsorbents, primarily carbon black, it is more appropriate to apply a thickness equation dedicated to carbons such as the so-called STSA equation suggested in the ASTM standard D-6556-01 [12] wherein the statistical layer thickness of carbon black is given by

$$t = 0.88(P/P_0)^2 + 6.45(P/P_0) + 2.98 \quad \text{Å} \tag{9.4}$$

Statistical film thickness data for the adsorption of nitrogen on carbon were also published by Kaneko [13a].

Differences between the shape of the experimental isotherm and the standard isotherm result in non-linear regions of the t-plot and positive or negative intercepts if the t-plot is extrapolated to $t = 0$. These deviations from the standard isotherm can be used to obtain information about the micropore volume and micropore surface area of the adsorbent.

Typical t-plots for both microporous and non-microporous samples are shown in Fig. 9.1: A, Standard type II isotherm; B, t-plot from type II isotherm; C, type II isotherm + microporous sample; D, t-plot from isotherm C. If the isotherm A is identical in shape to the standard isotherm of a nonporous sample (type II), the t-plot B will be a straight line passing through the origin, the slope of which is the surface area, according to equation (9.1). Using the slope, s, of the plot in Fig. 9.1b, equation (9.1) reduces to

$$S_t = s \times 15.47 \quad \text{m}^2/\text{g} \tag{9.5}$$

In the absence of micropores there is good agreement between the t-area, S_t, and the surface area determined by the BET method. When a relatively small fraction of micropores is present, the adsorption isotherm C will show an increased uptake of gas at low relative pressures. The t-plot D will be linear and when extrapolated to the adsorption axis will show a positive intercept, i, equivalent to the micropore volume V_{MP}:

$$V_{MP} = i \times 0.001547 \quad \text{cm}^3 \tag{9.6}$$

The external surface area S_{ext} of the microporous sample can be derived from the slope of the t-plot, D, (Fig 9.1d), i.e. the micropore surface area (S_{micro}) can be calculated from the relation $S_{micro} = S_{BET} - S_{ext}$ (see equation (9.8) and discussion of Fig. 9.1j).

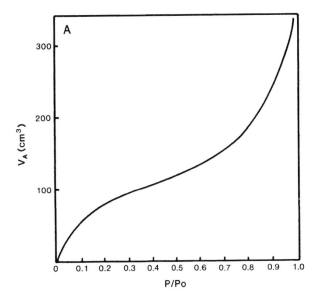

Figure 9.1a Standard Type II isotherm.

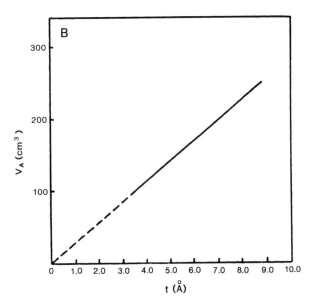

Figure 9.1b t-plot from Type II isotherm.

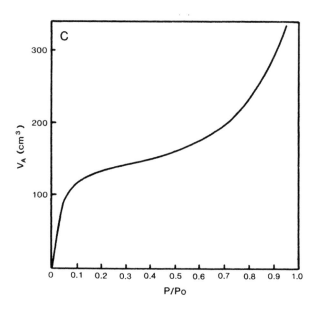

Figure 9.1c Type II isotherm + microporous sample.

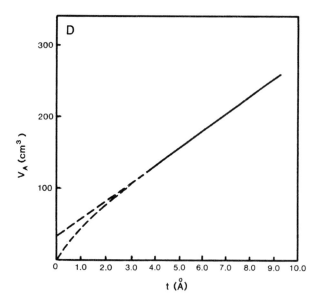

Figure 9.1d t-plot from isotherm C.

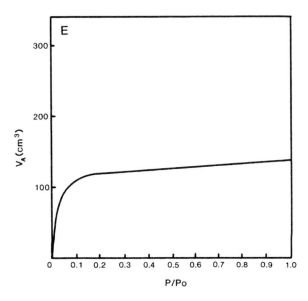

Figure 9.1e Isotherm of a microporous material.

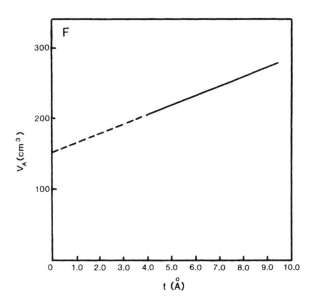

Figure 9.1f t-plot from E.

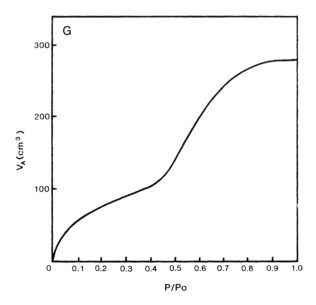

Figure 9.1g Type IV isotherm.

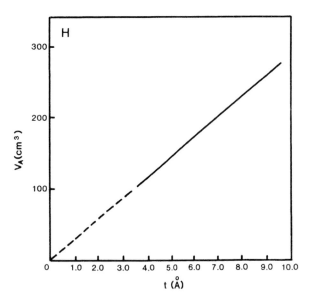

Figure 9.1h t-plot from G.

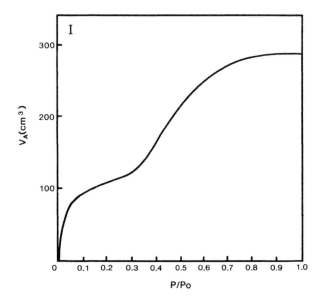

Figure 9.1i Type IV isotherm +microporous material.

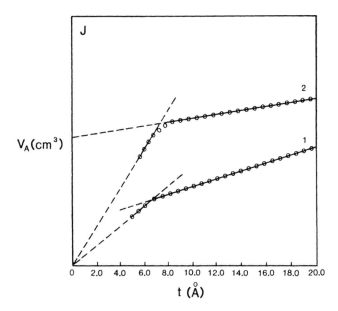

Figure 9.1j t-plots from isotherm in I.

The slope of the straight line in D is proportional to the external surface area since all micropores have been filled by micropore filling.

The adsorption isotherm E is typical of a sample having only micropores. The corresponding t-plot F is interpreted in the same manner as D. A typical type IV isotherm that indicates the presence of mesopores is depicted in G. It would result in the t-plot shown in H, provided no micropores are present.

If additional micropores are present the resulting isotherm would have the shape as shown I. The corresponding t-plots shown in Fig. 9.1j are illustrative of micropores in the presence of mesopores. J2 represents a material with greater micropore volume than J1. The initial slope of the V_A-t curve corresponds to small values of t, which represents an adsorbed film within large pores and complete filling of smaller pores. Those micropores smaller in diameter than the adsorptive molecule cannot contribute to this gas uptake. Therefore, from the initial slope of the V_A-t curve the total surface area of the sample can be obtained using equation (9.1). The surface area of the wide pores is similarly obtained from the slope of the upper linear portion of the t-plot. This area represents the build-up of a statistical thickness in all pores except the micropores, which are presumed filled at higher t values. Hence, the difference between these two surface areas is the surface area of the micropores only. The region between the two linear portions represents the transition that occurs as the micropores become filled while multilayer adsorption continues to occur in the larger pores. Thus, after converting the gas volume to the corresponding liquid volume, the micropore area can be calculated by applying equation (9.1) as follows:

$$S_{micro} = \left[\left(\frac{V_{liq}}{t} \right)_{lower} - \left(\frac{V_{liq}}{t} \right)_{upper} \right] \times 10^4 \qquad (9.7)$$

In the absence of sufficient data at low relative pressures, the total surface area from the lower linear portion of the t-plot cannot be calculated easily. Because the total area should be the same as the BET surface area, the micropore surface area can be calculated from

$$S_{micro} = S_{BET} - \left(V_{liq}/t \right)_{upper} \times 10^4 \qquad (9.8)$$

as already discussed in connection with Fig. 9.1d. The linear BET region for microporous materials generally occurs at relative pressures lower than 0.1. The linear t-plot range will be found at higher relative pressures and is dependent on the size distribution of micropores.

The abrupt break in the two linear parts of the t-plot J1 (Fig. 9.1j) indicates the presence of a group of micropores in a narrow pore size range, whereas the curvature between the two linear portions of J2 is an indication of a wider distribution of micropores.

9.2.3 The α_s-Method

Another method for estimating micropore volume and surface area without assuming a knowledge of the adsorbate statistical thickness is the α_s method, developed by Gregg and Sing [14a]. Hence, the construction of the α_s–plot does not require the monolayer capacity and allows therefore a more direct comparison between the test isotherm and the reference isotherm. The reference isotherm in this method is a plot of the amount of gas adsorbed, normalized by the amount of gas adsorbed at a fixed relative pressure, versus P/P_0. The referenced relative pressure is usually $P/P_0 = 0.4$, and the normalized term $V_{ads}/V_{ads}^{0.4}$ is α_s. Hence, the reduced isotherm for the non-porous reference adsorbent is called standard α_s –*curve*.

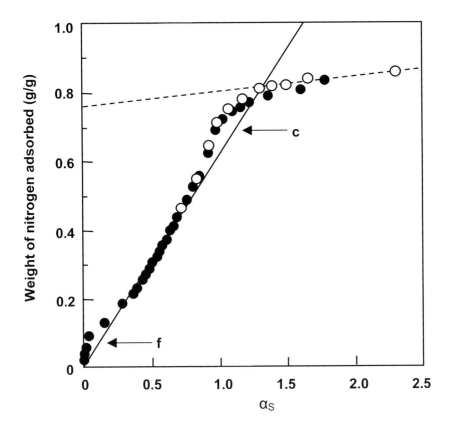

Figure 9.2 Example of high-resolution α_s-plot. From [13b]. The so-called swings at 'f' and 'c' are ascribed to filling and condensation respectively.

The α_s *plot* (Fig. 9.2) is obtained by plotting the volume of gas adsorbed by a test sample versus α_s, in the same way as one produces a t-plot. The estimation of micropore volume from an α_s plot, as in the t-method, involves extrapolation of the plots to the Y-axis Since the α_s-method does not assume any value for the thickness of an adsorbed layer, the calculation of the surface area is accomplished by relating the slope of the α_s-plot of the test sample to the slope of the corresponding plot for a standard sample of known surface area. In principle, the α_s-method can be used with any adsorptive gas and can be used to check the BET surface area, and to assess micro- and mesoporosity [14b].

Kaneko *et al.* [13c] introduced the high-resolution α_s analysis, based on a high-resolution standard isotherm [13a]. The high-resolution method makes particular use of the α_s plot below $\alpha_s = 0.4$ where, according to the texture of the adsorbent, the characteristic feature, f, can be observed. Such a detailed analysis of the α_s plot allows one to obtain more information concerning micro- and mesoporosity in the adsorbent.

9.3 THE MICROPORE ANALYSIS (MP) METHOD

Mikhail, Brunauer and Bodor [15] proposed an extension of de Boer's t-method that offers several advantages. These include the ability to obtain the micropore volume, surface area and pore size distributions from one experimental isotherm. Data for the MP (micropore analysis) method need not be measured at the very low pressures associated with the methods related to the Dubinin theory. The method assumes no site energy or adsorption volume distribution functions as required by the Kaganer and Dubinin theories (see chapter 4, §4.3). In addition, when the micropore analysis is completed, the MP method is applicable to adsorbents containing macropores, transitional pores (mesopores), and micropores with self-termination. When measuring mesopores, the statistical thickness, t, is used as a correction to allow for desorption from the adsorbed film. However, the value of t is much more critical when measuring micropores because in the MP method, t is the actual measure of the pore size.

In order to perform an accurate micropore analysis, a statistical thickness must be taken from a t versus P/P_0 curve that has approximately the same BET C value as the test sample. The unavailability of t versus P/P_0 plots on numerous surfaces with various C values is certainly a problem for the application of the MP equation. The calculation of t from equation (9.2) implies that the surface area can be accurately measured on microporous samples, which is also problematic as discussed in chapter 5.

To illustrate the MP method, consider the isotherm shown in Fig. 9.3. The volumes of adsorbed gas are converted to liquid volumes, from which t is calculated using equation (9.1). The V-t plot shown in Fig. 9.4 is then constructed using relative pressure intervals of 0.05. In the example shown, the t values were not calculated from equation (9.1) but were taken form a t versus P/P_0 plot prepared by de Boer *et al.* [16] on a sample with a similar BET C value. By choosing t values from a material with a similar but not identical C value, the surface area nevertheless agreed to within 1.4% of the BET measured area.

The slope of the linear portion of the curve, the origin through the first four points, is 0.0792; this gives a micropore surface area of 792 m²/g when calculated from equation (8.31). Straight-line #2, drawn tangentially to the curve between $t = 4$ and 4.5 Å, exhibits a slope of 0.0520. The area of all the pores remaining unfilled by adsorbate is 520 m²/g, and the surface area of pores in the range of thickness from 4 to 4.5 Å is 792 − 520 = 272 m²/g. The third line gives a slope of 0.0360 between t values of 4.5 and 5 Å, so the area of pores in this range is 520 − 360 = 160 m²/g. The calculation is continued in this manner until there is no further decrease in the slope of the V-t plot, which indicates that all the pores are filled.

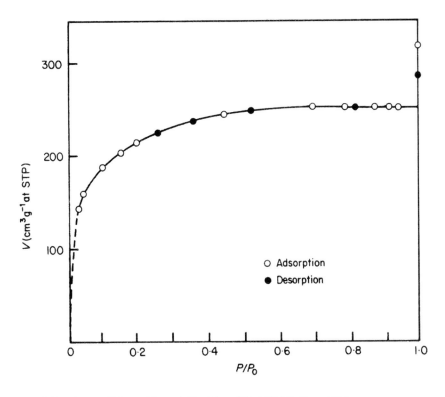

Figure 9.3 Isotherm of N₂ on silica gel (*Davidson 03*) at 77.3 K. From [15].

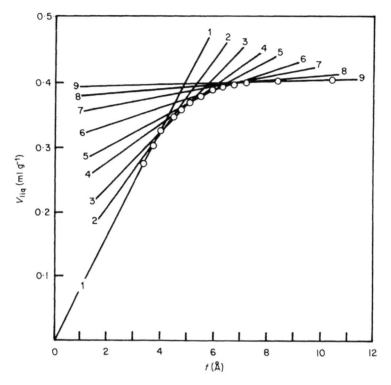

Figure 9.4 V-t curve from Fig. 9.5. From [15].

The calculation of pore volume is carried out in an equally straightforward manner. For example, the volume of pores is given by

$$V = 10^{-4}(S_1 - S_2)\frac{(t_1 + t_2)}{2} \text{ cm}^3/\text{g} \qquad (9.9)$$

Thus, for the first group of pores the volume is

$$V = 10^{-4}(792 - 520)\frac{(4.0 + 4.5)}{2} = 0.1156 \text{ cm}^3/\text{g} \qquad (9.10)$$

The micropore data for the isotherm shown in Fig. 9.3 are illustrated in Table 9.1, in which S_i is the total surface area. The exact pore shape is usually unknown and cylindrical pores are generally assumed. Mikhail, Brunauer and Bodor [15] show in their paper that equation (9.9) is equally valid for parallel plate or cylindrical pores. They also demonstrate that the hydraulic radius (r_h in Table 9.1) is the same as the plate separation or the cylinder radius.

Table 9.1. Isotherm data taken from Fig. 9.5. BET area = 793 m^2/g, V-t area = 782 m^2/g; total pore volume = 0.4034 cm^3/g; MP pore volume = 0.4088 cm^3/g. The authors attributed the difference between V-t and BET areas to surfaces that did not lie within pores.

Pore Group	S_{i+1} (m^2/g)	S_i - S_{i+1} (m^2/g)	Mean r_h (Å)	V_i (cm^3/g)
1	520	272	4.25	0.1156
2	360	160	4.75	0.0760
3	280	80	5.25	0.0420
4	200	80	5.75	0.0460
5	140	60	6.25	0.0375
6	80	60	6.75	0.0405
7	20	60	7.25	0.0435
8	10	10	7.75	0.0077
		ΣS_i = 782		

9.4 TOTAL MICROPORE VOLUME AND SURFACE AREA

Both de Boer's t-method and Brunauer's MP method are based on the assumption that the BET measured surface area is valid for micropores. Shields and Lowell [17], using this assumption, have proposed a method for the determination of the micropore surface area using mercury porosimetric data. The surface area of micropores is determined as the difference between the BET surface area and that obtained from mercury porosimetry (see chapter 10). Since mercury porosimetry is capable of measuring pore radii only as small as approximately 18 Å, this technique affords a means of calculating the surface area of all pores with radii between 3.5 Å (the approximate diameter of a nitrogen molecule) and 18 Å. Similarly, Shields and Lowell [17] suggested a method for the determination of the total micropore volume by a combination of techniques. The difference between the sample volumes measured by mercury porosimetry and helium pycnometry (see chapter 19) is the volume occupied by micropores.

9.5 THE DUBININ - RADUSHKEVICH (DR) METHOD

The basis of the DR theory was already discussed in §4.3. Dubinin and Radushkevich [18] put forward an equation based on Polanyi's potential theory, which allows the micropore volume to be calculated from the adsorption isotherm:

$$log\, W = log\left(\tilde{V}_0\,\rho\right) - \overline{K}\left[log\left(\frac{P_0}{P}\right)\right]^2 \qquad (9.11)$$

W and ρ are the weight adsorbed and the liquid adsorbate density, respectively; $\tilde{V}_0$ is the micropore volume. k is defined as

$$k = 2.303K\left(\frac{RT}{\beta}\right)^2 \qquad (9.12)$$

where β is the so-called affinity coefficient (for details see §4.3, and equation (4.48)), and K is a constant, determined by the shape of the pore size distribution. A plot of $log\, W$ versus $[log(P_0/P)]^2$ should give a straight line with an intercept of $log(\tilde{V}_0\rho)$, from which $\tilde{V}_0$, the micropore volume, can be calculated. Nikolayev and Dubinin [19] found linear plots using relative pressures ranging from 10^{-5} to 10^{-1} on a variety of microporous samples.

Kaganer [20] modified Dubinin's method in order to calculate the surface area within micropores. He obtained the following equation

$$log\, W = log\, W_m - \overline{K}\left[log\left(\frac{P_0}{P}\right)\right]^2 \qquad (9.13)$$

where

$$\overline{K} = 2.303\overline{k}\left(RT\right)^2 \qquad (9.14)$$

Equation (9.13) is similar to equation (9.11), Dubinin's equation. A plot of $log\, W$ versus $log\,[(P_0/P)]^2$ will yield a straight line with an intercept of $log\, W_m$ from which the surface area can be calculated by equation (4.13).

Some microporous carbons do give linear DR plots over a wide range of P/P_0, but there are many examples where DR plots exhibit only a very restricted range of linearity. The linear range of such plots is usually at very low pressures, $P/P_0 < 10^{-2}$. Linear DR plots over a large relative pressure range can be found for a number of microporous carbons, for many other adsorbents (zeolites are particularly problematic) the linear range is limited over a very narrow relative pressure range. The DR equation often fails to linearize the data when the microporous adsorbent is very heterogeneous with regard to surface chemistry and texture. To overcome such deficiencies of the original DR equation, a more general equation, known as the DA equation, was proposed by Dubinin and Astakhov [21] The DA equation (see also §4.3, equation (4.56)) is often applied in its linearized form

$$\ln W = \ln(\widetilde{V}_0, \rho) - K[\ln(P_0/P)]^n \qquad (9.15)$$

where K is again an empirical constant, and n is the so-called Dubinin-Astakhov parameter. For adsorbents with homogeneous micropore structure close to carbonaceous molecular sieves, parameter n is usually close to 2, but depending on the type of micropore system (and its heterogeneity), n may vary between 2 and 5 [22c]. The case where $n=2$ corresponds to the classical equation of Dubinin and Raduskevich (DR). Another generalization of the DR equation has been introduced by Stoeckli *et al.* [22], which also addressed the effect of heterogeneity in the micropore structure on the adsorption isotherm. Stoeckli's approach, which provides an altenative to the DA equation is based on the assumption that the original DR equation was compatible only for carbons which have a narrow range of micropore size. For strongly activated carbons with a heterogeneous collection of micropores it was assumed that the overall experimental sorption isotherm consists of the contributions from the different groups of pores. Under certain assumptions, both the total micropore volume and the range of pore size can be determined.

9.6 THE HORVATH-KAWAZOE (HK) APPROACH AND RELATED METHODS

Horvath and Kawazoe (HK) [23] described a semi-empirical, analytical method for the calculation of effective pore size distributions from nitrogen adsorption isotherms in microporous materials. The HK approach is based on a fundamental statistical analysis of a fluid confined to a slit-pore (e.g., applicable to carbon molecular sieves and active carbons).

An extension of the HK method for cylindrical pore geometry (e.g., applicable to some zeolites) was made by Saito and Foley [24] and for a spherical pore model by Cheng and Yang [25]. The spherical pore model is more suitable for cavity type zeolites (e.g., faujasite) whereas the cylindrical pore model is justified for channel type zeolites (e.g., ZSM-5) exhibiting cylindrical pores.

The HK method is based on the work of Everett and Powl [26]. They calculated the potential energy profiles for noble gas atoms adsorbed in a slit between two graphitized carbon layer planes. The separation between nuclei of the two layers is ℓ. The adsorbed fluid is considered as a bulk fluid influenced by a mean potential field, which is characteristic of the adsorbent-adsorbate interactions. The term *mean field* indicates that the potential interactions between an adsorbate molecule and the adsorbent, which may exhibit a strong spatial dependence, are replaced by an average, uniform potential field. Horvath and Kawazoe found by using thermodynamic arguments, that this average potential can be related to the free energy change of adsorption, yielding a relation between filling pressure and the effective pore width $d_p = \ell - d_a$, where d_a is the diameter of an adsorbent molecule:

$$\ln\left(\frac{P}{P_0}\right) = \frac{N_A}{RT} \times \frac{N_S A_S + N_a A_a}{\sigma^4 (l - 2d_0)} \times$$

$$\left[\frac{\sigma^4}{3(l-d_0)^3} - \frac{\sigma^{10}}{9(l-d_0)^9} - \frac{\sigma^4}{3(d_0)^3} + \frac{\sigma^{10}}{9(d_0)^3}\right] \quad (9.16)$$

The parameters d_0, σ, A_s and A_a can be calculated using the following equations:

$$d_0 = \frac{d_a + d_s}{2} \quad (9.17)$$

$$\sigma = \left(\frac{2}{5}\right)^{\frac{1}{6}} \cdot d_0 \quad (9.18)$$

$$A_s = \frac{6m_e c^2 \alpha_s \alpha_a}{\dfrac{\alpha_s}{\chi_s} + \dfrac{\alpha_a}{\chi_a}} \quad (9.19)$$

$$A_a = \frac{3}{2} m_e c^2 \alpha_a \chi_a \quad (9.20)$$

A_a = Kirkwood –Mueller constant of adsorptive
A_s = Kirkwood-Mueller constant of adsorbent
$d_0 = (d_s+d_a)/2$: distance between adsorptive and adsorbent molecules
d_a = diameter of an adsorptive molecule
N_a = number of atoms per unit area (m^2) of adsorbent
N_A = number of adsorptive molecules per unit area (m^2) of adsorbent
m = mass of an electron

c = speed of light
α_s = polarizability of adsorbent
α_a = polarizability of adsorbate
σ = distance between two molecules at zero interaction energy
χ_s = magnetic susceptibility of the adsorbent
χ_a = magnetic susceptibility of the adsorptive

According to equation (9.16), the filling of micropores of a given size and shape takes place at a characteristic relative pressure. This characteristic pressure is directly related to the adsorbent-adsorptive interaction energy.

Saito and Foley extended the HK method for the calculation of effective pore size distributions from argon adsorption isotherms at 87 K in zeolites [24]. They based their method, as did Horvath and Kawazoe, on the Everett and Powl potential equation, but for cylindrical pore geometry instead. Following the logic of the HK derivation, Saito and Foley derived a similar equation that relates the micropore filling pressure to the effective pore radius.

HK predictions for the pore filling pressures of nitrogen in slit-shaped carbon pores of different width are given in table 9.2. Table 9.3 shows the relation between the pore diameter D and the relative pressure P/P_0 where micropore filling of argon occurs at 87.27 K in cylindrical pore geometry according to the SF approach. The values in these tables were taken from the original papers of Horvath-Kawazoe [23] and Saito-Foley [24], respectively.

Table 9.2 Nitrogen (77 K) pore filling pressures for a slit-like carbon pore according to Horvath – Kawazoe [23].

D(nm)	0.4	0.6	0.8	1.1	1.5
P/P_0	1.47×10^{-7}	1.54×10^{-4}	2.95×10^{-3}	2.22×10^{-2}	7.59×10^{-2}

Table 9.3 Argon (87 K) pore filling pressures for a cylindrical zeolite pore according to the Saito-Foley theory [24b].

D(nm)	0.4	0.6	0.8	1.1	1.5
P/P_0	2.86×10^{-7}	1.03×10^{-4}	2.57×10^{-3}	2.29×10^{-2}	8.30×10^{-2}

It is important to mention that the pore width – relative pressure pairs given in the tables 9.2 and 9.3 are very sensitive with regard to the values of the magnetic susceptibility and polarizability chosen for the calculation (the parameter used to obtain the values in these tables are given in the original papers, [23, 24b]). Saito-and Foley [24b] have studied the effect of different values of the magnetic susceptibility of the zeolitic oxide ion (which was chosen as an adjustable parameter for adsorption interactions) on the pore filling pressures of argon in cylindrical pore model. The values given in table 9.2 were calculated using a value for the magnetic susceptibility that would correspond to zeolite Y (although zeolite Y does not exhibit a cylindrical pore geometry).

The semi-empirical HK and SF methods are widely used, and because they allow for the importance of the solid-fluid attractive forces in narrow pores, provide a better measure for micropore filling pressures than the macroscopic, classical methods of micropore analysis (e.g., the DR approach). However, the method still leads to an inaccurate micropore size analysis mainly because of these reasons: (i) the mechanism of micropore filling is not correctly described, i.e., this is thought to be a continuous process, but in the HK-related methods it is assumed that pore filling occurs discontinuously at a specific pressure characteristic of its' size; (ii) the assumption that the confined fluid behaves as a bulk fluid is questionable from a statistical thermodynamics point of view for small finite systems; and (iii) the HK-related methods do not take into account that the local density in the pore varies strongly with position due to fluid layering near the pore walls. The omission of this characteristic oscillating density profile leads to an underestimation of the pore size compared to exact pore sizes calculated for molecular simulation and density functional theory (see §9.7). Appropriate improvements of the HK method were recently put forward by Lastoskie et al. [27], and application of their modified HK-method for argon sorption at 77 K in carbon slit pores has led to a pore filling correlation (i.e., pore filling pressure versus pore size), which is in close agreement with the exact treatments from DFT.

It should be noted that the HK and SF methods are not applicable for mesopore size analysis. In contrast, statistical mechanical methods like the density functional theory (DFT) and computer simulation methods (Monte Carlo and Molecular Dynamics) provide a more realistic picture of micropore filling and can be applied for both micro- and mesopores size analysis. This will be discussed in the next section.

9.7 APPLICATION OF NLDFT: COMBINED MICRO/ MESOPORE ANALYSIS WITH A SINGLE METHOD

As indicated earlier, macroscopic, thermodynamic approaches like the DR-method and semi-empirical treatments such those of Horvath and Kawazoe

(HK), and Saito and Foley (SF) [23,24] do not give a realistic description of micropore filling. This leads to an underestimation of pore sizes for a given pore filling pressure as compared to methods like the Non-local Density Functional Theory (NLDFT) or methods of molecular simulation (Grand Canonical Monte Carlo simulation methods (GCMC), Molecular Dynamics methods (MD)). This situation is shown in Fig. 9.5.

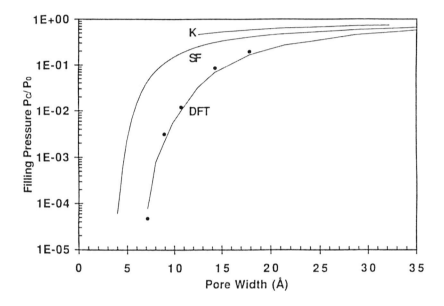

Figure 9.5 Pore filling pressures for nitrogen in cylindrical oxide pores at 77 K, as predicted, by Saito-Foley equation (SF), Kelvin equation (K), by NLDFT and Gibbs Ensemble Monte Carlo simulations (points). From [28].

where it is clearly shown that the Saito-Foley method significantly underestimates the pore size for given pore filling pressures of nitrogen in a cylindrical pore (of oxidic walls) compared to the predictions of the DFT method as well as Monte Carlo simulation.

Theoretical aspects of the DFT-method including information with regard to the calculation of a DFT based pore size distribution by applying the GAI equation was already described in chapter 4. We also discuss in chapter 8 that an accurate pore size analysis over the complete micro- and mesopore size range is only possible by applying microscopic methods based on statistical mechanics, such as the Non-Local-Density-Functional Theory (NLDFT) [e.g., 28,29,31-35] or Grand Canonical Monte Carlo simulation (GCMC) [e.g., 28, 30]. Pore size analysis data for micro- and

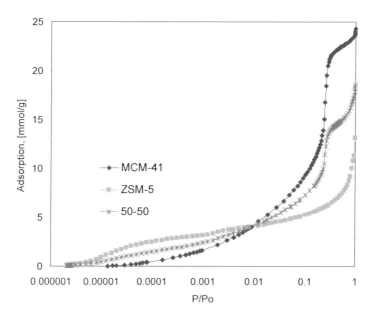

Figure 9.6 Argon sorption isotherms obtained at 87 K on ZSM 5, MCM 41, and on a mixture of ZSM5 and MCM 41. From [36].

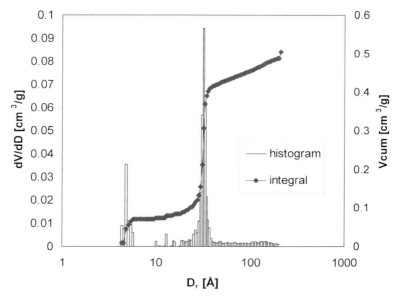

Figure 9.7 NLDFT pore size distribution curves for ZSM-5 and MCM-41 obtained from the argon sorption isotherms on the ZSM-5/MCM-41 mixture. From [36].

mesoporous molecular sieves (e.g., some zeolites, etc.) obtained with these methods agree very well with the results obtained from independent methods (based on XRD or TEM) [35, 36].

Hence, the application of microscopic methods leads to: (i) a much more accurate pore size analysis and (ii) allows in principle to perform a pore size analysis over the complete micro/mesopore size range. An example is shown in Figs. 9.6 and 9.7 where the NLDFT method was applied to calculate the pore size distribution curve of a 50:50 mixture of MCM-41 and ZSM-5 zeolite based on argon adsorption at 87.27 K.

Fig. 9.6 shows argon sorption isotherms obtained at 87 K on ZSM 5, a zeolite with cylindrical pore geometry, on MCM 41, a mesoporous material consisting of independent cylindrical-like pores, and on a mixture of ZSM5 and MCM 41, which resembles a combined micro/mesoporous material. Fig. 9.7 compares the experimental isotherm together with the theroetical DFT isotherm; the fit to the isotherm is excellent. The NLDFT pore size distribution of the "combined" material (also Fig. 9.7) shows two distinct groups of pores: micropores of the same size as in ZSM-5 and mesopores of the size as in MCM-41. It should be noted that the reported average pore diameter of ZSM-5 zeolite obtained from structural considerations is 5.1 - 5.5 Å, which agrees very well with the pore size distribution obtained from argon adsorption by the NLDFT method. The pore size obtained by independent methods (XRD) for the mesoporous MCM 41 is 3.2 nm, which is again in excellent agreement with the result obtained with the NLDFT method.

Meanwhile, many NLDFT and GCMC methods, applicable for various adsorptive/adsorbent systems, have been developed and commercialized [39]. Please note that the application of these advanced methods is useful (and therefore leads to accurate results) only if the given experimental adsorptive/adsorbent system is compatible with the NLDFT- or GCMC kernel available! A drawback of the currently available NLDFT and GCMC methods is certainly (as already stated in Chapter 8, §8.5), that they do not take sufficiently into account the chemical and geometrical heterogeneity of the pore walls, i.e., usually a structureless (i.e., chemically and geometrically smooth) pore wall model is assumed. This leads to small steps in the theoretical isotherms due to layering transition, which cannot be observed in experimental sorption isotherms obtained on adsorbents with chemical and geometrical heterogeneous surfaces. However, despite these deficiencies, methods for pore size analysis based on NLDFT and GCMC are meanwhile widely used and are considered the most accurate methods for micro- and mesopore size analysis.

9.8 ADSORPTIVES OTHER THAN NITROGEN FOR SUPER- AND ULTRAMICROPOROSIMETRY

Micropore analysis of microporous materials (e.g., activated carbon and zeolites) has mainly been performed by nitrogen adsorption at 77 K, but this is not satisfactory with regard to a quantitative assessment of the microporosity, especially in the range of ultramicropores (pore widths < 0.7 nm). A pore width of 0.7 nm corresponds to the bilayer thickness of the N_2 molecule. In addition, preadsorbed N_2 molecules near the entry of an ultramicropore may block further adsorption. The pore filling of such narrow pores occurs at relative pressures of 10^{-7} to 10^{-5}, where the rate of diffusion and adsorption equilibration is very slow. This leads to time-consuming measurements and may cause under-equilibration of measured adsorption isotherms, which will give erroneous results of the analysis.

For many microporous systems (in particular zeolites) the use of argon as adsorptive at its boiling temperature (87.27 K) appears to be helpful. Argon fills micropores of dimensions 0.4 nm – 0.8 nm in most cases at much higher relative pressures, (i.e., at least 1.5 decades higher in relative pressures) as compared to nitrogen, which leads to accelerated diffusion and equilibration processes and thus also to a reduction in analysis time.

The different pore filling ranges of argon adsorption at 87 K and nitrogen adsorption at 77 K is illustrated in Fig. 9.8 based on sorption data obtained on a faujasite-type zeolite. The much lower pore filling pressure for nitrogen compared to argon is still not completely understood, but clearly indicates that the attractive interactions of nitrogen molecules with the pore walls of the zeolite are much stronger as compared to argon. The possibility that this enhancement of the adsorption potential is related to specific quadrupole interactions is under discussion.

However, the micropores of some zeolites are too small to be characterized by nitrogen and argon adsorption at cryogenic temperatures. Here, the so-called *molecular probe method* offers a possible way of direct determination of the effective pore size. This method is based on the measurement of sorption rates and capacities using a series of sorbates of progressively increasing molecular diameter. There will be a very sharp sorption cutoff, when the molecule cannot enter anymore the micropore, and a good estimate of the effective pore size can be obtained in this way [37].

The problem of assessing ultramicroporosity appears in particular for microporous carbons, which often exhibit a wide distribution of pore sizes including ultramicropores. It has long been recognized [e.g., 38, 40, 41] that using CO_2 adsorption analysis at 273.15 K can eliminate problems of this type. At 273.15 K, CO_2 is still ca. 32 K below its critical temperature, and because the saturation pressure is very high (26200 torr), the relative pressure measurements necessary for the micropore analysis are achieved in

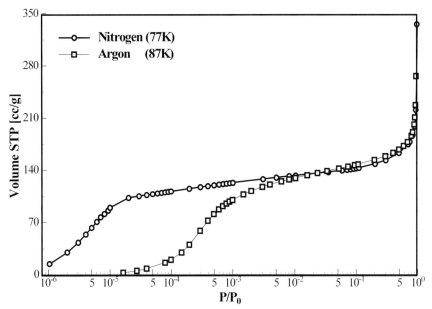

Figure 9.8 Nitrogen and argon adsorption at 77 K and 87 K, respectively, on a faujasite -type zeolite.

the range of moderate absolute pressures (1 – 760 torr). Hence, due to the relatively high absolute temperatures and pressures compared with nitrogen and argon adsorption at cryogenic temperatures, diffusion problems can be eliminated. Because of the higher diffusion rate, adsorption equilibrium is achieved faster, which allows completion of the adsorption isotherm in a significantly shorter time compared to nitrogen adsorption at 77 K. In addition, the range of analysis can be extended to pores of smaller sizes that are accessible to CO_2 molecules, but not to nitrogen and argon. However, if the analysis is performed with a conventional volumetric sorption analyzer, which can be used to pressures up to ca. 1 atm, the measurable pore size range is limited to pore sizes up to 1.5 nm. The CO_2 adsorption isotherms measured under such conditions can be analyzed using modern molecular models such as Density Functional Theory (DFT) or Monte Carlo simulations in order to obtain detailed information about the carbon micropore structure [38]. Such advanced methods for CO_2 analysis have been recently commercialized [39].

Gregg and Langford [42] suggested assessing microporosity by making use of the strong retention of *n*-nonane in narrow pores (i.e., the method of "nonane preadsorption"). The method was often applied to microporous carbon [43]. The aim is to fill all micropores with nonane while leaving the wider pores open. Because of the high physisorption energy, the preadsorbed nonane can only be removed at elevated temperatures. A possible procedure for the nonane preadsorption goes as follows: The sample is outgassed and a first nitrogen sorption isotherm is obtained. The sample is then exposed to nonane vapor. The sample, saturated with nonane is then outgassed again at room temperature, prior to the re-determination of the nitrogen sorption isotherm. In the following steps the sample is outgassed at increasingly higher temperatures. After each stage of outgassing, a nitrogen sorption isotherm is measured until the nonane has been completely removed. The difference in pore volumes and surface areas before and after nonane preadsorption is attributed to the fact that narrow micropores were completely blocked by preadsorbed nonane. However, the conclusions, which can be drawn from nonane preadsorption are not always straightforward (with regard to assessing the complete microporosity), mainly because of the pore size dependency of the adsorption potential, i.e. nonane molecules are more firmly trapped in the ultramicropores than in supermicropores. Another problem is associated with the blocking of wider pores due to the adsorption of nonane in the more narrow pores; this problem is in particular important in case the adsorbent consists of networks of pores with different sizes.

Kaneko *et al.* [44, 45] suggested helium as a good probe molecule for the assessment of ultramicropores, because helium is the smallest inert molecule and can possibly penetrate a narrow neck of a micropore. Within this context, Kaneko *et al.* determined the adsorption isotherms of He on activated carbon fibers at 4.2 K by a gravimetric method. The He adsorption isotherms were of typical type I and an analysis by the Dubinin-Radushkevich method gave micropore volumes greater than those obtained from N_2 adsorption at 77 K by 20 - 50%. The excess amount of He adsorption was ascribed to the presence of ultramicropores, which cannot be assessed by N_2 molecules. There are a couple of reasons that could explain the large difference in micropore volumes obtained by He and N_2 adsorption. The first reason is due to the fact that the He molecule is smaller and can enter narrower micropores than N_2. The second reason is associated with the packing efficiency; smaller molecules can fill more effectively the restricted micropore space. Helium adsorption at 4.2 K is considered to be a suitable tool to assess ultramicroporosity.

9.9 REFERENCES

1.   Sing K.S.W., Everett D.H., Haul R.A.W., Mouscou L., Pierotti R.A., Rouquerol J. and Siemieniewska T. (1985) *Pure & Appl. Chem.* **57** 603.
2.   Shull C.G. (1958) *J. Am. Chem. Soc.* **70**, 1405.
3.   Lippens B.C. and de Boer J.H. (1965) *J. Catal.* **4**, 319.
4.   Pierce C. (1953) *J. Phys. Chem.* **57**, 149.
5.   Harris M.R. and Sing K.S.W. (1959) *Chem Ind. (London)* 11.
6.   Cranston R.W. and Inkley F.A. (1957) *Adv. Catal.* **9**, 143.
7.   Halsey G.D. (1948) *J. Chem. Phys.* **16**, 931.

8. De Boer J.H., Lippens B.C., Linsen B.G., Broeckhoff J.C.P., van den Heuvel A. and Onsinga T.V. (1966) *J. Colloid Interface Sci.* **21**, 405.
9. Harkins W.D. and Jura G. (1944) *J. Am. Chem. Soc.* **66**, 1366.
10. Lecloux A.J. (1981) *Catalysis- Science and Technology, Vol.2* (J.R. Anderson and P. Boudard, eds) Springer Verlag, Berlin p171.
11. Hudec P., Smiešková A., Židek Z, Schneider P. and Šolcová O. (2002) *Stud. Surf. Sci. Catal.* **142**, 1587.
12. *Standard Test Method for Carbon Black-Total and External Surface Area by Nitrogen Adsorption.* D-6556-01, ASTM International, West Conshohocken, Pennsylvania.
13. [a]Kaneko K. (1994) *J. Membrane Sci.* **96**, 59; [b]Ohkubo T., Miyawaki J. and Kaneko K. (2002) *J. Phys. Chem. B* **106**, 6523; [c]Kaneko K., Ishii C., Ruike M. and Kuwabara H. (1992) *Carbon* **30**, 1075.
14. [a]Gregg S.J. and Sing K.S.W. (1982) *Adsorption, Surface Area and Porosity*, 2nd edn, Academic Press, London; [b]Carrott P.J.M. and Sing K.S.W. (1988) *Stud. Surf. Sci. Catal.* **39**, 77.
15. Mikhail R.H., Brunauer S. and Bodor E.E. (1968) *J. Colloid Interface. Sci.* **26**, 45.
16. de Boer J.H., Linsen B.G., van der Plas T. and Zondervan G.J. (1965) *J. Catal.* **4**, 649.
17. Shields J.E. and Lowell S. (1983) *Powder. Technol.* **36**, 1.
18. Dubinin M.M. and Radushkevich L.V. (1947) *Proc. Acad. Sci. USSR*, **55**, 331.
19. Nikolayev K.M. and Dubinin M.M. (1958) *Izv. Akad. Nauk. SSSR, Otd. Tekn. Nauk.*, 1165.
20. Kaganer M.G. (1959) *Zh. Fiz. Khim.* **32**, 2209.
21. Dubinin M.M. and Astahakov V.A. (1971) *Adv. Chem. Soc.* **102**, 69.
22 [a]Stoeckli H.F. (1977) *J. Colloid Interface Sci.* **59**, 184; [b]Dubinin, M. M. and Stoeckli H.F. (1980) *J. Colloid Interface Sci.* **75**, 34; [c]Stoeckli H.F., Kraehenbuehl F., Ballerini L. and De Bernandini S. (1982) *Carbon* **27**, 125.
23. Horvath G. and Kawazoe K. (1983) *J. Chem. Eng. Jpn.* **16**, 474.
24. [a]Saito A. and Foley H.C. (1991) *Am. Inst. Chem. Eng.* **J. 37**, 429; [b]Saito A. and Foley H.C (1995) *Microporous Mater.* **3**, 531.
25. Cheng L.S. and Yang R.T. (1994) *Chem. Eng. Sci.* **49**, 2599.
26. Everett D.H. and Powl J.C. (1976) *J. Chem. Soc. Faraday Trans.* **I 72**, 619.
27. Lastoskie M.L. (2000) *Stud. Surf. Sci. Catal.* **128**, 475.
28. [a]Gubbins K.E. (1997) In *Physical Adsorption: Experiment, Theory and Applications* (J. Fraissard, ed.) Kluwer, Dordrecht; [b] Peterson B.K., Gubbins K.E., Heffelfinger G.S., Marconi U.M.B. and van Swol F. (1988) *J. Chem. Phys.* **88**, 6487.
29. [a]Evans R., Marconi U.M.B. and Tarazona P. (1986) *J. Chem. Soc. Faraday Trans.* **82**, 1763; [b]Tarazona P., Marconi U.M.B. and Evans R. (1987) *Mol. Phys.* **60**, 573; [c]Evans R. (1990) In *Liquides aux Interfaces, Les Houches* 1988, Session XLVIII, Course 1 (J. Charvolin, J. Joanny and J.F. Zinn-Justin, eds.) North-Holland, Amsterdam.
30. Walton J.P.R.B. and Quirke N. (1989) *Mol. Simulation* **2**, 361.
31. Seaton N.A., Walton J.R.P.B. and Quirke N. (1989) *Carbon* **27**, 853.
32. Lastoskie C., Gubbins K. and Quirke N. (1993) *J. Phys. Chem.* **97**, 4786.
33. Olivier J.P., Conklin W.B. and von Szombathely M. (1994) In *Characterization of Porous Solids* III (K.S.W. Sing, ed.) Elsevier, Amsterdam.
34. Neimark A.V. (1995) *Langmuir* **11**, 4183.
35. [a]Ravikovitch P.I., Vishnyakov A. and Neimark A.V. (2001) *Phys. Rev. E*, **64**, 011602. [b]Neimark A.V., Ravikovitch P.I. and Vishnyakov A. (2003) *J. Phys. Condens. Matter* **1**, 347.
36. [a]Ravikovitch P.I. and Neimark A.V. (2001) personal communication; [b]Thommes M. (2001) In *Particle Technology News* **4**, Quantachrome, Florida.
37. Ruthven D.M. (1984) *Principles of Adsorption and Adsorption Processes*, Wiley Interscience, New York.
38. [a]Ravikovitch P.I. and Neimark A.V. (2000) *Langmuir* **16**, 2311; [b]Vishnyakov A.,

Ravikovitch P.I. and Neimark A.V. (1999) *Langmuir* **15**, 8736.

39. Quantachrome Instruments, Boynton Beach, Florida, USA.
40. Garrido J., Linares-Solano A., Martin-Martinez J.M., Molina-Sabio M., Rodriguez-Reinoso F. and Torregosa R. (1987) *Langmuir* **3**, 76.
41. Cazorla-Amoros D., Alcaniz-Monje J. and Linares-Solano A. (1996) *Langmuir* **12**, 2820.
42. Gregg S.J. and Langford J.F. (1969) *Trans. Faraday Soc.* **65**, 1394.
43. Carrott P.J.M., Drummond F.C., Roberts R.A. and Sing K.S.W. (1988) *Chem. Ind. (London)* **6**, 372.
44. Kuwabara H., Suzuki T. and Kaneko K. (1991) *J. Chem. Soc. Faraday Trans.* **87**, 1915.
45. Kaneko K., Setoyama N., Suzuki T. and Kuwabara H. (1992) In *Proc. IVth Int. Conf. on Fundamentals of Adsorption*, Kyoto.

10 Mercury Porosimetry: Non-wetting Liquid Penetration

10.1 INTRODUCTION

The method of mercury porosimetry for the determination of the porous properties of solids is dependent on several variables. One of these is the wetting or contact angle between mercury and the surface of the solid.

When a liquid is placed in contact with a surface of a porous solid, the question arises as to whether it will penetrate into the pores. The answer must be pursued in the realm of capillarity, which deals with the equilibrium geometries of liquid-solid interfaces and the angle of contact between the liquid and the pore wall.

In the absence of gravity or other external forces, a liquid will assume a spherical shape, i.e. one that possesses the minimum area to volume ratio of all geometric forms. If any external force distorts the sphere, molecules must be brought from the interior to the surface in order to provide for the necessarily increased surface area. This process will require that work be expended in order to raise the potential energy of a molecule when the number of stabilizing interactions with neighboring molecules is reduced upon reaching the surface. The work expended will raise the free energy G of the liquid. That part of the total free energy change ΔG that is altered is the free surface energy G^s. Therefore, the change of free surface energy ΔG^s is the net work required to alter the surface area of a substance. Since spontaneous processes are associated with a decrease in free energy, in the absence of external forces, liquids will spontaneously assume a spherical shape in order to minimize their exposed surface area and thereby their free surface energy. The spontaneous coalescence of two similar liquid droplets into one large drop when brought into contact is a dramatic demonstration of the free surface energy decrease brought about by the decrease in total surface area by the formation of a single larger drop.

The surface tension, γ, of a substance is identical to the free surface energy G^s per unit area and is the work required to alter the surface area by one square centimeter. Therefore, γ_S has the dimensions of energy per unit area.

According to the preceding argument, a bubble should collapse in order to minimize its free surface energy. In the process of shrinking, however, the gas pressure within the bubble will increase thereby preventing any further reduction in radius. When the bubble radius decreases from r to $r–dr$, the free energy decreases by

$$-\,dG = 8\pi r\gamma dr \tag{10.1}$$

When the bubble shrinks, the volume change is $4\pi r^2 dr$. The gas within the bubble undergoes compression while the external atmosphere undergoes expansion. The net work associated with the compression and expansion is given by

$$-\,dW = \left(P_{int} - P_{ext}\right)4\pi r^2 dr \tag{10.2}$$

The internal pressure, P_{int}, is greater than the external pressure, P_{ext}, since it supports both the external pressure and the tendency for the area of the surface film to decrease. Since the work performed is equal to the free energy difference, one obtains from equations (10.1) and (10.2) at mechanical equilibrium

$$\left(P_{int} - P_{ext}\right) = \frac{2\gamma}{r} \tag{10.3}$$

or

$$\Delta P = \frac{2\gamma}{r} \tag{10.4}$$

Equation (10.4) dictates that the smaller the bubble radius the greater will be the pressure difference across the wall. Thus, a large and small bubble each exposed to the same external pressure will result in a greater pressure within the smaller bubble. A vivid demonstration of this occurs when a balloon is inflated. The lung pressure required decreases rapidly as the balloon radius increases until the elastic limit is approached.

10.2 YOUNG-LAPLACE EQUATION

Equation (10.4) is a special case of a more general concept represented by the Young [1] and Laplace [2] equation. A sphere possesses a constant radius of curvature. For an area element belonging to a non-spherical curved surface, there can exist two radii of curvature (r_1 and r_2). If the two radii of curvature are maintained constant while an element of the surface is stretched along the x-axis from x to $x + dx$ and along the y-axis from y to $y + dy$ the work W_1 performed will be

$$W_1 = \gamma\left[\left(x + dx\right)\left(y + dy\right) - xy\right] \tag{10.5}$$

which, ignoring the product of differentials, reduces to

$$W_1 = \gamma(x\,dy + y\,dx) \tag{10.6}$$

If the area element is stretched due to an increase in internal pressure relative to the external pressure, there will also be displacement along the z-axis as the surface expands. The work performed will be

$$W_2 = \Delta P (x + dx)(y + dy)dz \tag{10.7}$$

Again, neglecting the product of differentials, equation (10.7) reduces to

$$W_2 = \Delta P x y dz \tag{10.8}$$

Because the two radii of curvature r_1 and r_2 remain unchanged, one can write

$$\frac{(x + dx)}{(r_1 + dz)} = \frac{x}{r_1} \tag{10.9}$$

and

$$\frac{(y + dy)}{(r_2 + dz)} = \frac{y}{r_2} \tag{10.10}$$

Equations (10.9) and (10.10) reduce to

$$dx = \frac{x\,dz}{r_1} \tag{10.11}$$

and

$$dy = \frac{y\,dz}{r_2} \tag{10.12}$$

Substitution of the values in equations (10.11) and (10.12) for dx and dy into equation (10.6) yields

$$W_1 = \gamma \left(\frac{1}{r_1} + \frac{1}{r_2} \right) xy\,dz \qquad\qquad (10.13)$$

At mechanical equilibrium, W_1 must equal W_2, leading to

$$\Delta P = \gamma \left(\frac{1}{r_1} + \frac{1}{r_2} \right) \qquad\qquad (10.14)$$

The Young-Laplace equation, equation (10.14), reduces to equation (10.4) for the special case of a sphere with r_1 equal to r_2. For a bubble, the right hand side of equations (10.14) and (10.4) should be multiplied by two to allow for the fact that there are two surface being stretched, the interior and the exterior.

10.3 CONTACT ANGLES AND WETTING

The affinity of a liquid for a solid surface is usually described as wetting. When a liquid spreads spontaneously across a solid surface it is said to wet the surface because, in such a case, the *adhesive* forces exceed the *cohesive* forces. If the liquid in the form of a drop remains stationary and appears pseudo-spherical, then it is non-wetting because the cohesive forces exceed the adhesive forces. Fig. 10.1 illustrates wetting and non-wetting liquids on a solid surface. A measure of the degree of wetting is given by the wetting, or *contact* angle, θ, the measurement of which is discussed in Chapter 18. The wetting angle is greater or less than 90° for non-wetting and wetting liquids, respectively.

A drop of liquid at rest on a solid surface is under the influence of three forces or tension. As shown in Fig. 10.2, the circumference of the area of contact of a circular drop is drawn toward the center of the drop by the solid-liquid interfacial tension, γ_{SL}. The equilibrium vapor pressure of the liquid produces an adsorbed layer on the solid surface that causes the circumference to move away from the drop center and is equivalent to a solid-vapor interfacial tension, γ_{SV}. The interfacial tension between the liquid and vapor, γ_{LV}, essentially equivalent to the surface tension γ of the liquid, acts tangentially to the contact angle θ, drawing the liquid toward the drop center with its component $\gamma_{LV}\cos\theta$.

At mechanical equilibrium, the tensions or forces cancel to yield

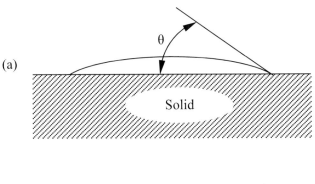

(a)

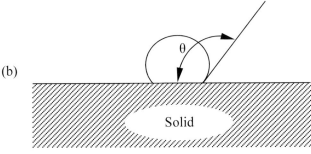

(b)

Figure 10.1 Wetting and nonwetting liquids on a solid surface: (a) wetting, $\theta < 90°$, (b) nonwetting, $\theta > 90°$.

$$\gamma_{SV} = \gamma_{SL} + \gamma_{LV} \cos\theta \qquad (10.15)$$

Equation (10.15), derived by Young [3] and Dupré [4], establishes the criteria for wettability. The contact angle is determined from the values of the three interfacial tensions.

$$\cos\theta = \frac{\gamma_{SV} - \gamma_{SL}}{\gamma_{LV}} \qquad (10.16)$$

When γ_{SV} exceeds γ_{SL} then $\cos\theta$ must be positive with $\theta < 90°$. When γ_{SL} exceeds γ_{SV} the $\cos\theta$ term is negative and $\theta > 90°$. Complete wetting of a surface occurs at a contact angle of $0°$ and total nonwetting at $180°$. Since wetting occurs when adhesive forces, F_A, exceed cohesive forces, F_C, and non-wetting when cohesion exceeds adhesion, from 10.16 one can write

$$F_A = \frac{\gamma_{SV}}{\gamma_{LV}} \qquad (10.17)$$

and

$$F_C = \frac{\gamma_{SL}}{\gamma_{LV}} \qquad (10.18)$$

10.4 CAPILLARITY

When a capillary is immersed under the surface of a liquid, the liquid will either rise in the capillary (Fig. 10.3a) or be depressed below the surface of the external liquid (Fig. 10.3b).

   As illustrated in Fig. 10.3a, the pressure above the meniscus within a capillary of radius r at point A is the same as that at point C at the surface of the liquid level external to the capillary. The small pressure difference due to the gravitational difference in gas densities is neglected. According to equation (10.4), the pressure on the concave side of a curved surface at point A must be in excess of that on the convex side at point B. Because the pressure at point B is less than that at A the liquid will rise in the capillary

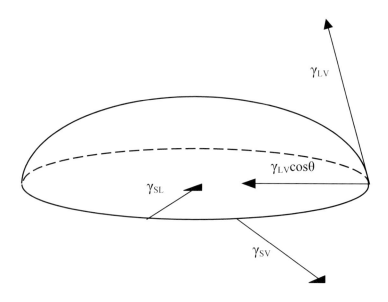

Figure 10.2 Interfacial tensions.

until the sum of the pressure at point B and the pressure head induced by the liquid rising to height h is equal to the external pressure. At equilibrium then

$$P_A = P_B + hg(\rho_1 - \rho_g) \tag{10.19}$$

Where P_A and P_B are the pressures at points A and B, ρ_1 and ρ_g are the densities of the liquid and the gas that it displaces in the capillary, and g is the gravitational constant. Equation (10.19) is more conveniently expressed as

$$P_A - P_B = \Delta P = hg(\rho_1 - \rho_g) \tag{10.20}$$

Combining equations (10.20) and (10.4) yields

$$\frac{2\gamma}{r} = hg(\rho_1 - \rho_g) \tag{10.21}$$

which is often used to describe the surface tension of a liquid by the capillary rise method.

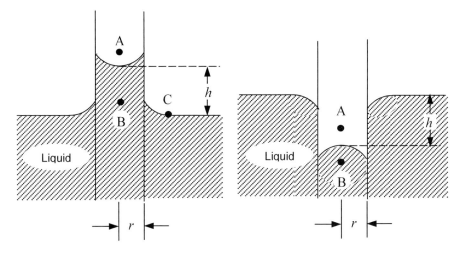

Figure 10.3 (a) Capillary rise, $\theta > 90°$; (b) Capillary depression, $\theta < 90°$.

When a liquid does not wet the walls of a capillary, as shown in Fig. 10.3b, the concave side of the meniscus is within the liquid at point B, which is at a higher pressure than the gas immediately above the surface. As described by equation (10.21), the liquid is depressed a distance h below the level of the external liquid. Equation (10.21) was derived for capillary rise or depression assuming complete wetting, that is $\theta = 0°$. In the case of contact angles greater than $0°$ and less than $180°$, equation (10.21) must be modified. As liquid moves up the capillary during capillary rise the solid-vapor interface disappears and the solid-liquid interface appears. The work required for this process is

$$W = (\gamma_{SL} - \gamma_{SV})\Delta A \qquad (10.22)$$

Where ΔA is the area of the capillary wall covered by a liquid as it rises. From the Young-Dupré equation, equation (10.15), and equation (10.22), it can be seen that

$$W = -(\gamma_{LV} \cos \theta)\Delta A \qquad (10.23)$$

in which γ_{LV} is the surface tension γ of the liquid. The work required to raise a column of liquid a height h in a capillary of radius r is identical to the work that must be performed to force it out of the capillary. When a volume V of liquid is forced out of the capillary with gas at a constant pressure above ambient, ΔP_{gas}, the work is given by

$$W = V\Delta P_{gas} \qquad (10.24)$$

Combining equations (10.23) and (10.24) yields

$$\Delta P_{gas} V = -(\gamma \cos \theta)\Delta A \qquad (10.25)$$

Where ΔP_{gas} is the excess pressure above that of the external pressure and V is the volume of liquid displaced. If the capillary is circular in cross-section, the terms V and ΔA are given by $\pi r^2 \ell$ and $2\pi r\ell$, respectively. Substituting these values into equation (10.22) gives

$$\Delta Pr = -2\gamma \cos \theta \qquad (10.26)$$

10.5 THE WASHBURN EQUATION

Equation (10.26) was first presented by Washburn [5] and is the operating equation in mercury porosimetry. For wetting angles less than $90°$ (as is the condition for capillary rise), $\cos\theta$ is positive and ΔP is negative, indicating that pressure greater than ambient must be applied to the top of a liquid column to force the liquid out. When θ is greater than $90°$, resulting in capillary depression, $\cos\theta$ is negative and pressure greater than ambient must be applied *to the reservoir* in which the capillary is immersed, to force the liquid *into* the capillary.

According to the Washburn equation (10.26) a capillary of sufficiently small radius will require more than one atmosphere of pressure differential in order for a non-wetting liquid to enter the capillary filled with ambient atmosphere. The method of mercury porosimetry requires the evacuation of the sample and subsequent pressurization to force mercury into the pores. Since the pressure difference across the mercury interface is then the absolute applied pressure, equation (10.26) reduces to:

$$Pr = -2\gamma \cos\theta \tag{10.27}$$

An alternative derivation of the Washburn equation can be pursued as follows. For a pore of circular cross-section with radius r, the surface tension acts to force a non-wetting liquid out of the pore. The force developed, F, due to interfacial tensions is the product of the surface tension of the liquid, γ, and the circumference, $2\pi r$, of the pore, that is:

$$F = 2\pi r\gamma \tag{10.28}$$

Since the interfacial tension acts tangentially to the contact angle θ, the component of force driving mercury out of the pore becomes

$$F_{out} = 2\pi r\gamma \cos\theta \tag{10.29}$$

The force driving mercury into the pore can be expressed as the product of the cross-sectional area of the pore and the applied pressure P directed against F_{out}, that is

$$-F_{in} = P\pi r^2 \tag{10.30}$$

Equating (10.29) and (10.30) at equilibrium gives

$$Pr = -2\gamma \cos\theta \tag{cf. 10.27}$$

Another approach to the Washburn equation involves the work required to force mercury into a pore. Because of its high surface tension, mercury tends not to wet most surfaces and must be forced to enter a pore. When forced under pressure into a pore of radius r and length ℓ an amount of work W is required which is proportional to the increased surface exposed by the mercury at the pore wall. Therefore, assuming cylindrical pore geometry

$$W = 2\pi r \ell \gamma \tag{10.31}$$

Since mercury exhibits a wetting angle greater than 90° and less than 180° on all surfaces with which it does not amalgamate, the work required is reduced by cosθ and equation (10.31) becomes

$$W_1 = 2\pi r \ell \gamma \cos\theta \tag{10.32}$$

When a volume of mercury, ΔV, is forced into a pore under external pressure P an amount of work W_2, given as follows, is performed

$$W_2 = -P\Delta V = -2\pi r^2 \ell \tag{10.33}$$

where the negative sign implies a decreasing volume. At equilibrium $W_1=W_2$ and since $\Delta V = \pi r^2 \ell$, equations (10.32) and (10.33) combine to give

$$Pr = -2\gamma \cos\theta \tag{cf. 10.27}$$

Because the product Pr is constant, assuming constancy of γ and θ, equation (10.27) dictates that as the pressure increases mercury will intrude into progressively narrower pores.

10.6 INTRUSION – EXTRUSION CURVES

A plot of the intruded (or extruded) volume of mercury versus pressure is called a porosimetry curve, sometimes a porogram. The authors will use the terms "intrusion curve" to denote the volume change with increasing pressure and "extrusion curve" to indicate the volume change with decreasing pressure. Fig. 10.4 shows a typical porosimetry curve of cumulative volume plotted versus both pressure (bottom abscissa) and radius (top abscissa). The same data plotted on semi log scale is illustrated in Fig. 10.5.

As shown in Fig. 10.4 the initial intrusion at very low pressure is due to penetration into the interparticle voids when the sample is a powder. The slight positive slope between points A and B on the intrusion curve results from a continuous filling of the toroidal volume [6-10] between the contacting particles. As the pressure is increased, mercury will penetrate deeper into the narrowing cavities between the particles. Depending on the size, size distribution, shape and packing geometry of the particles, there will exist some interparticle voids of various dimensions and shapes that will progressively fill as the pressure is increased. Between points B and C on the intrusion curve in Fig. 10.4, intrusion commences into a new range of cavities, which, if cylindrical in shape, would possess circular openings of about 500Å to 75Å radius. This range of intrusion occurs at a pressure substantially greater than was required to fill the interparticle voids. Therefore, mercury is intruding into small pores within the particles. At point D, intrusion commences into a range of pores with even smaller radii (29 Å) after which no further intrusion takes place up to the maximum pressure corresponding to point E on the curve. The slight slope between points C and D indicates a small amount of porosity in the range of about 30Å to 75Å.

During depressurization, pores that commenced filling at point D begin to empty at point F and at lower pressures pores that filled between points B and C begin to empty at point G. At point H on the extrusion curve, the cycle is terminated. The intrusion – extrusion cycle does not close when the initial pressure is reached indicating that some mercury has been permanently entrapped by the sample. Often, at the completion of an intrusion-extrusion cycle mercury will slowly continue to extrude, sometimes for hours.

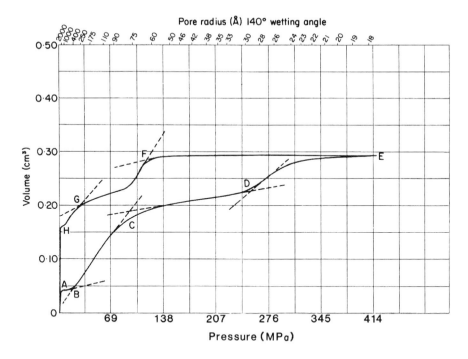

Figure 10.4 Example of mercury porosimetry data (alumina-I).

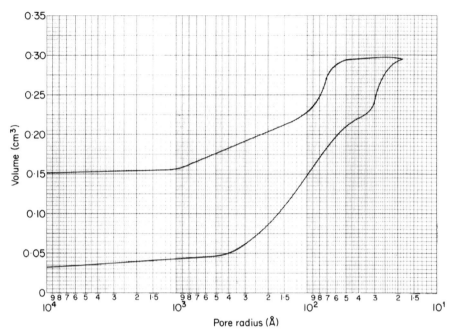

Figure 10.5 Plot of volume versus log radius (c.f. Fig. 10.4).

Figs 10.4 and 10.5 illustrate mercury porosimetry data of a bimodal size distribution. However, other types of less typical curves are often encountered. For example, samples of controlled porous glass exhibit intrusion-extrusion curves illustrated by Fig. 10.6, in which all pores are essentially of one radius.

On the other hand, some samples possess a wide and continuous range of pore sizes, resulting in the porosimetry curve shown in Fig. 10.7. Still another type of porosimetry curve has been reported [11]. Fig. 10.8 illustrates mercury intrusion into distinct pore sizes in crystals of calcium hydrogen phosphate. The vertical steps on the curve indicate very narrow bands of pore radii between which essentially no pores exist. Therefore, pores do not always exist over a continuum of radii.

The authors have found similar stepwise intrusion on other materials. The low-pressure (0.003 to 0.1 MPa) intrusion curve in Fig. 10.8 was obtained using a scanning porosimeter [12] that continuously recorded the pressure and corresponding intruded volume. Only in this continuous manner was it possible for the exact position and magnitude of each intrusion step to be fully determined.

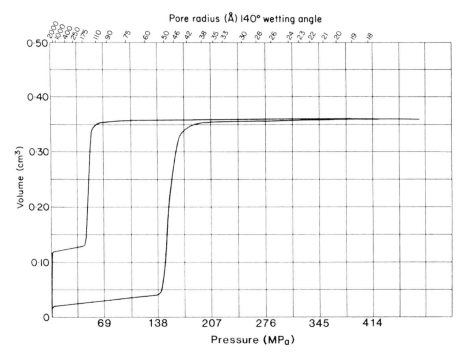

Figure 10.6 Intrusion-extrusion curves of porous glass.

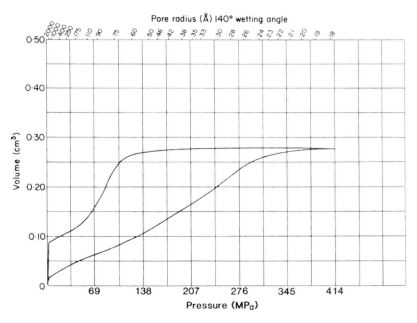

Figure 10.7 Intrusion-extrusion curves of broad pore size range (alumina-II).

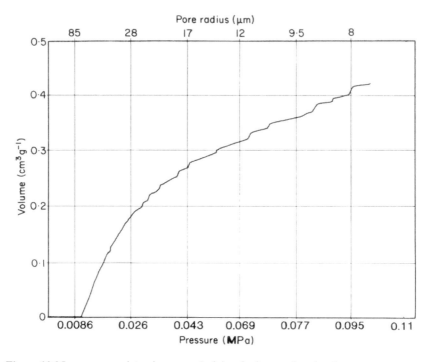

Figure 10.8 Low-pressure intrusion curve (calcium hydrogen phosphate).

10.7 COMMON FEATURES OF POROSIMETRY CURVES

All mercury intrusion-extrusion curves have certain characteristics in common. These include:

1. Powders may rearrange (A) followed by intrusion at relatively low pressures as mercury penetrates the large interparticle voids (B).
2. Intrusion occurs at higher pressures into pores *within* the particles (C) which, for susceptible materials, is followed by (more or less) reversible compression (D).
3. All porosimetry curves exhibit hysteresis, that is, the path followed by the extrusion curve (E) is not the same as the intrusion path. At a given pressure the volume indicated on the extrusion curve is greater than that on the intrusion curve and for a given volume the pressure indicated on the intrusion curve is greater than that on the extrusion curve.
4. Upon completion of a first intrusion-extrusion cycle, some mercury is always retained by the sample, thereby preventing the loop from closing (F).

5. Intrusion-extrusion cycles after the first will continue showing hysteresis (G) but eventually the loop will close, showing that further entrapment of mercury eventually ceases. On most samples, the loop closes after just the second cycle.
6. Each step on an intrusion curve has a corresponding step on the extrusion curve at a lower pressure, – except powder rearrangement. The corresponding extrusion step will also be absent from the completed analysis if it would have occurred below the finishing pressure (H, usually slightly above ambient atmospheric pressure).

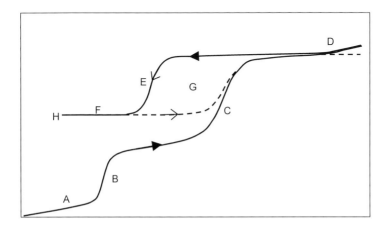

Figure 10.9 Features of porosimetry curves

10.8 HYSTERESIS, ENTRAPMENT AND CONTACT ANGLE

Cumulative intrusion curves generated by intrusion mercury into porous samples are not followed as the pressure is lowered and the mercury extrudes out of the pores. In all cases, the depressurization curve lies above the pressurization curve and the hysteresis loop does not close even when the pressure is returned to zero, indicating that some mercury is entrapped in the pores. Usually after the sample has been subjected to a first pressure cycle, no additional entrapment occurs during subsequent cycles. In some cases however, a third or even fourth cycle is required before entrapment ceases.

The pressure-volume ($P \cdot V$) work associated with an initial intrusion-extrusion cycle can, with reference to Fig. 10.10, be calculated from the areas above curves A and C to the maximum intruded volume, indicated by the horizontal dotted line. This $P \cdot V$ work, $\oint dW_1$, can be expressed as

$$\oint dW_1 = \int_0^V P_i dV_i + \int_V^{V'} P_e dV_e > 0 \qquad (10.34)$$

where P_i and P_e are intrusion and extrusion pressures and V_i and V_e are intrusion and extrusion volumes respectively. The limit V' is used in the extrusion integral in equation (10.34) to indicate that, at the completion of a cycle, some mercury has been retained by the sample.

Even in the absence of mercury entrapment, second or subsequent intrusion-extrusion cycles, curves B and C of Fig. 10.10, continue to show hysteresis with the $P \cdot V$ work, $\oint dW_2$, through a cycle given by

$$\oint dW_2 = \int_0^V P_i dV_i + \int_V^0 P_e dV_e > 0 \qquad (10.35)$$

As illustrated in Fig. 10.10 the $P \cdot V$ work through a second cycle is always les than that through a first cycle since the area between curves B and C is less than that between A and C. Then,

$$\oint dW_1 > \oint dW_2 \qquad (10.36)$$

The difference between cyclic integrals $\oint dW_1$ and $\oint dW_2$ is the work associated with entrapment of mercury and can be evaluated from the area between curves A and B.

Equations (10.34) and (10.35) indicate that the work of intrusion always exceeds the work of intrusion. Therefore, the surroundings must experience a decrease in potential energy in order to provide the required work difference around an intrusion-extrusion cycle. After many cycles, the system either must store a boundless amount of energy or must convert the energy into heat as irreversible entropy production. Since neither of these phenomena is observed, one must be able to show that around any intrusion-extrusion cycle, after entrapment has ceased, work is conserved, that is,

$$\oint dW = 0 \qquad (10.37)$$

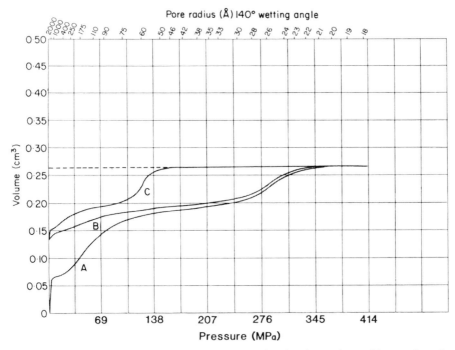

Figure 10.10 Cumulative volume versus pressure plot showing regions of hysteresis and entrapment

Inspection of a plot of volume versus radius for an intrusion-extrusion cycle (Fig. 10.10) would lead one to the conclusion that, at a given volume, mercury intrudes into pores of one size, but extrudes from pores of a smaller size. This anomaly and the question of energy conservation will be considered in the following sections of this Chapter.

10.9 CONTACT ANGLE CHANGES

The contact angle of a liquid on a solid might be expected to vary as the liquid moves over the surface, i.e. as it advances over a "dry" surface or recedes from a wetted surface [13]. For example, a contact angle hysteresis of 22° (124° to 146°) has been reported for mercury on stainless steel [14]. The higher, *advancing* contact angle is associated with penetration of, or intrusion into, pores and the *receding* angle with extrusion from pores. Smithwick [15] equates this hysteresis in mercury porosimetry to the irreversibility of the "immersion-emersion" work cycle, i.e. frictional loss, but nevertheless quantifies the irreversibility in terms of the change in cosine of the wetting angle, and found extrusion contact angles near 100° for an intrusion angle of 130° [16]. Lowell and Shields [17,18] have shown that superimposition of the intrusion and extrusion curves, when plotted as volume versus radius, can be achieved if the contact angle θ is adjusted from $θ_i$, the intrusion contact angle, to $θ_e$, the extrusion angle. Fig. 10.11 illustrates the resulting curves when the contact angles are properly adjusted. Curve A is the first intrusion curve using the intrusion contact angle and is identical to Curve A in Fig. 10.4. Curve B is the second intrusion curve using $θ_i$ and is identical to curve B in Fig. 10.4. The difference between curves A and B in both figures along the volume axis reflects the quantity of entrapped mercury after completion of the first intrusion-extrusion cycle.

When the contact angle is changed from $θ_i$ to the smaller value $θ_e$, the pore radius calculated from the Washburn equation will decrease by the ration of $r(\cos θ_e/\cos θ_i)$. This does not imply that intrusion occurs into pores of the smaller size since the actual intrusion contact angle is applicable. However, when the correction is made from $θ_i$ to $θ_e$ at the start of extrusion, the pore radius corresponding to the maximum pressure will be $r(\cos θ_e/\cos θ_i)$. Thus, when depressurization commences, shown in Fig. 10.11 as curve C, no extrusion occurs between $r(\cos θ_e/\cos θ_i)$ and r since no mercury has intruded into pores in this range. In Fig. 10.11 the extrusion curve is superimposed on the second intrusion curve B so that no hysteresis is exhibited. If mercury exhibits two or more different intrusion contact angles with a given material, superimposition of curves cannot be achieved.

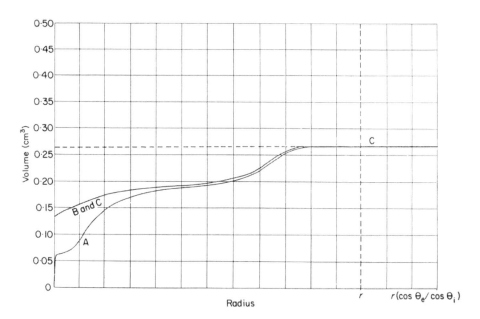

Figure 10.11 Cumulative volume versus radius plot after adjustment of extrusion contact angle.

Fig. 10.11 indicates clearly, that changing the extrusion contact angle from 140° to 104.5° eliminates hysteresis. In this case, hysteresis is entirely due to intrinsic effects and network effects do not play a significant role; hysteresis can be completely explained by a single pore mechanism. Fig. 10.11 also shows that a change in contact angle does not affect the entrapment phenomenon at all.

If the hysteresis loop can be eliminated by just changing an intrinsic parameter, as for instance the contact angle, one can conclude that networking or pore blocking effects are not affecting the intrusion data, and neither, therefore, the appropriate calculation of the pore size distribution. By inference therefore, if hysteresis cannot be accounted for in this manner, then some other (structure-related) factor is involved [19].

10.10 POROSIMETRIC WORK

It has been demonstrated [18] that first and subsequent intrusion-extrusion curves can be viewed as consisting of three significant areas when the extrusion contact angles are not corrected, as shown in Fig. 10.4.

The total area above the first intrusion curve, A in Fig. 10.4, to the maximum intruded volume indicated by the horizontal dotted line, corresponds to the $P \cdot V$ work of intrusion, W_i. This work term consists of three parts:

- The first is the work of entrapment, W_t, corresponding to the area between curves A and B.
- The second is the work $W_{\Delta\theta}$ associated with the contact angle change from θ_i to θ_e, which corresponds to the area between curves B and C.
- The third and final contribution to the work of intrusion is the area between curve C and the maximum intruded volume and which corresponds to the work of extrusion using the incorrect or intrusion contact angle.

W_t, $W_{\Delta\theta}$ and W_e can be evaluated [18] either graphically from Fig. 10.4 or calculated from the equations below.

The work associated with entrapment of mercury is given by

$$W_t = \gamma \left| \cos \theta_i \right| A_t \tag{10.38}$$

where A_t is the area of mercury entrapped in all the pores.

The work associated with the contact angle change from θ_i to θ_e is

$$W_{\Delta\theta} = \gamma \left(|cos\ \theta_i| - |cos\ \theta_e| \right) A_{\Delta\theta} \tag{10.39}$$

in which $A_{\Delta\theta}$ is the area of mercury undergoing a change in contact angle and is the same as A_e, the area of mercury which extrudes from the pores. The work of extrusion using the incorrect or intrusion contact angle is given by

$$W'_e = \gamma |cos\ \theta_i| A'_e \tag{10.40}$$

When the correct extrusion angle is employed, equation (10.40) becomes

$$W_e = \gamma |cos\ \theta_i| A_e \tag{10.41}$$

It would appear from equations (10.40) and (10.41) there are two different values for W_e, the work of extrusion. This is not the case. When equation (10.40) is used to calculate the work of extrusion, A_e is determined from Fig. 10.10 where the intrusion contact angle θ_i was used to obtain the extrusion curve. In fact, A_e' is a smaller area than that of the pores actually emptied. When equation (10.41) is employed, A_e represents the correct pore area and is larger than A_e' by the factor $cos\theta_i/cos\theta_e$. Thus, regardless of whether equation (10.40) or (10.41) is used, the calculated value of W_e will be the same.

Equations (10.38) through (10.41) assume no specific pore shape. By assuming cylindrical pore geometry the validity of these equations can be established. For a cylindrical pore the area in any pore interval is given by

$$A = \frac{2}{\bar{r}} V \tag{10.42}$$

where $\bar{r}$ is the mean pore radius in a narrow pore interval and V is the pore volume in that radius interval. Using equation (10.42) and expressing V as the volumetric difference between the appropriate curves less the volume difference for the previous interval, W_t, $W_{\Delta\theta}$ and W_e can be evaluated from

$$W_t = 2\gamma |cos\ \theta_i| \sum_1^n \left(\frac{V_t}{\bar{r}} \right)_n \tag{10.43}$$

$$W_{\Delta\theta} = 2\gamma \left(|cos\ \theta_i| - |cos\ \theta_e| \right) \sum_1^n \left(\frac{V_{\Delta\theta}}{\bar{r}} \right)_n \tag{10.44}$$

and

$$W'_e = 2\gamma |cos\ \theta_i| \sum_1^n \left(\frac{V_{\Delta\theta}}{\overline{r}}\right)_n \tag{10.45}$$

The values of the work terms for numerous samples, calculated from equations (10.43), (10.44) and (10.45), show excellent agreement [14] with the corresponding work terms obtained by graphical integration of the areas in the intrusion-extrusion cycles.

10.12 THEORY OF POROSIMETRY HYSTERESIS

From the previous discussion, it can be shown that complete intrusion-extrusion cycle is composed of the following steps: Intrusion with contact angle θ_i into pores with area A_i, for which the corresponding work, W_i, is

$$W_i = \gamma |cos\ \theta_i| A_i \tag{cf. 10.23}$$

The change in contact angle from θ_i to θ_e as extrusion commences, for which the corresponding work $W_{\Delta\theta}$, is

$$W_{\Delta\theta} = \gamma \left(|cos\ \theta_i| - |cos\ \theta_e|\right) A_e \tag{cf.10.39}$$

By adding the area of entrapped mercury, A_t, to A_e in the above equation, the work associated with the contact angle change, $W'_{\Delta\theta}$, becomes the work necessary to alter the contact angle over the entire length of the pore, not only that portion which extrudes. This $W'_{\Delta\theta}$ can be expressed as

$$W'_{\Delta\theta} = \gamma \left(|cos\ \theta_i| - |cos\ \theta_e|\right) A_i \tag{10.46}$$

Extrusion with contact angle θ_e is accompanied by an amount of work, W_e, given by

$$W_e = \gamma |cos\ \theta_e| A_e \tag{cf.10.41}$$

After completion of extrusion and at the start of another intrusion process the entrapped mercury undergoes a contact angle change from θ_e back to θ_i, for which the corresponding work, $W_{-\Delta\theta}$, is

$$W_{-\Delta\theta} = \gamma \left(\left| \cos\theta_i \right| - \left| \cos\theta_e \right| \right) A_t \qquad (10.47)$$

The work, W_t, associated with mercury entrapment in the pores with contact angle θ_i is given by

$$W_t = \gamma \left| \cos\theta_i \right| A_t \qquad (cf.10.38)$$

Summing all of the above work terms, with positive work being done *on* the system and the negative sign indicating work done *by* the system, leads to the total work $\oint dW_1$ around a first intrusion-extrusion cycle. That is,

$$\oint dW_1 = W_i + W_{-\Delta\theta} - W'_{\Delta\theta} - W_e - W_t = 0 \qquad (10.48)$$

For second or subsequent cycles, when no entrapment occurs, W_t vanishes and the total work through the cycle becomes

$$\oint dW_2 = W_t = 0 \qquad (10.49)$$

It is evident that the thermodynamic processes associated with mercury porosimetry are far more complex than just the consideration of $P \cdot V$ work.
In volume versus radius plots for second cycles, during which no entrapment occurs, hysteresis vanishes when the correct contact angles are employed, as shown in Fig. 10.11. Residual hysteresis exhibited in a first intrusion-extrusion cycle, when the correct contact angles are employed, is due solely to entrapment of mercury.
The $P \cdot V$ work differences between intrusion and extrusion, corresponding to the area between curves A and C of Fig. 10.10 can be expressed as

$$W_i - W_e = P_i V_i - P_e V_e = \gamma \left(\left| \cos\theta_i \right| A_i - \left| \cos\theta_e \right| \right) A_e \qquad (10.50)$$

When the additional work terms W_t, $W'_{\Delta\theta}$ and $W_{-\Delta\theta}$ are considered the difference in the work of intrusion and extrusion is written as

$$W_i - W_e = \left(P_i V_i + W_{-\Delta\theta} \right) - \left(P_e V_e + W'_{\Delta\theta} + W_t \right) = 0 \qquad (10.51)$$

Equation (10.51) establishes that the hysteresis energy, W_H, the energy associated with the area in the hysteresis region of an intrusion-extrusion cycle is given by

$$W_H = W'_{\Delta\theta} + W_t - W_{-\Delta\theta} = \gamma\left(\left|cos\,\theta_i\right|A_i - \left|cos\,\theta_e\right|A_e\right) \qquad (10.52)$$

Since $\oint dW = 0$ and since no work difference exists between W_i and W_e for any element of the intrusion-extrusion curve, it follows that hysteresis on volume-pressure plots results only as a consequence of ignoring the additional energy terms: $W'_{\Delta\theta}$, W_t and $W_{-\Delta\theta}$.

10.13 PORE POTENTIAL

The fact that hysteresis is attributable to the $W'_{\Delta\theta}$, W_t and $W_{-\Delta\theta}$ terms as shown in the preceding section, does not, however, explain the processes which lead to these terms. Lowell and Shields [20] postulated that, in volume versus pressure plots, hysteresis is due to the influence of a pore potential.

The concept of a pore potential is generally accepted in gas sorption theory to account for capillary condensation at pressures well below the expected values. Gregg and Sing [21] described the intensification of the attractive forces acting on an adsorbate molecule by overlapping fields from the pore walls. Adamson [22] has pointed out that evidence exists for changes induced in liquids by capillary walls over distances in the order of a micrometer. The Polanyi potential theory [23] postulates that molecules can "fall" into the potential well at the surface of a solid, a phenomenon that would be greatly enhanced in a narrow pore.

In mercury porosimetry, it was proposed [20] that the pore potential prevents extrusion of mercury from a pore until a pressure less than the nominal extrusion pressure is reached. Similarly, the pore potential, when applied to gas sorption, is used to explain desorption at a lower relative pressures than adsorption for a given quantity of condensed gas.

In mercury porosimetry, the pore potential can be derived as follows: The force, F, required for intrusion into a cylindrical pore is given by

$$F = 2\pi r\gamma\left|cos\,\theta_i\right| = P_i\pi r^2 \qquad (10.53)$$

If the pore potential, U, is the difference between the interaction of the mercury along the total length of all pores with radius r when the pores become filled at pressure P_i and when partially emptied at P_e, U can be expressed as

$$U = \int_0^{\ell_i} F d\ell_i - \int_0^{\ell_e} F d\ell_e \qquad (10.54)$$

where ℓ_i represents the total length of mercury columns in all the pores when filled and ℓ_e is the total length of the mercury which extruded from these pores.

Expressing force as pressure times the area of pores in a radius interval with mean radius $\bar{r}$ and mean length $\bar{\ell}$, equation (10.54) can be rewritten as

$$U = \pi \bar{r}^2 \left(P_i \bar{\ell}_i - P_e \bar{\ell}_e \right) \qquad (10.55)$$

In terms of the intruded and extruded volumes in a radius interval,

$$U = P_i V_i - P_e V_e \qquad (10.56)$$

Thus, the expression for the pore potential is identical to the pressure-volume work differences between intrusion and extrusion and can be expressed as the hysteresis energy, that is,

$$-U = W'_{\Delta\theta} + W_t - W_{-\Delta\theta} = W_H \qquad \text{(cf. 10.52)}$$

Equation (10.52) predicts that changes in U, the pore potential, will affect the quantity of entrapped mercury and/or the difference in contact angle between intrusion and extrusion. Hence, changes in the pore potential will alter the size of the hysteresis loop.

The model described by equation (10.52) is that mercury, when intruded into a pore with contact angle θ_i, acquires an increased interfacial free energy. As the pressure decreases, mercury will commence extruding from the pore at pressure P_e, reducing the interfacial area and simultaneously the contact angle, thereby spontaneously decreasing the interfacial free energy. Mercury will continue leaving the pores with the extrusion contact angle θ_e, until the interfacial free energy equals the pore potential at which point extrusion ceases. That is, upon depressurization, mercury separates from the pore with contact angle θ_e, leaving the trapped portion of mercury near the pore opening rather than at the base of the pore. This is consistent with the observation that penetrated samples show discoloration due to finely divided mercury at the pore entrances.

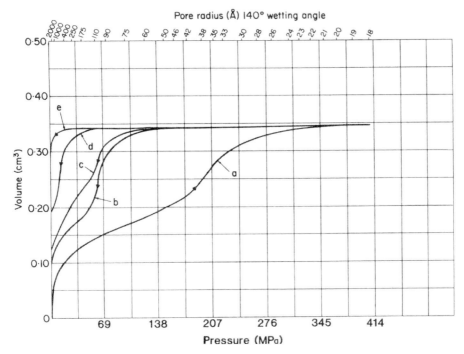

Figure 10.12 Intrusion-extrusion curves on an alumina sample coated with various amounts of copper sulfate: (a) Intrusion curves for all samples; (b) Extrusion curve for untraeated alumina; (c) Extrusion curve for alumina treated with 0.5% $CuSO_4$; (d) Extrusion curve for alumina treated with 2% $CuSO_4$; (e) extrusion curve for alumina treated with 40% $CuSO_4$.

Evidence for the pore potential was experimentally obtained [20] from mercury intrusion-extrusion data by impregnating various samples with both polar and non-polar materials. Fig. 10.12 is typical mercury porosimetry data of a sample coated with various amounts of polar material.

It is evident from Fig. 10.12 that, as the copper sulfate concentration in the sample is increased, the hysteresis increases, that is the difference between P_i and P_e increases while, at the same time, the extrusion contact angle decreases. Similarly, the work of entrapment, W_t, increases as the salt concentration is raised, as evidenced by the quantities of mercury entrapped in the sample. It can be seen in Fig. 10.12 that the intrusion curves for both treated and untreated samples are virtually identical indicating that impregnation does not significantly alter the radius of the pore opening. However, in all cases the volume of mercury intruded decreased with increasing salt concentration, an indication that precipitation of the salt occurred near the base of the pores. These pore volume differences were eliminated in Fig. 10.12 by normalizing the data to the maximum volume intruded into the untreated sample.

The effect of pore impregnation with non-polar material was studied by treating samples with dichlorodimethylsilane (DCDMS). In each case, a decrease in hysteresis area, compared to untreated material, was observed after coating samples with DCDMS. The increase in the extrusion contact angle, with DCDMS compared to untreated sample, resulted in decreases in $W_{\Delta\theta}$. In some cases, impregnation with DCDMNS led to greater mercury retention or an increase in W_t over the untreated material. However, this was always accompanied by a larger decrease in $W_{\Delta\theta}$ and thus a decrease in the pore potential.

In some instances, it is possible to plot mercury intrusion curves corresponding to intrusion into pores within samples with which mercury amalgamates. Presumably, a thin oxide film or some activation process inhibits the rate of amalgamation so that intrusion can occur as the pressure is increased before the unfilled pores are degraded. In these cases, the depressurization curve is always a straight horizontal line indicating that no extrusion occurs. Thus, the hysteresis energy W_H is a maximum corresponding to the largest possible value of W_t, namely $W_t = W_i$, and corresponding to the largest possible value of $W_{\Delta\theta}$ since the extrusion angle would be that of a wetting liquid or near zero degrees.

10.14 OTHER HYSTERESIS THEORIES (THROAT-PORE RATIO NETWORK MODEL)

In addition to contact angle hysteresis, networking effects (e.g., pore blocking) may contribute to the occurrence of hysteresis. These effects can be assessed by applying network models. A graphical representation of such a three-dimensional pore-throat ratio model [24, 25] is shown in Fig. 10.13.

One inference from network models is the relationship between mercury intrusion and extrusion curves and the shape of pores and throats [26]. Based on geometrical considerations, the intrusion curve (intrusion pressures P_i) describes the distribution of pore constrictions (throats, of diameter D_i), whereas the mercury extrusion curves (extrusion pressure P_e) could be related to the shapes of cavities beyond the constrictions (pores, D_e). The throat/pore size ratio $R_{TP,j}$ is calculated as a function of the void fraction filled with mercury, ϕ_j, where

$$R_{TP,j} = \left(\frac{D_i}{D_e} \right) = \left[\left(\frac{P_e}{P_i} \right) \left(\frac{f_e}{f_i} \right) \left(\frac{\cos\theta_i}{\cos\theta_e} \right) \right]_j \qquad (10.57)$$

and

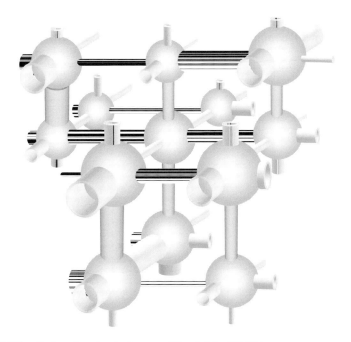

Figure 10.13 Visualization of a network of pores and throats (after [24, 25])

$$\phi_j = \frac{V_a}{V_c} \qquad\qquad (10.58)$$

where D, P, f and θ are the throat or pore size, the applied pressure, the pore shape factor and the mercury contact angle, respectively, with the subscripts i and e referring to intrusion and extrusion curves, and where V_a is the mercury volume intruded at any given pressure and V_c is the volume of mercury intruded at the maximum experimental pressure.

  If the shape factors are assumed to be the same for throats and pores, and the intrusion and extrusion contact angles are also assumed to be equal, R_{TP} is simply given by the ratio of extrusion to intrusion pressures, P_e/P_i at any given ϕ_j. The variations in P_e/P_i ratios with ϕ_j have been correlated with pore shape characteristics of a variety of agglomerated microparticles [26] and as shown in Fig. 10.14, [27]. Similarly, pore/throat ratios have been used to characterize the textural properties of colloidal SiC castings [28], pharmaceutical excipients and tablets [28, 29], and paper [29].

  However, a rigorous *3-D* reconstruction of a porous material cannot be achieved solely from an analysis of mercury intrusion/extrusion curves by applying a simple network model that consists of pore bodies interconnected by throats. Additional structural information, stored in the so-called *pore-pore correlation function*, is needed for this task. The application of complementary techniques such as neutron scattering [30], SEM techniques [31], X-ray microtomography [32, 33], etc is needed.

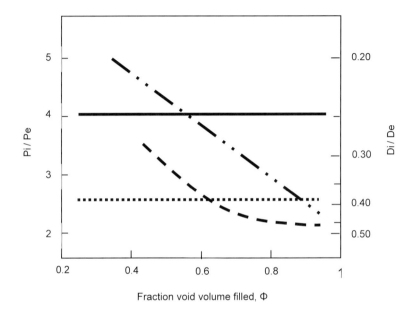

Figure 10.14 Ratio of intrusion to extrusion pressure as a function of the fraction of filled volume between particles of different shape: ▬▬ rods, ▪▪▪▪▪▪ spheres, ▬ ▪ ▪ needles, ▬ ▬ plates (after [22])

10.15 EQUIVALENCY OF MERCURY POROSIMETRY AND GAS SORPTION

Lowell and Shields [34] have shown that vapor condensation-evaporation and mercury intrusion-extrusion into and out of pores are thermodynamically equivalent processes.

A vapor will condense into pores of radius r according to the Kelvin equation

$$\ln \frac{P_V}{P_0} = \frac{-2\gamma \overline{V} \cos \theta}{rRT} \qquad \text{(cf 8.1)}$$

The molar free energy change associated with the isothermal vapor pressure change from P_0 to P_V is given by

$$\Delta G = RT \, ln\frac{P_V}{P_0} = \frac{-2\gamma\overline{V}\,cos\,\theta}{r} \tag{10.59}$$

The process of mercury intrusion requires the application of hydraulic pressure, P_h, to force mercury into pores for which the molar free energy change is given by

$$\Delta G = \overline{V}\int_0^{P_h} dP_h = \overline{V}P_h \tag{10.60}$$

Combining equations (10.59) and (10.60) results in the Washburn equation, that is,

$$P_h r = -2\gamma \, cos\,\theta \tag{10.61}$$

Equating the molar free energy terms in (10.59) and (10.60) affords an expression which relates the hydraulic pressure, P_h, required to force mercury into pores, to the relative pressure, P_V/P_0, exerted by the liquid with radius of curvature, r. That is,

$$P_h = \frac{RT}{\overline{V}} \, ln\frac{P_V}{P_0} \tag{10.62}$$

Porosimetry isotherms, corresponding to condensation and evaporation, have been constructed [23] by conversion of the hydraulic pressure to the corresponding relative pressure using equation (10.62). A typical isotherm from mercury intrusion-extrusion data (Fig. 10.15) is shown in Fig. 10.16 for an alumina sample.

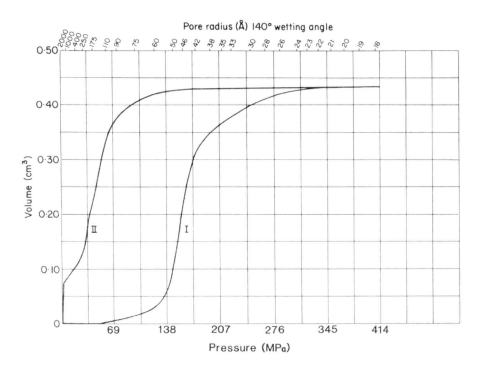

Figure 10.15 Mercury intrusion (I) and extrusion (II) curves of an alumina sample.

Following the convention in gas adsorption-desorption isotherms, the mercury isotherm, illustrated in Fig. 10.16, is plotted as volume versus relative pressure such that the radius increases from left to right. Curve I in Fig.10.16 represents the condensation isotherm from the extrusion curve and curve II is the evaporation isotherm from the intrusion data. Since no adsorption takes place on the pore walls prior to the filling of pores in mercury porosimetry, as occurs in gas adsorption, the usual "knee" of the isotherm is absent. However, condensation-evaporation isotherms from mercury porosimetry are strikingly similar to adsorption-desorption isotherms at relative pressure above the "knee". The maximum volume of the intrusion curve I in Fig. 10.15, is plotted as zero volume on the isotherm (Fig. 10.16) to conform to the requirement that the volume increases with increasing pore radius.

An advantage of isotherms constructed from mercury intrusion-extrusion curves is the capability of extending the isotherm well beyond the limits of vapor adsorption-desorption isotherms. Intruded and extruded volumes can be measured for pores of several hundred micrometers in diameter at pressures below one psia.

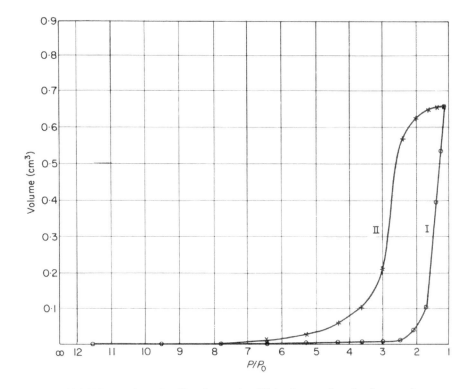

Figure 10.16 Mercury intrusion (I) and extrusion (II) isotherms of an alumina sample.

A significant aspect of the equivalency of mercury porosimetry and gas sorption is the implication can fill pores by either liquid or vapor transport.

10.16 REFERENCES

1. Young T. (1855) *Miscellaneous Works*, Vol 1(Peacock G., ed) J. Murray, London, p418.
2. de Laplace P.S. (1806) *Traité de Mécanique Céleste* (Supplement to Book 10).
3. Young T. (1805) *Phil. Trans. Roy. Soc. London* **95**, 65.
4. Dupré A. (1869) *Théorie Mécanique de la Chaleur*, Gauthier-Villars, Paris.
5. Washburn E.W. (1921) *Proc. Natl. Acad. Sci. USA*, **7**, 115.
6. Frevel L.K. and Kressley L.J. (1963) *Anal. Chem.* **35**, 1492.
7. Mayer R.P. and Stowe R.A. (1965) *J Colloid. Interface Sci.* **20**, 893.
8. Mayer R.P. and Stowe R.A. (1966) *J Phys Chem.* **70**, 3867.
9. Lee J.A. and Maskell W.C. (1974) *Powder Technol.* **9**, 165.
10. Stanley-Wood N.G. (1979) *Analyst* **104**, 79.
11. Lowell S. and Shields J.E. (1981) *Powder Technol.* **28**, 201.
12. Autoscan Filling Apparatus, *Quantachrome Corporation*, Syosset, NY, USA.
13. Adam N.K. (1948) *Trans. Faraday Soc.* **44**, 5.
14. Hubert C. and Swanson D. (2001) *GSFC Flight Mechanics Symposium*, NASA.
15. Smithwick R.W. (1986) *Powder Technol.* **48**, 233.

16. Smithwick R.W. and Fuller E.L. (1984) *Powder Technol.* **38**, 165.
17. Lowell S. and Shields J.E. (1981) *J Colloid. Interface Sci.* **80**, 192.
18. Lowell S. and Shields J.E. (1981) *J Colloid. Interface Sci.* **83**, 273.
19. Conner W.C., Lane A.M. and Hoffman A.J. (1984) *J Colloid. Interface Sci.* **100**, 185.
20. Lowell S. and Shields J.E. (1981) *J Colloid. Interface Sci.* **90**, 203.
21. Gregg S.J. and Sing K.S.W. (1982) *Adsorption Surface Area and Porosity*, 2nd edn., Academic Press, New York, p195.
22. Adamson A.W. (1982) *Physical Chemistry of Surfaces*, 4th edn., Wiley International, New York, p242.
23. Polanyi M. (1914) *Verh. Deut. Phys. Ges.*, **16**, 1012.
24. Conner W.C. and Lane A.M. (1984) *J. Catal.* **89**, 217.
25. Conner W.C., Cevallos-Candau J.F., Weist E.L., Pajares J., Mendioroz S. and Cortés A. (1986) *Langmuir*, **2**, 151.
26. Conner W.C., Blanco C., Coyne K., Neil J., Mendioroz S. and Pajares J. (1988) *Stud. Surf. Sci. Catal.* **39**, 273.
27. Zgrablich G., Mendiorez S., Daza L., Pajares J., Mayagoitia V., Rojas F. and Conner W.C. (1991) *Langmuir* **7** 779.
28. von Both H., Oberacker R., Hoffmann M.J. and Thommes M. (1999) *Proceedings of Euromat, Symposium F 1.4*, Munich.
29. León y León C.A. (1998) *Adv. Colloid Interface Sci.* **76-77**, 341.
30. Eschricht N., Hoinkis E., Maedler F. and Schubert-Bischoff P. (2002) *Proceedings of the 6th International Symposium on the Characterization of Porous Solids*, Alicante, Spain, May 8-11.
31. Talukdar M.S., Torsæter O. and Ioannidis M. (2002) *J. Colloid. Interface Sci.* **248**, 419.
32. Lin C.L. and Miller J.D. (1999) *1st World Congress on Industrial Process Tomography*, , Buxton, Derbyshire, U.K., April 14-17.
33. Lindquist W.B., Venkatarangan A., Dunsmuir J. and Wong T-F. (2000) *J. Geophys. Res. – Solid Earth* **105**, 21509.
34. Lowell S. and Shields J.E. (1981) *Powder Technol.* **29**, 225.

11 Pore Size and Surface Characteristics of Porous Solids by Mercury Porosimetry

11.1 APPLICATION OF THE WASHBURN EQUATION

The experimental method employed in mercury porosimetry is presented in detail in chapter 18. It involves filling an evacuated sample holder with mercury and then applying pressure to force the mercury into interparticle voids and intraparticle pores. Both applied pressure and intruded volume are recorded.

Ritter and Drake [1] measured the contact angle, θ, between mercury and a variety of materials and found them to be between 135° and 142°, with an average value of 140°. Using this value for θ and 0.480 N/m (480 dyne cm^{-1}) for γ, the surface tension of liquid mercury at room temperature, in the Washburn equation (10.24) one obtains the following expression

$$r = \frac{0.736}{P} \tag{11.1}$$

where P, the applied pressure, is in MPa and r, the pore radius, is in micrometers. When P is expressed in kg cm^{-2}, the relationship becomes

$$r = \frac{1530}{P} \tag{11.2}$$

and when P is expressed in psia, the relationship becomes

$$r = \frac{106.7}{P} \tag{11.3}$$

Therefore, the range of pore sizes that can be measured by mercury intrusion is extremely wide, from approximately 200 micrometers radius down to less than 1.8nm (18Å). Note that the actual size limits depend on the value of the contact angle used in the calculation. Engineering limitations preclude measurement of micropores (< 1nm radius) by mercury intrusion by virtue of the required pressure (736 MPa or 107,000 psia) being impractical.

11.2 PORE SIZE AND PORE SIZE DISTRIBUTION FROM MERCURY POROSIMETRY

The primary purpose of measuring mercury intrusion (and extrusion) curves is to determine pore size and pore volume. There exists a number of ways to express the pore size/volume relationship, i.e. the distribution of pore volume or pore area with respect to pore size. The most commonly encountered are described below.

11.2.1 Linear Pore Volume Distribution

When the radius of a cylindrical pore is changed from r to $r - dr$ the corresponding decremented change in the pore volume V is

$$dV = -2n\pi r \ell dr \tag{11.4}$$

where n is the number of pores with radius r and length ℓ. However, when pores are filled according to the Washburn equation, the volumetric change with decreasing radius does not necessarily decrease since it corresponds to the filling of a new group of pores. Thus, when the pore radius into which intrusion occurs changes from r to $r - dr$ the corresponding volume change is given by

$$dV = -D_V(r)dr \tag{11.5}$$

where $D_V(r)$ is the volume pore size distribution function, defined as the pore volume per unit interval of pore radius as follows. Differentiation of the Washburn equation (10.24), assuming constancy of γ and θ, yields

$$Pdr + rdP = 0 \tag{11.6}$$

Combining equations (11.5) and (11.6) gives

$$dV = D_V(r)\frac{r}{P}dP \tag{11.7}$$

Rearranging yields

$$D_V(r) = \frac{P}{r}\left(\frac{dV}{dP}\right) \tag{11.8}$$

Table 11.1 Data and calculations for the cumulative (Fig. 11.1) and distribution (Fig 11.2) curves.

P (MPa)	ΔP (MPa)	r × 10^3 (μm)	V (cm^3)	ΔV (cm^3)	$\Delta V/\Delta P$ × 10^3 (cm^3 MPa^{-1})	$D_v(r)$ (cm^3 μm^{-1})	$D_s(r)$ × 10^3 (m^2 μm^{-1})
0.345	0.345	213.0	0.085	0.085	0.246	0.039	0.000
1.38	1.035	533.0	0.142	0.057	0.0551	0.142	0.000
3.45	2.070	213.0	0.251	0/109	0.0527	0.852	0.001
6.90	3.45	106.0	0.330	0.079	0.0229	1.482	0.003
34.5	27.6	21.3	0.461	0.131	0.0047	7.602	0.071
69.0	34.5	10.6	0.633	0.172	0.0050	32.364	0.605
103.5	34.5	7.11	0.772	0.139	0.0040	58.228	1.638
138.0	34.5	5.33	0.825	0.053	0.0015	38.837	1.457
207.0	69.0	3.5	0.860	0.035	0.0005	29.073	1.633
276.0	69.0	2.6	0.862	0.002	0.00003	3.101	0.232
345.0	69.0	2.13	0.863	0.001	0.00001	1.620	0.152
414.0	69.0	1.7	0.863	0.000			

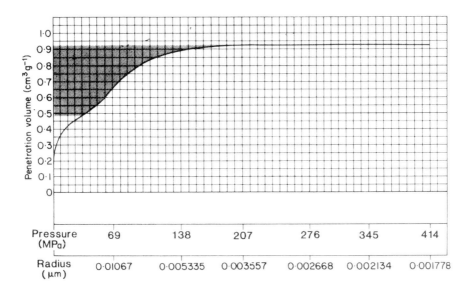

Figure 11.1 Cumulative pore volume plot.

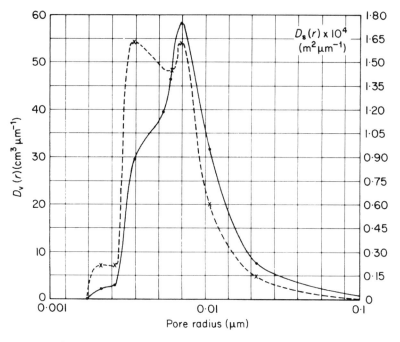

Figure 11.2 Distribution curves for volume $D_V(r)$ ─●─ and area $D_S(r)$ ─✕─

Equation (11.8) represents a convenient means of reducing the cumulative curve to the distribution curve, which gives the pore volume per unit radius interval. Each value of $\Delta V/\Delta P$ is multiplied by the pressure at the upper end of the interval and divided by the corresponding pore radius. Alternatively, the mean pressure and radius in an interval may be used to calculate $D_V(r)$. Unlike surface area calculations, the volume distribution function is based on the model of cylindrical pore geometry.

11.2.2 Logarithmic Pore Volume Distribution
Another useful function often used in place of the linear volume distribution function is the logarithmic volume distribution function, $D_V(\ln r)$, which can be expressed as

$$D_V(\ln r) = \frac{dV}{d\ln r} \tag{11.9}$$

Since

$$d\ln r = \frac{1}{r}dr \tag{11.10}$$

substitution of (11.10) in (11.9) gives

$$D_V (\ln r) = r\left(\frac{dV}{dr}\right) = rD_V (r) \tag{11.11}$$

Substituting for Dv(r) from (11.8) gives

$$D_V (\ln r) = P\left(\frac{dV}{dP}\right) \tag{11.12}$$

and since

$$d \ln P = \frac{1}{P} dP \tag{11.13}$$

it follows that

$$D_V (\ln r) = \frac{dV}{d \ln P} \tag{11.14}$$

The logarithmic volume distribution function serves to reduce the wide apparent disparity of derivative values that the linear volume distribution function can create (as a function of pore size). For example, if the linear volume distribution function *ratio* is, for example, 1000:1 for intrusion into pores of 5 nm and 1000 nm

i.e.

$$\frac{D_V (r) \text{ at } 5 \text{ nm}}{D_V (r) \text{ at } 1000 \text{ nm}} = 10^3 \tag{11.15}$$

then the ratio of the logarithmic volume functions, from (11.11), would be 5:1.

$$\frac{D_V (\ln r) \text{ at } 5 \text{ nm}}{D_V (\ln r) \text{ at } 1000 \text{ nm}} = \frac{5}{1000} \times 10^3 = 5 \tag{11.16}$$

The effect that this has on the appearance of a "pore size distribution" plot is clearly evident in Fig. 11.3.

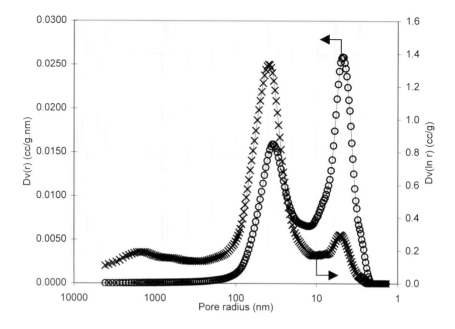

Figure 11.3 Comparison of linear derivative $D_V(r)$ —O— and logarithmic derivative $D_V(\ln r)$ —✕— .

11.2.3 Pore Number Distributions
The linear volume distribution function $D_V(r)$ represents the volumetric uptake in a unit interval of pore radii, irrespective of the variation in the number or the length of the pores. However, two related distribution functions, pore length distribution and pore population distribution, can be derived from cylindrical pore geometry.

11.2.4 Pore Length Distribution
When $D_V(r)$ is divided by πr^2, the mean cross-sectional area in the radius interval, the new function reflects only variations in the pore length and is called the length distribution function, $D_\ell(r)$. Thus

$$D_\ell(r) = \frac{D_V(r)}{\pi r^2} \tag{11.17}$$

This function effectively assigns the variation in the volumetric uptake in a radius interval to differences in the pore lengths and assumes constancy of pore population in each interval.

11.2.5 Pore Population (Number Distribution)

When a volume ΔV is intruded into a narrow pore radius range of Δr, centered about a unit radius r, one can write

$$\frac{\Delta V}{\pi r^2} = n\ell \qquad (11.18)$$

where n is the number of pores of average length, ℓ. Since, for sufficiently small values of Δr

$$\frac{\Delta V}{\Delta r} = D_V(r) \qquad (11.19)$$

it follows that

$$\frac{D_V(r)\Delta r}{\pi r^2} = n\ell \qquad (11.20)$$

For equal, small pore radius intervals, assuming equal pore lengths, equation (11.20) can be used to obtain relative pore populations, viz.,

$$\frac{\dfrac{D_V(r_1)}{r_1^2}}{\dfrac{D_V(r_2)}{r_2^2}} = \frac{n_1}{n_2} \qquad (11.21)$$

Since

$$D_V(r) = \frac{2D_S(r)}{r} \qquad (11.22)$$

it follows that

$$\frac{n_1}{n_2} = \frac{\dfrac{D_V(r_1)}{r_1^2}}{\dfrac{D_V(r_2)}{r_2^2}} = \frac{\dfrac{D_S(r_1)}{r_1}}{\dfrac{D_S(r_2)}{r_2}} \qquad (11.23)$$

11.2.6 Surface Area and Surface Area Distribution from Intrusion Curves

The surface area of all pores and voids filled up to a given pressure P can be obtained from mercury intrusion data. Fig. 11.1 is a cumulative pore volume intrusion curve that shows the summation of volume intruded into the pores and interparticle voids plotted versus the applied pressure.

Recognizing that the increase in interfacial area ΔA from equation (10.25) is effectively the pore and void surface area S, this equation can be rewritten as

$$S\gamma|\cos\theta| = -P\Delta V \qquad\qquad\qquad \text{(cf.10.25)}$$

Then, pores that take up volume dV will possess area dS as given by

$$dS = -\frac{PdV}{\gamma|\cos\theta|} \qquad\qquad\qquad (11.24)$$

The area of pores filled up to pressure P is given by integration of the preceding equation which, assuming constancy of surface tension and wetting angle, is given by

$$S = -\frac{1}{\gamma|\cos\theta|}\int_0^V PdV \qquad\qquad\qquad (11.25)$$

The surface area calculated by the above method is modelless and assumes no specific pore geometry. The surface area value obtained depends strongly on the chosen contact angle. Therefore, one should not necessarily expect excellent agreement between the surface area calculated from mercury intrusion data and that from gas adsorption and the BET method. However, for many materials with simple pore structures general consensus between the two methods can be obtained within ten percent or so, assuming contact angles between 130° and 140°. Indeed, forcing the mercury intrusion area to agree with the BET area by judicious adjustment of the contact angle has been proposed as a method of determining contact angle. But, the mercury intrusion area can only be considered a reasonable estimate of the true surface area if all pores have been intruded. This is not true for those materials containing pores with diameters smaller than 3.6 nm (using the Washburn equation, a maximum pressure of about 414 MPa and assuming a contact angle of 140°).

11.2.7 Pore Area Distributions

The pore surface distribution $D_s(r)$ is the surface area per unit pore radius. By chain differentials, one can write

$$D_s(r) = \left(\frac{dS}{dV}\right)\left(\frac{dV}{dr}\right) \tag{11.26}$$

Assuming cylindrical pore geometry, then

$$\frac{dS}{dV} = \frac{2}{r} \tag{11.27}$$

and

$$D_S(r) = \frac{2}{r} D_V(r) \tag{11.28}$$

11.3 PORE SHAPE FROM HYSTERESIS

A detailed discussion is given in Chapter 10 as to the origins of hysteresis and interpretation. However, for reference or comparative purposes, a characteristic throat/pore ratio, R_{TP}, can be defined as

$$R_{TP} = \frac{|D|_{max,I}}{|D|_{max,E}} \tag{11.29}$$

where $|D|_{max,I}$ and $|D|_{max,E}$ are the modal throat and pore sizes respectively, the subscripts I and E referring to intrusion and extrusion respectively.

11.4 FRACTAL DIMENSION

The fractal dimension D of a solid is a parameter that can characterize the degree of roughness of its surface. Perfectly smooth surfaces expose areas that can be calculated as a function of a characteristic dimension, e.g., $4\pi r^2$ for a nonporous sphere of radius r. Any surface roughness or porosity would increase the sphere's surface area up to an extreme point in which the sphere would be so porous that its entire volume could be occupied by pore walls. At this hypothetical point, the surface area would be proportional to the volume of the sphere, i.e., $4/3\pi r^3$. Real solids expose areas which, following fractal arguments, are proportional to r^D, with their fractal dimension D ranging between 2 for flat surfaces and about 3 for extremely rough surfaces.

For pore wall surfaces in general, it was shown [2,3] that the pore size distribution function, -dV/dr, could be expressed as

$$\frac{-dV}{dr} = k_1 r^{(2-D)} \tag{11.30}$$

where k_1 is a proportionality constant, r is the pore radius, and D is the fractal dimension. Now, combining (11.5) and (11.8) gives

$$\frac{dV}{dr} = \frac{P}{r}\left(\frac{dV}{dP}\right) \tag{11.31}$$

Equating (11.30) with (11.31) yields

$$\frac{P}{r}\left(\frac{dV}{dP}\right) = k_1 r^{(2-D)} \tag{11.32}$$

From the Washburn equation (10.24) one can write the general expression

$$r = kP^{-1} \tag{11.33}$$

Substituting for r in (11.32) and collecting proportionality constants gives

$$P^2\left(\frac{dV}{dP}\right) = k_2 P^{(D-2)} \tag{11.34}$$

That is

$$\left|\frac{dV}{dP}\right| = k_2 P^{(D-2)} P^{-2} \tag{11.35}$$

It follows therefore that fractal dimensions can be derived from mercury porosimetry data according to

$$\frac{-dV}{dP} = k_2 P^{(D-4)} \tag{11.36}$$

Taking logarithms on both sides of the expression yields

$$log\left(\frac{dV}{dP}\right) = log(k_2) + (D-4)log\,P \qquad (11.37)$$

Hence, values of D can be derived from the slope of log (dV/dP) versus log$P$ plots. Smooth surfaces (D = 2) would present slopes close to -2, whereas rougher surfaces (for which D approaches 3) would present slopes approaching -1. Inspection of typical mercury porosimetry curves reveals that constant slopes in the appropriate range are often encountered at pressure regions that coincide with the filling or emptying of pores of distinct sizes. Hence, each step of a cumulative intrusion or extrusion volume versus pressure plot can yield a unique fractal dimension that characterizes the particular range and type of pores being filled or emptied at given pressure ranges.

Once pores of a distinct size become filled or emptied, the mercury volume ceases to change significantly and dV/dP decreases, thereby complicating fractal calculations in the transition regions between pore ranges of materials with multimodal pore size distributions.

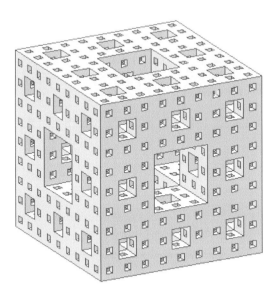

Figure 11.4 A Menger[1] sponge, analogous to a porous medium, has fractal dimension 2.73 ([1] Named after the Austrian mathematician Karl Menger, 1902-1985.)

11.5 PERMEABILITY

Numerous attempts have been made to relate the permeability, k, of a solid to intrinsic and more readily measurable properties, such as porosity and pore diameters. One such approach models the flow of fluids across straight cylindrical channels in a bed of powder by combining Darcy's[4] and Poiseuille's laws [5]; equations (11.38) and (11.45) respectively.

       The volume flow rate, q, as originally defined for a vertical sand bed, [4] is a function of hydrostatic head, Δh, hydraulic conductivity, K, and tube cross sectional area, A, and length, ℓ:

$$q = K \frac{A\Delta h}{\ell} \tag{11.38}$$

The hydrostatic head, h, also known as the *pressure potential*, has the familiar relationship with resulting pressure, P, for a fluid of density ρ, where g is the acceleration due to gravity:

$$h = \frac{P}{\rho g} \tag{11.39}$$

Substitution yields

$$q = K \frac{A\Delta P}{\rho g \ell} \tag{11.40}$$

The interstitial velocity, v, (the speed with which the fluid penetrates the pores) is given by

$$v = \frac{q}{A\phi} \tag{11.41}$$

where ϕ is the system porosity. Substitution for q yields

$$v = K \frac{\Delta P}{\phi \rho g \ell} \tag{11.42}$$

The fluid independent, or intrinsic medium, permeability, k, was derived [6] as a function of Darcy's conductivity (K) and fluid viscosity, μ:

$$K = \frac{k\rho g}{\mu} \tag{11.43}$$

Therefore, substituting k/μ for $K/\rho g$ in (11.42) gives

$$v = k\frac{\Delta P}{\mu\phi\ell} \tag{11.44}$$

Poiseuille [5] developed a description of fluid flow in small tubes. The relationship between volumetric flow rate (flux), q, tube radius, r, pressure drop ΔP, tube length, ℓ, and fluid viscosity, μ, is also known as the Hagen-Poiseuille equation [7]:

$$q = \frac{\pi r^4}{8\mu}\frac{\Delta P}{\ell} \tag{11.45}$$

It follows that the velocity within a tube (pore of radius r) of circular cross section (πr^2) is

$$v = \frac{r^2}{8\mu}\frac{\Delta P}{\ell} \tag{11.46}$$

Equating (11.44) and (11.46)

$$k\frac{\Delta P}{\mu\phi\ell} = \frac{r^2}{8\mu}\frac{\Delta P}{\ell} \tag{11.47}$$

Therefore, by elimination

$$k = \frac{\phi\bar{r}^2}{8} = \frac{\phi\bar{d}^2}{32} \tag{11.48}$$

where ϕ is the sample or powder bed porosity and $\bar{r}$ and $\bar{d}$ are the average (mean volume) radius and diameter respectively of the pore distribution.

Eqn (11.48) has a form similar to the Kozeny-Carman relationship [8,9] which describes permeability corrected in terms of the parameter *tortuosity factor*, τ (see below), in addition to the terms already given. For a porous medium constructed of bundled cylinders [10], the Kozeny- Carman permeability is given by

$$k = \frac{\phi^3 d^2}{32\tau(1-\phi)^2}$$ (11.49)

and when constructed of packed spheres[11]

$$k = \frac{\phi^3 d^2}{72\tau(1-\phi)^2}$$ (11.50)

Assigning an arbitrary pore shape factor, f, one can write

$$k = \frac{\phi^3 d^2}{16 f \tau(1-\phi)^2}$$ (11.51)

At the median porosity for all porous systems (50%), i.e. $\phi_{0.5}$,

$$\frac{\phi_{0.5}^3}{(1-\phi_{0.5})^2} = \phi_{0.5}$$ (11.52)

So, on average

$$k = \frac{\phi d^2}{16 f \tau}$$ (11.53)

such that $f = 2$ for cylindrical pores of circular cross section, i.e. $\tau = 1$ (cf 11.55)

11.6 TORTUOSITY

When modeling the diffusion of fluids in porous solids it is usually stated that the effective (or measured) diffusivity, D_{eff}, differs from the theoretical (or bulk fluid) diffusivity, D_b, by a factor related to the structure of the solid as follows:

$$D_{eff} = \frac{D_b \phi}{\tau} \qquad (11.54)$$

Where ϕ is the pore volume fraction (porosity) and τ is the tortuosity factor. The tortuosity factor lumps all deviations from straight diffusion paths into a single dimensionless parameter. Simply put,

$$\tau = \frac{L_{eff}}{L} \qquad (11.55)$$

where L_{eff} is the traveled distance through the porous medium and L is the simple orthogonal distance across the medium. Tortuosity factor values are expected to range from the theoretically lower limit of 1 (for straight non-intersecting pores *at infinite porosity*) to more than 2 for pore volume fractions below 0.5. Some workers prefer to use $(L_{eff}/L)^2$ and refer to *this* quantity as *tortuosity* (*cf* tortuosity factor). The compound structure parameter $f\,\tau^2$ (the Carman-Kozeny coefficient) typically attains values around 5 [12], which, assuming a shape factor of 2, suggests tortuosity factor (τ) values of approximately 1.6.

Using Fick's first law [13] to describe fluid diffusion through cylindrical paths Carniglia [14] derived the following empirical expression:

$$\tau = 2.23 - (1.13 V\rho) \qquad (11.56)$$

where V is the total specific pore volume (which can be approximated by the mercury volume intruded at the maximum experimental pressure attained) and ρ is the bulk or particle density of the solid – their common product being porosity, ϕ.

For simplicity (11.56) can be rewritten

$$\tau = \frac{2.23}{1.13} - \phi \;\cong 2 - \phi \qquad (11.57)$$

Carniglia pointed out that this relationship was generally applicable within $0.05 \le \phi \le 0.95$, so that if cylindrical pores truly prevailed in a material its tortuosity must lie in the range $1 < \tau > 2$. However, experimental values significantly larger than that led Carniglia to expand (11.57) into a more generalized form:

$$\tau = (2.23 - 1.13 V\rho)(0.92\gamma)^{1+\varepsilon} \qquad (11.58)$$

where γ is a pore shape factor given by

$$\gamma = \frac{4V}{A\bar{d}} = \frac{S}{A} \tag{11.59}$$

A = total surface area by BET gas adsorption, V = total pore volume, $\bar{d}$ = average pore diameter, S = total pore area (by equation 11.25) and ε = pore shape factor *exponent* ($0 < \varepsilon > 1$).

Carniglia reported "remarkable"(sic) agreement between tortuosity factors computed with the above model and those calculated from experimental diffusion measurements for a wide variety of metal oxides and catalysts, with few values of ε falling below 1.0.

11.7 PARTICLE SIZE DISTRIBUTION

The notion that mercury intrusion curves yield information about pores between particles has led several researchers to postulate that the same intrusion curves also contain structural information about the particles themselves. Kruyer [15] studied with mercury the filling and emptying of spaces between packed spheres of known size and predicted hysteresis between mercury penetration (intrusion) and retraction (extrusion). Two models in particular for deducing particle size from intrusion curves have found general acceptance in the literature: the simple mercury breakthrough theory of Mayer and Stowe [16,17] and the integral patch-wise approach of Smith and Stermer [18,19,20].

11.7.1 Mayer & Stowe Approach
The manner in which mercury penetrates a bed of uniform spherical particles was examined in detail by Mayer and Stowe [16]. They postulated that the breakthrough pressure P_b required to force mercury to penetrate the void spaces between packed spheres is a function of the geometry of the system (in addition to contact angle θ and surface tension γ), and is given by

$$P_b = \gamma \frac{L_{L,V} - \cos\theta L_{L,S}}{A} \tag{11.60}$$

where L is the perimeter length of the incipient mercury "lobe", A its cross-sectional area and where subscripts L,V and S refer to liquid, vapor and solid respectively. Note that for a circular opening or radius r, in which the mercury perimeter is solely in contact with the solid surface, $L_{L,V}$ is zero and substituting the usual terms $2\pi r$ and πr^2 for circumference and area respectively, (11.61) reduces to

$$P_b = \frac{-2\gamma\cos\theta}{r} \qquad\qquad \text{(cf 10.61)}$$

Pospěch and Schneider [21] combined contact angle and the spatial relationship between spheres of diameter D into a single proportionality constant, K, such that

$$P_b = \frac{K\gamma}{D} \qquad\qquad (11.62)$$

Using interparticle porosity, ϕ, as a measure of packing arrangement, it can be shown that for randomly close-packed spheres ($\phi = 37.7\%$ [22]) and a typical mercury contact angle of $\theta = 140^\circ$, $K \approx 10.7$. In general, K was found to increase with θ and to decrease with ϕ. Interparticle porosity does change with packing type, but cannot be used as a unique measure of the arrangement of particles. For example, it is possible for two arrangements of spheres to exhibit the same porosity but have different coordination numbers [15]. This leads to different-size passages with different breakthrough pressures filling (effectively) identically sized cavities. However, independent work [23] confirmed that the average particle coordination number (N_C) could be estimated from

$$N_C = \frac{\pi}{\phi} \qquad\qquad (11.63)$$

The interparticle porosity ϕ can be conveniently determined from the bulk density, ρ_b, and particle (envelope) density, ρ_e.

$$N_C = \frac{\pi}{\left(1 - \dfrac{\rho_b}{\rho_e}\right)} \qquad\qquad (11.64)$$

If the true (skeletal) density ρ_t, is known from e.g. helium pycnometry, and the specific *intra*pore volume, V_p, is measured directly by mercury porosimetry, then

$$N_C = \frac{\pi}{1 + \left(\rho_b\left[V_p - \dfrac{1}{\rho_t}\right]\right)} \qquad\qquad (11.65)$$

The validity of the MS theory has been confirmed experimentally by quantitative comparison with particle size distributions derived from independent techniques such as X-ray sedimentation and electron microscopy and was used successfully to characterize a series of reference carbon blacks [24] as illustrated in Figure 11.Y and oxides [25]. The best agreement among different techniques was found for solids with narrow, monomodal particle size distributions and for materials with relatively few interparticle voids.

Particle shape is thought to play a minor role in the results [26], although dimensionless intrusion (or extrusion) particle shape factors, f, have occasionally been introduced in the MS expression as follows:

$$P_b = \frac{fK\gamma}{D} \qquad (11.66)$$

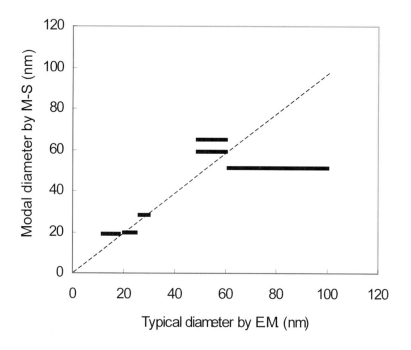

Figure 11.5 Agreement between Mayer-Stowe method (M-S) and Electron Microscopy (EM) for determination of carbon black particle size [24].

11.7.2 Smith & Stermer Approach

Smith and Stermer [18] applied a correction to the shape of the mercury meniscus as it fills the toroidal space between two contacting spheres, and recognized that while this volume is relatively small to the total pore volume, it is only completely filled at pressures tens or hundreds times that at breakthrough.

Upon applying Mayer and Stowe's approach, Smith *et al* observed that even for narrow particle size distributions mercury intrusion curves generally do not exhibit unique breakthrough pressures [19] for random sphere packings. These deviations were attributed to the fact that particles of a given size do not pack in separate regions of a powder bed, as was implied by others, and because of resulting network effects.

Smith and Stermer [20] corrected partly Mayer and Stowe's approach by postulating that the total volume of mercury V_i intruded into a bed of particles of different sizes at any given pressure P_i is the sum of volumes intruded between particles of each size D according to the Fredholm integral

$$V_i = \int_{D_{min}}^{D_{max}} K(P_i, D) F(D) dD \qquad (11.67)$$

where $K(P_i, D)$ is a "kernel" which describes mercury intrusion between "sets" of particles of fixed diameter, D, as a function of pressure, P, and $F(D)$ is the desired particle size distribution function. Using their previously published experimental volume, V_i, and pressure, P_i, values and a generalized kernel function, Smith and Stermer adopted a numerical approach in order to solve equation (11.67) for $F(D)$. Their numerical approach called for the division of the expected particle size range into discrete intervals within which the distribution function $F(D_j)$ for its average pore size (D_j) could be evaluated:

$$V_i = \sum_{j=1}^{N} K(P_i, D_j) F(D_j) \Delta D_j \qquad (11.68)$$

If the summation is applied in the interval $1 \le j \le N$, and the equation is solved for $1 \le i \le M$, evaluating $F(D)$ involves solving simultaneously a matrix of M by N equations subject to a non-negativity constraint such as Lawson and Hanson's Non-Negative Least Squares (NNLS) approach [27], i.e., making the function

$$E = \int_{D_{min}}^{D_{max}} \left[V_{exp}(P) - V_{theor}(P) \right]^2 dD \qquad (11.69)$$

fall within a minimum accuracy, E. To minimize the oscillatory behavior of the calculated distribution function a "damping" term [28, 29], commonly called smoothing (not to be confused with smoothing by averaging), can be added as follows:

$$E' = E + \lambda \int_{D_{min}}^{D_{max}} F(R)^2 \, dD \qquad (11.70)$$

where λ is the so-called regularization parameter. In practice a wide range of values of λ (three orders of magnitude or more) are employed to minimize the number of possible solutions, and in order to effectively vary λ, .i e. exponentially, it is not uncommon to use a regularization *factor*, α, in the form

$$\alpha = log_{10} \lambda \quad i.e. \quad \lambda = 10^{\alpha} \qquad (11.71)$$

To speed up the convergence of the iterative solution of the above summations, a relaxation parameter, ω, can also be incorporated by correcting successive iteration values as follows:

$$F(D_{k+1}) = F(D_k) + \omega[F'(D_{k+1}) - F(D_k)] \qquad (11.72)$$

where $F'(D_{k+1})$ is the iteration value before applying the relaxation approach. Using $0 < \omega < 1$ (under relaxation) slows convergence, and $0 > \omega > 1$ (over relaxation) can speed convergence.

Smith and Stermer concluded that meaningful particle size information on powder compacts could be generated from mercury intrusion data. Their method yields much narrower (more realistic) pore size distributions than straightforward application of the M-S method. However, the S-S method is constrained to narrow monodisperse distributions of a rather restricted porosity range (on the order of 0.37). In addition, the S-S is more computationally demanding and generally yields results of lower resolution than the M-S method.

11.8 COMPARISON OF POROSIMETRY AND GAS SORPTION

The useful range of the Kelvin equation (and its derivative methods e.g. BJH, HK) is limited at the narrow pore end by the question of its applicability (see Chapter 8, §8.5) and at the wide pore end measurements are limited by the rapid (logarithmic) change of the core radius with relative pressure. Modern methods of calculation from gas sorption data (e.g. DFT) confidently extend accurate pore size determination well into the micropore region. Pore diameters in excess of 500nm are rarely reported from gas sorption data.

Mercury porosimetry has parallel constraints at the narrow pore end of its range, in that questions arise regarding the constancy of surface tension and wetting angle for mercury, not to mention the practical difficulties associated with very high pressure generation. Consequently, mercury porosimetry has a lower limit (~3.5nm pore diameter for mercury intrusion at 414 MPa) that does not approach the upper limit for micropores (2nm). However, at the large pore end, mercury porosimetry does not have the limitations of the Kelvin equation and, for example, at 0.0069 MPa (1psi) pore volumes can be measured in pores of approximately 107-micrometer radius or 1.07×10^5 nm.

Comparisons have been made in the range where the two methods overlap and they generally show reasonable agreement. Zweitering [30] obtained distribution curves from mercury porosimetry and the nitrogen isotherms on chromium oxide-iron oxide catalysts. Each curve showed a narrow distribution with the peak near 15nm, the nitrogen curve being slightly narrower and higher. Using nitrogen, Joyner, Barrett and Skjold [31] obtained good agreement between the two methods. By adjusting the wetting angle for mercury of charcoals to values between 130° to 140°, they were able to closely match the curve produced from the nitrogen isotherm. Cochran and Cosgrove [32] using n-butane, reported total pore volume of 0.458 cm^3g^{-1} and 0.373cm^3g^{-1} with adsorption and porosimetry, respectively. The maximum pore radius was under 50nm and they attributed the difference to n-butane entering pores narrower than 3nm. Dubinin et al [33] found that nitrogen and benzene both gave type I isotherms with hysteresis on several activated carbons. The pore distributions based on the Kelvin equation agreed with porosimetry measurements.

Comparisons have been made between surface areas measured by porosimetry and gas adsorption [34] as well as by permeametry [35] with results ranging from excellent to poor.

The most definitive surface area measurements are probably those made by nitrogen adsorption using the BET theory. Neither the Brunauer, Emmett and Teller (BET) theory, nor equation (11.25) used to calculate surface area from mercury intrusion data, makes any assumptions regarding pore shape for surface area determinations. When these two methods are compared, there is often surprisingly good agreement. When the two methods do not agree, it does not imply the theoretical failure of either one. Indeed the differences between results from the two methods can be used to deduce meaningful information that neither alone can supply. For example, when the BET value is large compared to the area measured by porosimetry, the implication is that there is substantial volume of pores smaller than that penetrated by mercury at the maximum pressure.

If the porosimetry can generate 414MPa (60,000 psi) of hydraulic pressure, the minimum diameter into which the intrusion can occur will be about 3.5nm. Assuming that pores centered about 1.5nm are present and have a volume of $0.02cm^3$, an approximation of their surface area can be made by assuming cylindrical geometry. Thus,

$$S = \frac{4V}{d} = \frac{4 \times 0.02}{15 \times 10^{-8}} \times 10^{-4} = 53.3\,m^2 \qquad (11.73)$$

Therefore, the area measured by porosimetry would be approximately $53.3m^2$ less than that measured by the BET method.

Another factor that can lead to BET areas slightly higher than those from porosimetry is pore wall roughness. Slight surface roughness will not alter the porosimetry surface area since it is calculated from the pore volume while the same roughness *will* be measured by gas adsorption.

Cases that lead to porosimetry-measured surface areas exceed those from nitrogen adsorption can result from inkbottle shaped pores having a narrow entrance with a wide inner body. Intrusion into the wide inner body will not occur until sufficient pressure is applied to force the mercury into the narrow entrance. It will appear, therefore, as if a large volume intruded into narrow pores, generating an excessively high, calculated surface area.

11.9 SOLID COMPRESSIBILITY

Mercury intrusion involves subjecting samples to hydrostatic pressures that are applied equally in all directions. This means that upon intrusion the walls of all pores penetrated by mercury at any given pressure are uniformly affected by similar stresses. Hence, a collapse of the pore walls before or as they are filled with mercury is in general unlikely, though not impossible [36]. On the other hand, solid samples could, in principle, compress after the pores are filled, and thus generate an additional volume change for apparent mercury intrusion to take place.

$$\beta = \frac{1}{V_s}\left(\frac{dV_s}{dP}\right) = \rho_B\left(\frac{dV_s}{dP}\right) \qquad (11.74)$$

The influence of sample compressibility on mercury porosimetry data can be assessed by means of a solid compressibility factor, β, defined above as the fractional change in solid volume V_s per unit of pressure and ρ_B is the solid's bulk density.

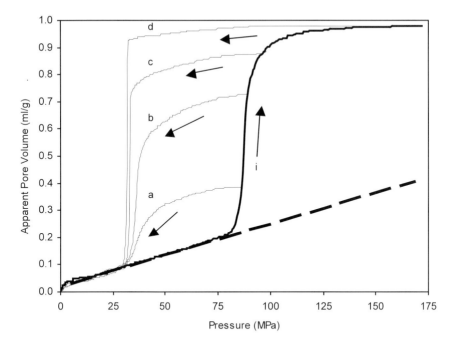

Figure 11.6 A silica gel undergoes reproducible elastic compression (shown by slope of dashed line) prior to intrusion (i) of mercury, regardless of degree of intrusion (after [36]). Extrusion curves (a)-(d) represent pressurization-depressurization cycles to increasing turn-round pressures. Arrows added for clarity.

Most solids exhibit very low compressibilities (typically of the order of 10^{-11} m^2/N), which vary fairly linearly with pressure. This implies that a typical sample could compress by about 0.5% of its original volume when subjected to an applied pressure of 60,000 PSI. In agreement with these observations, mercury porosimetry curves often exhibit small yet finite dV/dP slopes at pressures higher than those required filling all accessible pores in many solids. However, since solid compressibility tends to decrease (the solid stiffens) at some higher pressure, many materials exhibit no compressibility whatsoever. Nevertheless, the compressibility factor β can be estimated from slopes, when present, of the linear portions of high-pressure mercury intrusion or extrusion curves. Similarly, a bulk modulus of elasticity ($= 1/\beta$) can be derived as a correlation of the elastic (reversible deformation) [37] or plastic (irreversible deformation) behavior of solids in general.

11.10 REFERENCES

1. Ritter H.L. and Drake L.C. (1945) *Ind. Eng. Chem. Anal. Ed.* **17,** 782.
2. Pfeifer P. and Avnir D. (1983) *J. Phys. Chem.* **79,** 3558.
3. Neimark A.V. (1990) *Russ.J.Phys.Chem.(Engl.Transl.)* 1990, **64**(10) 1397.
4. Darcy H. (1856) *Les Fontaines Publiques De La Ville De Dijon*, Appendix Note D, Victor Dalmont, Paris.
5. [a]Poiseuille J.L. (1840) *Comptes Rendus, Acad. des Sci.*, **11,** 961; [b]Poiseuille J.L. (1841) **12,** 112.
6. Hubbert M.K. (1956) *AIME Transact.* **207,** 222.
7. Hagen G. (1839) *Pogg. Ann. Phys. Chem*, **46,** 423.
8. Kozeny J. (1927) *Akad. Wiss. Wien*, **136,** 271.
9. Carman P.C. (1937) *Trans. Inst. Chem. Eng.*, **15,** 150.
10. Scheidegger A.E. (1974) *The Physics of Flow Through Porous Media*, University of Toronto Press, Toronto.
11. Panda M.N. and Lake L.W. (1994) *AAPG Bulletin* **79,** 431.
12. Dullien F.A.L. (1992) *Porous Media – Fluid Transport and Pore Structure*, 2nd edn, Academic Press, San Diego/New York/Boston/ London/Sydney/Tokyo/Toronto, p237.
13. Fick's first law states that the rate of diffusion of a fluid is proportional to its concentration gradient; Fick A. (1855) *Ann. Physik.* **94,** 59.
14. Carniglia S.C. (1986) *J. Catal.* **102,** 401.
15. Kruyer S. (1958) *Trans. Faraday Soc.* **54,** 1758.
16. Mayer R.P. and Stowe R.A. (1965) *J. Colloid Interface Sci.* **20,** 893.
17. Mayer R.P. and Stowe R.A. (1966) *J. Phys. Chem.* **70,** 3867.
18. Smith D.M. and Stermer D.L. (1986) *J. Colloid Interface Sci.* **111,** 160.
19. Smith D.M., Gallegos D.P. and Stermer D.L. (1987) *Powder Technol.* **53,** 11.
20. Smith D.M. and Stermer D.L. (1987) *Powder Technol.* **53,** 23.
21. Pospěch R. and Schneider P. (1989) *Powder Technol.* **59,** 163.
22. Frevel L.K. and Kressley L.J. (1963) *Anal. Chem.* **35,** 1492.
23. Ben Aim R. and Le Goff P. (1968) *Powder Technol.* **2,** 1.
24. León y León C.A. (1998) *Adv. Colloid Interface Sci.* **76-77,** 341.
25. León y León C.A., Thomas M.A. and Espindola-Miron B. (1997) *G.I.T. Laboratory Journal* **1,** 101.
26. Svatá M. and Zábranský Z. (1970) *Powder Technol.* **3,** 296.
27. Lawson C.L. and Hanson R.J. (1995) *Solving Least-Squares Problems (Classics in Applied Mathematics, No.15)* Society for Industrial and Applied Mathematics.
28. Tikhonov A.N. and Arsenin V.Y. (1977) *Solutions of Ill-Posed Problems*, Wiley, New York.
29. Press W.H., Teukolsky S.A., Vetterling W.Y. and Flannery B.P. (1992) *Numerical Recipes in C: The Art of Scientific Computing, 2nd Ed.*, Cambridge University Press, chapter 18.
30. Zweitering P. (1958) *The Structure and Properties of Porous Solids*, Butterworths, London, p287.
31. Joyner L.G., Barrett E.P. and Skold R. (1951) *J. Am. Chem. Soc.* **73,** 3158.
32. Cochran C.N. and Cosgrove L.A. (1957) *J. Phys. Chem.* **61,** 1417.
33. Dubinin M.M., Vishnayakova E.G., Zhukovskaya E.A., Leontev E.A., Lukyanovich V.M. and Saraknov A.I. (1960) *Russ. J. Phys. Chem.* **34,** 959.
34. Rootare H.M. and Prenzlow C.F. (1967) *J. Phys. Chem.* **71,** 2733.
35. Henrion P.N., Geenen F. and Leurs A. (1977) *Powder Technol.* **16,** 167.
36. Thomas M.A. and Coleman N.J. (2001) *Colloids Surf. A* **187-188,** 123.
37. Suuberg E.M., Deevi S.C., and Yun Y. (1995) *Fuel* **74,** 1522.

12 Chemisorption: Site Specific Gas Adsorption

12.1 CHEMICAL ADSORPTION

When the interaction between a surface and an adsorbate is relatively weak, only physisorption takes place via dispersion and coulombic forces (see Chapter 2). However, surface atoms often possess electrons or electron pairs that are available for chemical bond formation. Resulting chemical adsorption or chemisorption has been defined by IUPAC [1] as "adsorption in which the forces involved are valence forces of the same kind as those operating in the formation of chemical compounds" and as "adsorption which results from chemical bond formation (strong interaction) between the adsorbent and the adsorbate in a monolayer on the surface" [2].

It is often found to occur at temperatures far above the critical temperature of the adsorbate. This often-irreversible adsorption, at least under mild conditions, is characterized by large interaction potentials that lead to high heats of adsorption. However, to distinguish between physical and chemical adsorption based on heat of adsorption alone is not entirely satisfactory. For example, some experimentally determined enthalpy values for physical adsorption in zeolites [3] have been reported to be higher than 30 kcal mol^{-1} and almost 50 kcal mol^{-1} for nitrogen and carbon dioxide, respectively. In comparison, heats of chemisorption range from over 100 to *less than* 20 kcal mol^{-1}.

As is true for most chemical reactions, chemisorption is often associated with an *activation energy*, which means that adsorptive molecules attracted to a surface must go through an energy barrier before they become strongly bonded to the surface. In the following Figs. 12.1 – 12.3 curves C and P represent potential energy (plotted as a function of internuclear separation) for chemisorption and physisorption respectively. The minimum of curve P is equal to the heat of physisorption, ΔHp, and the minimum in curve C is equal to the heat of chemisorption, ΔHc. The minimum of curve C occurs at a smaller internuclear separation than that of curve P since chemical bonding, which involves orbital overlap, brings nuclei closer together than can the less energetic physical adsorption forces.

It should be noted that an adsorptive molecule first approaches the surface along the low energy curve P. It is important to recognize therefore that in almost all cases physisorption occurs as a pre-cursor to chemisorption. The adsorption process can be said to transition from physical to chemical at the cross over point, A. In Fig. 12.1, the chemisorption process indicated proceeds without the hindrance of an

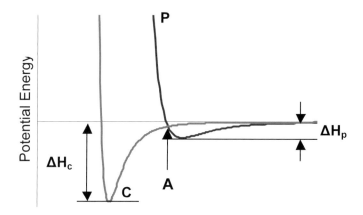

Figure 12.1. Potential energy curves for molecular (non-dissociative) adsorption

activation barrier since the chemisorption potential energy curve approaches from zero. This chemisorption curve is that which might be expected for non-dissociative, or molecular, adsorption. The crossover point must lie below zero potential and such adsorption is expected to have no activation energy (though in reality a small configurational or orientation barrier may exist). Examples include adsorption of oxygen on certain crystal faces of silver [4], iodine monochloride (ICl) on aluminum [5] molecular hydrogen, oxygen and chlorine on clean carbon *in vacuo* [6] and ethylene on silver {4,1,0} [7].

In many cases, the molecule to be adsorbed undergoes dissociation. This "*dissociative adsorption*" is defined by IUPAC [8] as "adsorption with dissociation into two or more fragments, both or all of which are bound to the surface of the adsorbent". Dissociative adsorption may be non-activated (typical for platinum group metals) or activated.

Dissociative adsorption is non-activated if not only the bonding, but also the antibonding molecular orbitals (MO) hybridize with the metal's d-band, leading to a partial occupation of the antibonding states and a resultant weakening of the molecular bond [9].

The dissociative adsorption of hydrogen on many metals, often non-activated, is said to be *homolytic* because a single bond, i.e. the electron pair between the two hydrogens in the molecule is divided.

$$H_2 \xrightarrow[catalyst]{} 2H_{ads} \qquad\qquad (12.1)$$

e.g. spontaneous hydrogen adsorption on palladium [10-12], H_2 on Rh [9,13], H_2 on W [14]. Note however, hydrogen adsorption is not unactivated

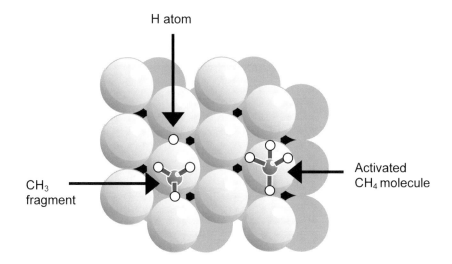

Figure 12.2 Dissociative adsorption of CH₄ on Iridium (after Henkelman [25]). Distortion of the metal lattice not shown.

on all metals, even occasionally platinum [15]; the dissociation at the copper surface [16-18] for example, is regarded as a classic case of activated dissociative adsorption. Iron [19], cobalt [20] and ZnO [21] are other examples.

If the bond splits asymmetrically, the dissociative adsorption is said to be *heterolytic*. Not surprisingly, this is most likely in three- (or more) center adsorptives. Such cases include water adsorption on MgO [22,23]:

$$MgO + H_2O \xrightarrow[catalyst]{} Mg(OH)_2 \qquad (12.2)$$

H₂S on alumina-rich zeolite [24]:

$$H_2S \xrightarrow[catalyst]{} HS^- + H^+ \qquad (12.3)$$

$$H^+ + O^{2-} \rightarrow OH^- \qquad (12.4)$$

Methane on Ir [25], Ni [26],Pt [27] and MgO [28]:

$$CH_4 \xrightarrow[catalyst]{} CH^{3-} + H^+ \qquad (12.5)$$

or hydrogen adsorbed on metal oxide surfaces (MgO [29,30], ZrO2 [31]), in which surface ions are of lower coordination than in the bulk.

The difference between activated and non-activated adsorption can be gleaned from Figs. 12.2 and 12.3 respectively.

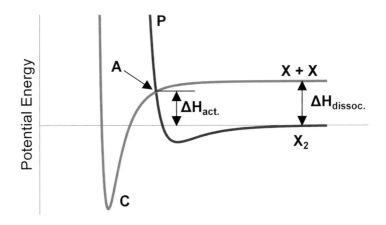

Figure 12.3 Potential energy curves for activated adsorption.

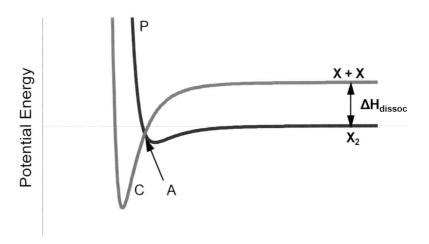

Figure 12.4 Potential energy curves for non-activated adsorption.

In both cases the dissociation of a diatomic species is characterized by a heat of dissociation ΔH_{dissoc} for $X_2 \rightarrow X + X$. The relative position of the two potential energy curves P and C results in a cross over point lying either above (activated) zero, or below zero (non-activated). At low temperatures most of the net heat of adsorption is due to physisorption whereas at higher temperatures most of the heat of adsorption is due to chemisorption since there is sufficient thermal energy to overcome the activation energy barrier.

Note, the rate of adsorption frequently decreases at higher temperatures indicating that the adsorption mechanism undergoes a transition to an *activated* process. A resulting observation is that the quantity adsorbed varies with temperature as shown in Fig. 12.5(c). The initial decrease is due to thermal desorption of physically adsorbed gas – along isobar 12.5(a). Subsequently, the quantity adsorbed increases with increasing temperature due to commencement of activated chemisorption. Finally, the curve slopes downward gradually when sufficiently high temperature is reached to desorb the chemisorbed state thermally – along isobar 12.5(b).

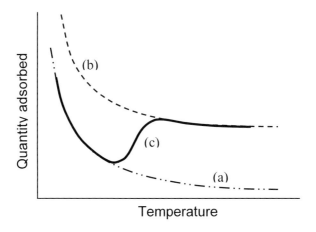

Figure 12.5 Isobaric variation in quantity adsorbed with temperature. Physisorption isobar (a) represents lower heat of adsorption than chemisorption isobar (b).

Because chemisorption involves a chemical bond between adsorbate and adsorbent, unlike physisorption, only a single layer of chemisorbed species can be realized on localized *active* sites such as those found in *heterogeneous catalysts*. However, further physical adsorption on top of the chemisorbed layer and diffusion of the chemisorbed species into the bulk solid can obscure the fact that truly chemisorbed material can be only one layer in depth.

The methods available to distinguish between strong chemisorption and weaker adsorption will be discussed.

12.2 QUANTITATIVE MEASUREMENTS

Because the formation of a chemical bond takes place between an adsorbate molecule and a localized, or specific, site on the surface of the adsorbent, the number of *active sites* on *catalysts* can be determined simply by measuring the quantity of chemisorbed gas. The active site of catalytic importance in many cases is an exposed metal atom, zero valence, i.e. in its elemental state. Classical examples include nickel for hydrogenation of unsaturated carbon-carbon bonds. However, a number of important oxide and other non-metallic catalysts do exist. Some catalysts are essentially pure metals (e.g. iron, albeit with added promoters, for ammonia synthesis and platinum gauze used for ammonia oxidation,) but it is more likely that the catalyst exists as a collection of metal atoms distributed over an inert, often refractory, support material such as alumina. At the atomic level, it is normal that these atoms are assembled into "islands" [32-34] or "clusters" [35-39] on the surface of the support.

Figure 12.6 *Metal* atoms form nanoparticles on the support surface.

Since these islands vary in size due to both the intrinsic nature of the metal and the support beneath, plus the method of manufacture, more or less of the metal atoms in the whole sample are actually exposed at the surface (Fig 12.6). It is evident therefore that the method of gas adsorption is perfectly suited to the determination of exposed active sites, whereas a simple chemical determination of the entire metal content of the sample is not.

12.3 STOICHIOMETRY

A simple association of one gas molecule with one exposed metal atom yields the most straightforward determination of the quantity of active sites. However, it is important to recognize that many polyatomic gas molecules do *not* adsorb in a ratio of unity with each active site. Consider the example of the reaction between a hydrogen molecule (H_2) and a surface constructed of platinum atoms. It is generally recognized that the hydrogen is *dissociatively* adsorbed, that is, the hydrogen splits into two atoms each of which reacts with a single metal atom. Thus, *one* gas molecule has bound *two* metal atoms. Therefore, the stoichiometry is said to be two for this surface reaction. Alternatively, one adsorbate molecule might be associated with more than one metal atom without being dissociated (see Fig. 12.7 below). For example, carbon monoxide, which is normally expected to bond in a ratio of one to one, might "bridge" between two metal atoms. This situation would also result in a stoichiometry of two. Excess adsorption beyond a ratio of unity might even lead to compound formation: hydride in the case of hydrogen and carbonyl for carbon monoxide. Such behavior yields stiochiometries below unity, and should be avoided.

> *The gas-sorption stoichiometry is defined as the average number of metal atoms with which each gas molecule reacts.*

Since, in the gas adsorption experiment to determine the quantity of active sites in a catalyst sample, it is the quantity of adsorbed gas that is actually measured, the knowledge of (or at least a reasonably sound assumption of) the stoichiometry involved is essential in meaningful active site determinations. This information is most easily obtained from the catalyst literature, or possibly by using one of the calculation methods described below, the Langmuir method, depending on the exact nature of the sorption process.

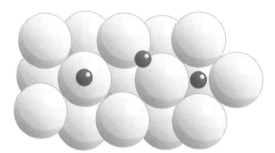

Figure 12.7 There may exist a number of different adsorption sites that involve different numbers of metal atoms per adsorbate molecule.

12.4 MONOLAYER COVERAGE

Once armed with the experimentally determined quantity of gas adsorbed by the sample, or *isotherm*, one can set about calculating the number of active sites from the monolayer capacity, V_m. There are a number of graphical and numerical methods available for this, and those most commonly used are described below.

12.4.1 Extrapolation

This method involves measuring three or more data points of a single isotherm. The points should be measured at a pressure high enough such that the plateau, or low-slope, region of the isotherm has been reached. In this region, the surface has been saturated with chemisorbate and the formation of the monolayer assured. Further increase in pressure leads only to additional physical adsorption. The contribution from this physisorption can be accounted for by assuming that it is zero at zero pressure. A straight line constructed through the linear region of the isotherm, extrapolated back to zero pressure, yields the chemisorption only monolayer as the intercept on the y-axis. This value of V_m represents the total amount of chemisorbed gas, regardless of the exact nature of the type of bonding, be it strong or weak. The type of isotherm used is commonly referred to, therefore, as the *combined* (strong plus weak) isotherm.

12.4.2 Irreversible Isotherm and Bracketing

Some applications require that only strong chemisorption sites be determined and that the contribution by weak chemisorption, in addition to the gas uptake due to physisorption, be excluded. In those cases, it is not sufficient to merely back-extrapolate the combined isotherm. Where weak chemisorption might be expected, a second repeat isotherm must be obtained. After the acquisition of the first (combined) isotherm the sample is evacuated at analysis temperature to cause the desorption of weakly bonded gas molecules. Strongly adsorbed molecules remain bonded to the active sites at the sample surface. The second isotherm consisting of weak chemisorption plus physisorption is then collected in the same way as the first. The difference between the two isotherms at any given pressure represents the amount of strong chemisorption at that pressure – the so-called "bracketing method".

Alternatively, the plateau region of the entire difference plot, or *irreversible isotherm*, can be extrapolated to zero pressure to determine V_m graphically.

12.4.3 Langmuir Theory

The Langmuir theory [40-42] seeks to bridge the gap between the kinetics and the thermodynamics of any gas sorption process. Kinetic principles allow the description of the rate of any adsorption process as:

$$\text{rate of adsorption} = k_a P(1 - \theta) \qquad (12.6)$$

where k_a is a constant (the *adsorption rate constant*), P is the adsorptive pressure, and θ is the fraction of the surface already covered with adsorbate, i.e.,

$$\theta = \frac{V}{V_m} \qquad (12.7)$$

where V is the volume adsorbed and V_m is the volume of gas required to complete a monomolecular layer, or monolayer, of adsorbed gas. The fraction θ is commonly termed *surface coverage*, or simply *coverage*. Similarly, the rate of desorption can be simply described as:

$$\text{rate of desorption} = k_d \theta \qquad (12.8)$$

where k_d is the *desorption rate constant*, In other words, while the adsorption process depends on both gas pressure and the availability of uncovered surface (1-θ), the desorption process depends only on the amount of gas already adsorbed on the surface – i.e. the coverage, θ.

At equilibrium, the adsorptive pressure remains relatively constant. That means that any gas that does leave the surface is immediately replaced by an equal amount of gas being adsorbed. It can be postulated therefore that, at equilibrium the rate of adsorption = rate of desorption. Therefore, combining (12.6) and (12.8) yields

$$k_a P(1 - \theta) = k_d \theta \qquad (12.9)$$

Defining k_a/k_d as the Langmuir constant K, and rearranging (12.9) gives

$$KP = \frac{\theta}{1 - \theta} \qquad (12.10)$$

Substituting for θ from (12.7) and rearranging yields

$$\theta = \frac{V}{V_m} = \frac{KP}{1+KP} \qquad (12.11)$$

In its linear form, the above equation can be expressed as:

$$\frac{1}{V} = \frac{1}{V_m} + \frac{1}{V_m KP} \qquad (12.12)$$

Hence, a plot of $1/V$ against $1/P$ often yields a straight line with slope $1/(V_m K)$ and y-intercept $1/V_m$, so that V_m can be simply computed thus:

$$V_m = \frac{1}{\text{intercept}} \qquad (12.13)$$

However, since chemical adsorption, unlike physisorption, is specific, it can only occur on certain types of sites. Consider the adsorption of a typical adsorbate, hydrogen, on a typical catalyst, platinum. Adsorption of the diatomic hydrogen molecule results in its breakup into two hydrogen atoms and it is these that are actually chemisorbed. This requires that the molecule fall on *two* empty sites, i.e. two neighboring platinum atoms. Therefore, the rate of chemical adsorption and desorption become, respectively:

$$\text{rate of chemical adsorption} = k_{ca} P(1-\theta)^2 \qquad (12.14)$$

and

$$\text{rate of chemical desorption} = k_{cd}\theta^2 \qquad (12.15)$$

where k_{ca} and k_{cd} are constants. It can be easily confirmed that, at equilibrium, the above equations yield the following linearized form of the Langmuir equation:

$$\frac{1}{V} = \frac{1}{V_m} + \frac{1}{V_m K_c P^{1/2}} \qquad (12.16)$$

where $K_c = k_{ca}/k_{cd}$. Since different gas molecules require different types and number of available chemisorption sites, the above expression can be generalized as follows:

$$\frac{1}{V} = \frac{1}{V_m} + \frac{1}{V_m K_c P^{1/S}}$$ (12.17)

where S is the adsorption stoichiometry. The generalized Langmuir equation yields N_m from the intercept of a plot of $1/V$ vs. $1/P^{1/S}$. If the value of S is not known, nor available from accepted literature values, then it can be estimated by applying the Langmuir equation to the experimental isotherm data. Various Langmuir plots are constructed using different values of S until a reasonably linear relationship is found. Of course, this approach requires that the adsorption really is "Langmuir-like" in nature, which can be recognized as a type I isotherm using the BDDT classification, and is therefore normally limited to strong chemisorption only.

12.4.4 Temperature Dependent Models

The theories developed by Temkin [43] and Freundlich [44] differ from that of Langmuir in the way they treat the parameter K of the Langmuir expression; from (12.10)

$$K = \frac{k_a}{k_d} = \frac{\theta}{P(1-\theta)}$$ (12.18)

All three theories can be represented by the above equation if K is replaced by the generalized Arrhenius expression

$$K = K_0 \, exp\left(\frac{q}{RT}\right)$$ (12.19)

where K_0 is a constant, R is the universal gas constant, T is the adsorption temperature and q is the *heat of adsorption; i.e. K* varies as a function of q. Langmuir assumed that the parameter K is constant and therefore that q is also constant at all degrees of coverage. However, q can reasonably be expected to vary as a function of coverage, θ, due to the presence of previously adsorbed molecules.

12.4.5 Temkin Method

Temkin assumed that q decreases linearly with increasing coverage, that is,

$$q = q_0 (1 - \lambda\theta)$$ (12.20)

Where q_o is a constant equal to the heat of adsorption at zero coverage ($\theta = 0$) and λ is a proportionality constant. In practice, this behavior occurs when the adsorption of gas is gradually slowed down as the coverage increases because it becomes increasingly difficult for impinging gas molecules to find the specific adsorption sites they seek. Combining equations (12.18), (12.19) and (12.20) yields:

$$K_0 \, exp \, q_0 \left(\frac{1 - \lambda\theta}{RT} \right) = \frac{\theta}{P(1 - \theta)} \tag{12.21}$$

Taking natural logarithms of both sides gives

$$ln \, K_0 + q_0 \left(\frac{1 - \lambda\theta}{RT} \right) = ln \left(\frac{\theta}{P(1 - \theta)} \right) \tag{12.22}$$

Expanding both sides

$$ln \, K_0 + \frac{q_0}{RT} - \frac{q_0\lambda\theta}{RT} = ln \left(\frac{\theta}{1 - \theta} \right) - ln \, P \tag{12.23}$$

Multiplying by $\dfrac{RT}{q_0\lambda}$ and rearranging yields

$$\theta + \left(\frac{RT}{q_0\lambda} \right) ln \left(\frac{\theta}{1 - \theta} \right) = \left(\frac{RT}{q_0\lambda} \right) ln \, P + \left(\frac{RT}{q_0\lambda} \right) ln \, K_0 + \frac{1}{\lambda} \tag{12.24}$$

or

$$\theta + A \, ln \left(\frac{\theta}{1 - \theta} \right) = A \, ln \, P + B \tag{12.25}$$

where A and B are dimensionless constants given by

$$A = \frac{RT}{q_0\lambda} \tag{12.26}$$

and

$$B = A \, ln \, K_0 + \frac{1}{\lambda} \tag{12.27}$$

In above equations, the factor $A ln(\theta/1-\theta)$ can be neglected if A is very small and especially when θ approaches 0.5. Since the value of A often falls in the range 0.01 to 0.05, it is generally acceptable to neglect the aforementioned factor even at coverages far away from $\theta = 0.5$. In such cases, equation (12.25) becomes

$$\theta = A \ln P + B \qquad (12.28)$$

or, since $\theta = V/V_m$

$$V = V_m A \ln P + V_m B \qquad (12.29)$$

The Temkin method predicts that plotting V versus lnP yields a straight line (with slope $= V_m A$ and y-intercept $= V_m B$) at intermediate coverages, say between 0.2 and 0.8.

The above equations show that the values of A and B are temperature dependent. On the other hand, for relatively narrow temperature ranges, V_m should be constant. Therefore, one can determine V_m from isotherms at two or more temperatures and the *intersection* of the resulting Temkin plots. This approach is illustrated graphically in Fig.12.8.

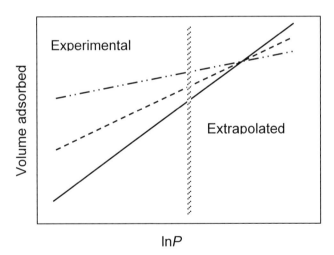

Figure 12.8 Illustrative Temkin plot for three isotherms at low — ·· — ·· , intermediate − − − − − − , and high ——————— temperatures.

12.4.6 Freundlich Method

Freundlich also postulated that the heat of adsorption, q, varies with surface coverage, θ. However, unlike the Temkin method, in which q is assumed to vary linearly with coverage, the Freundlich method infers a more complex dependence of q on θ. Essentially, it is assumed that the heat of adsorption decreases exponentially with increasing coverage, θ. This assumption leads to the following relationship:

$$q = -q_m \, ln\,\theta \tag{12.30}$$

Where q_m is a constant equal to the heat of adsorption at $\theta = 0.3679$. Combining equations (12.18), (12.19) and (12. 30) yields

$$K_0 \, exp\left(-q_m \frac{ln\,\theta}{RT}\right) = \frac{\theta}{P(1-\theta)} \tag{12.31}$$

Taking natural logarithms of both sides gives:

$$ln\,K_0 - q_m \frac{ln\,\theta}{RT} = ln\left(\frac{\theta}{1-\theta}\right) - ln\,P \tag{12.32}$$

Multiplying by $\dfrac{RT}{q_m}$ and rearranging yields

$$ln\,\theta + \frac{RT}{q_m} ln\left(\frac{\theta}{1-\theta}\right) = \frac{RT}{q_m} ln\,P + \frac{RT}{q_m} ln\,K_0 \tag{12.33}$$

or

$$ln\,\theta + C\,ln\left(\frac{\theta}{1-\theta}\right) = C\,ln\,P + D \tag{12.34}$$

where C and D are constants given by

$$C = \frac{RT}{q_m} \tag{12.35}$$

and

$$D = C\,ln\,K_0 \tag{12.36}$$

The magnitude of C is often small. Indeed, it is similar in magnitude to that of A in the Temkin method. In addition, as with the Temkin method, the factor $ln(\theta/1-\theta)$ approaches zero at intermediate coverages, say 0.2 to 0.8. Hence, in general, the factor $Cln(\theta/1-\theta)$ can be neglected, and thus the above equation becomes

$$ln\,\theta = C\,ln\,P + D \tag{12.37}$$

If the adsorption behavior is correctly described by the above relationship, a plot of lnV versus lnP should yield a straight line. The above equations also show that values of C, and hence D, are temperature dependent. As in the Temkin method, one can assume that V_m is constant for a relatively narrow temperature range, and so V_m is computed from the intersection of the Freundlich plots of isotherms collected at two or more temperatures. Whenever two or more isotherms at different temperatures are available for the same material, one can employ either the Temkin or Freundlich methods. Normally, one or the other, but not both methods, will be found to be appropriate.

12.4.7 Isotherm Subtraction – Accessing Spillover

If, by application of one or more of the above methods, it is suspected that the apparent chemisorption behavior is not simple, it might be appropriate to consider interference by the support material. The irreversible isotherm approach is often successful in eliminating adsorption on the inert material, but by its very nature also dismisses adsorption on the support that is caused by the presence of the active sites themselves. It is typical, for example, for hydrogen that is dissociatively adsorbed on platinum to migrate to the support surface. This type of adsorption is termed *spillover*.

If it is important to include spillover in the determination of adsorption capacity then two isotherms should be measured: one being of the supported catalyst and the other being of support material only - without active catalyst. The first isotherm yields adsorption data consisting of strong chemisorption on the active sites, weaker chemisorption, physisorption on active sites and exposed support surface, plus spillover in the immediate vicinity of active sites. The second isotherm consists merely of physisorption on the support. By subtracting the second data set from the first, the net chemisorption amount, *including* spillover, can be easily computed.

12.4.8 Surface Titration

Often it is sufficient to measure the total number of strongly adsorbing surface sites, to calculate crystallite size for example, without recourse to the entire isotherm. Adsorption data acquired by the flowing method yields quantitative strong chemisorption values of V_m directly. Since the adsorptive is presented to the material in a pulses carried in a flow of a second, inert, non-adsorbing gas any weakly chemisorbed and physisorbed molecules are removed by the continuous flow of carrier gas after the pulse of adsorptive has passed by the sample. Multiple pulses of adsorptive may have to be introduced to the sample before all active sites have been reacted. The surface is then said to be *saturated*. The cumulative amount of adsorbate retained by the surface is equal to V_m, the monolayer capacity. Chapter 17, section 17.5, in the Experimental Section details the methodology and exact calculations necessary.

12.5 ACTIVE METAL AREA

To calculate the equivalent area of exposed active sites, or active metal area (A_m) one must first know the following: the number of adsorbed gas molecules in the monolayer, N_m, the adsorption stoichiometry, s, and the cross-sectional area occupied by each active surface atom, A_X. N_m is calculated from the experimental determination of V_m thus:

$$N_m = \left(\frac{V_m}{22414} \right) L_{Av} \tag{12.38}$$

Where V_m is expressed in mL at standard temperature and pressure (STP), 22414 being the volume, in mL, occupied by one mole of gas at STP and L_{Av} is Avogadro's number or 6.022×10^{23}. The number of metal atoms, N_a, is given by multiplying N_m, the number of gas molecules, by S, the stoichiometry:

$$N_a = N_m \cdot S \tag{12.39}$$

The *total* sample active metal area is simply the product of the number of exposed metal atoms, N_a, and the cross-sectional area of each atom, A_X. It is normal to express the result in terms of *specific* metal area, that is the active metal area per unit mass, W:

$$A_m = \frac{N_a A_x}{W} \tag{12.40}$$

12.6 DISPERSION

In the case of supported metal catalysts, it is important to know what fraction of the active metal atoms is exposed and available to catalyze a surface reaction. Those atoms that are located inside metal particles do not participate in surface reactions, and are therefore wasted. Dispersion is defined as the percentage of all metal atoms in the sample that are exposed. The total amount of metal in the sample is termed the *loading*, χ , as a percentage of the total sample mass, and is known from chemical analysis of the sample. The dispersion, δ, is calculated from:

$$\delta = M \frac{N_a}{L_{Av}} \times \frac{100}{W_S \chi} \times 100\% \qquad (12.41)$$

Where M is the molecular weight of the metal and W_S is the mass of the sample.

12.7 CRYSTALLITE (NANOPARTICLE) SIZE

If both the mass of metal in the sample and its density (ρ) are known, the volume of metal can be estimated. Since the surface area of the metal (A_m) has already been calculated, the equivalent particle diameter, d, can be estimated assuming a particle shape factor, f

$$d = \frac{\chi}{100} \frac{f}{A_m \rho} \qquad (12.42)$$

where χ is the *percent* loading.

Consider a cube of six sides (faces) each of edge-length d. Its specific volume, V, is given by

$$V = d^3 \qquad (12.43)$$

and its specific metal surface area, A_m, is given by

$$A_m = 6d^2 \qquad (12.44)$$

Dividing equation (12.42) by (12.43) yields.

$$\frac{V}{A_m} = \frac{d}{6} \tag{12.45}$$

Substituting $\dfrac{1}{\rho}$ for V, and rearranging results in

$$d = \frac{6}{A_m \rho} \tag{12.46}$$

For a supported metal, it is necessary to consider the percent loading, P,

$$d = \frac{P}{100}\frac{6}{A_m \rho} \tag{12.47}$$

In this example, the number 6 is the shape factor, f. For a particle of unit length it is equal to the surface area-to-volume ratio. Other geometries can be treated in a similar fashion. For example, a sphere of unit diameter also yields a shape factor equal to 6.

12.8 HEATS OF ADSORPTION AND ACTIVATION ENERGY

Whenever a gas molecule adsorbs on a surface, heat is (generally) released, i.e. the process of adsorption is exothermic. This heat comes mostly from the loss of molecular motion associated with the change from a 3-dimensional gas phase to a 2-dimensional adsorbed phase. Heats of adsorption provide information about the chemical affinity and the heterogeneity of a surface, with larger amounts of heat denoting stronger adsorbate-adsorbent bonds. There are at least two ways to quantify the amount of heat released upon adsorption: in terms of (i) *differential* heats, q, and (ii) *integral* heat, Q, as described below.

12.8.1 Differential Heats of Adsorption
The differential heat of adsorption, q, is defined as the heat released upon adding a small *increment* of adsorbate to the surface. The value of q depends on (i) the strength of the bonds formed upon adsorption and (ii) the degree to which a surface is already covered with adsorbate. Hence, q is most often expressed in terms of its variation with surface coverage, θ. A plot of q versus θ provides a characteristic curve illustrating the energetic heterogeneity of the solid surface. This curve not only serves to fingerprint

the surface energetics of a particular sample, but it also allows a direct test of the validity of any V_m evaluation method used (see earlier) since each method assumes a different relationship between q and θ, that is,

> (i) the *Langmuir* method assumes that q remains *constant* with increasing θ,
> (ii) the *Temkin* method assumes that q decreases *linearly* with increasing θ and
> (iii) the *Freundlich* method assumes that q decreases *exponentially* with increasing θ.

Since q can, and most often does, vary with θ, it is convenient to express it as an *isosteric* heat of adsorption, that is, at *equal* surface coverage for *different* temperatures. Thus, in order to evaluate this q, one must obtain two or more isotherms at different temperatures. By inspecting plots of coverage (where $\theta = V/V_m$) against pressure, one can determine those pressures corresponding to equal coverage at the different temperatures. These pressures and temperatures are then used to construct an Arrhenius plot of the natural logarithm of pressure ($\ln P$) versus the reciprocal of absolute temperature ($1/T$). Values for q at any given coverage, θ, can be calculated from the slope, m, of each Arrhenius plot using the following equation:

$$q = -mR \qquad\qquad (12.48)$$

where m = d $\ln P$/d($1/T$) and R is the universal gas constant.

12.8.2 Integral Heat of Adsorption

This is simply defined as the *total* amount of heat released, Q, when one gram of adsorbent takes up X grams of adsorbate. It is equivalent to the sum, or integral, of q over the adsorption range considered, that is:

$$Q = \frac{V_m}{22414} \int_{\theta_{min}}^{\theta_{max}} q\,d\theta \qquad\qquad (12.49)$$

where V_m is expressed in mL at STP, and θ ideally ranges from $\theta_{min} = 0$ to θ_{max} = maximum coverage attained experimentally.

12.8.3 Activation Energy

A series of temperature-programmed studies (TPR, TPO, TPD), each performed on a fresh sample of the material under test, yields sufficient information to permit the estimation of activation energy. The peak maximum temperature is not only related to the activation energy for the process being monitored, but it is also a function of the heating rate. Therefore, the shift in peak maximum (T) as a function of heating rate (β) can be used to calculate the activation energy (E_a) using the Kissinger equation [45-48]:

$$ln\left(\frac{\beta}{T^2}\right) = \frac{E_a}{RT} + ln\left(\frac{AR}{E_a}\right) + C \tag{12.50}$$

where R is the universal gas constant and A and C are constants. A plot of $ln(\beta/T^2)$ versus $1/T$ yields a straight line with slope $= -Ea/R$ from which the activation energy, E_a, is easily computed.

12.9 REFERENCES

1. Everett D.H. (1972) *Pure Appl. Chem.* **31**, 579.
2. [a]*IUPAC Compendium of Chemical Terminology 2nd Edition* (1997); [b]*Pure Appl. Chem.* (1990) **62**, 2179.
3. Shen D., Bulow M., Siperstein F., Engelhard M. and Myers A.L. (2000) *Adsorption* **6**, 275.
4. Vattuone L., Burghaus U., Valbusa U. and Rocca M. (1998) *Surf. Sci.* **408**, L693.
5. Pettus K.A., Taylor P.R. and Kummel A.C. (2000) *Faraday Discuss.* **117**, 321.
6. Russell Jr J.N., Butler J.E., Wang G.T., Bent S.F., Hovis J.S., Hamers R.J and D'Evelyn M.P. (2001) *Mat. Chem. Phys.* **72**, 147.
7. Rocca M, Savio L and Vattuone L. (2002) *Surf. Sci.* **502-503**, 331.
8. Burwell R.L. (1976) *Pure Appl. Chem.* **46**, 71.
9. Eichler A., Hafner J., Groß A. and Scheffler M. (1999) *Phys. Rev. B* **59**, 297.
10. Wilke S. and M. Scheffler M. (1996) *Phys. Rev. B,* **53**, 4926.
11. Lischka M, and Groß A. (2002) *Phys. Rev., B* **65**, 075420.
12. Kroes G-J., Gross A., Baerends E-J., Scheffler M. and McCormack D.A. (2002) *Acc. Chem. Res.* **35**, 193.
13. Hafner J. (1998) Presented at: *"Fundamentals Aspects of Surface Science: Elementary Processes in Surface Reactions"* ERC, June 20-25, Acquafredda di Maratea, Italy.
14. Darling G.R., Kay M. and Holloway S. (1997) *Phys. Rev. Lett.* **78**, 1731.
15. Pennemann B., Oster K. and Wandelt K. (1991) *Surf. Sci.* **251/252**, 877.
16. Clarke L. (1937) *J. Am. Chem. Soc.* **59**, 1389.
17. Michelsen H.A., Rettner C.T. and Auerbach D.J. (1992) In *Surface Reactions* (R.J. Madix, ed.) Springer-Verlag, Berlin, chapter 6.
18. McCormack D.A. and Kroes G.J. (1998) *Chem. Phys. Lett.* **296**, 515.
19. Weatherbee G.D., Rankin J.L., and Bartholomew C.H. (1984) *Appl. Catal.* **11**, 73.
20. Zowtiak J.M., Weatherbee G.D. and Bartholomew C.H. (1983) *J. Catal.* **82**, 230.
21. Burwell R.L. and Taylor H.S. (1936) *J. Am. Chem. Soc.* **58**, 1753.
22. Johnson M.A., Stevanovich E.V. and Truong T.N. (1998) *J. Phys. Chem., B* **102**, 6391.
23. Johnson M.A., Stevanovich E.V., Truong T.N., Günster J. and Goodman D.W. (1999) *J. Phys. Chem., B* **103**, 3391.
24. Yeom Y.H. and Kim Y. (1996) *J. Phys. Chem.,* **100** 8373.
25. Henkelman G. and Jónsson H. (2001) *Phys. Rev. Lett.* **86**, 664.
26. Wonchoba S.E. and Truhlar D.G. (1998) *J. Phys. Chem., B* **102**, 6842.
27. Gee A.T., Hayden B.E., Mormiche C., Kleyn A.W. and Riedmüller B. (2003) *J. Chem. Phys.* **118**, 3334.

28. Onal I. and Senkan S. (1997) *Ind. Eng. Chem. Res.* **36,** 4028.
29. Paganini M.C., Chiesa M., Giamello E., Coluccia S., Martra G., Murphy D.M. and Pacchioni G. (1999) *Surf. Sci.* **421,** 246.
30. Cavalleri M., Pelmenschikov A.G., Morosi G., Gamba A., Coluccia S. and Martra G. (2001) *Stud. Surf. Sci. Catal.* **140,** 131.
31. Hong Z., Fogash K.B. and Dumesivc J.A. (1999) *Catal. Today* **51,** 269.
32. Gavioli L., Kimberlin K., Tringides M.C., Wendelken J.F. and Zhang Z. (1999) *Phys. Rev. Lett.* **82,** 129.
33. Legare P., Madani B., Cabeza G.F. and Castellani N.J. (2001) *Int. J. Mol. Sci.* **2,** 246.
34. Campbell C.T. and Starr D.E. (2002) *J. Am. Chem. Soc.* **124,** 9212.
35. Hansen K.H., Worren T., Stempel S., Lægsgaard E., Bäumer M., Freund H-J., Besenbacher F., and Stensgaard I. (1999) *Phys. Rev. Lett.* **83,** 4120.
36. Dulub A.O., Hebenstreit W. and Diebold U. (2000) *Phys. Rev. Lett.* **84,** 3636.
37. Argo A.M., Odzak J.F., Lai F.S. and Gates B.C. (2002) *Nature* **415,** 623.
38. Aizawa M., Lee S. and Anderson S.L. (2002) *J. Chem. Phys.* **117,** 5001.
39. Kim Y.D. and Goodman D.W. (2002) *"Toward an understanding of catalysis by supported metal nanoclusters"* Presented at International Workshop on Nanochemistry, COST D19, Sept 26-28.
40. Langmuir I. (1916) *J. Am. Chem. Soc.* **38,** 2221.
41. Langmuir I. (1917) *J. Am. Chem. Soc.* **39,** 1848.
42. Langmuir I. (1918) *J. Am. Chem. Soc.* **40,** 1361.
43. Temkin M.I. (1941) *J. Phys. Chem. (USSR),* **15,** 296.
44. Freundlich H. (1906) *Z. Phys. Chem. A,* **57,** 385.
45. Kissinger H.E. (1956) *J. Res. Natl. Bur. Stand.* **57,** 217.
46. Kissinger H.E. (1957) *Anal. Chem.* **29,** 1702.
47. Redhead P.A. (1962) *Vacuum* **12**, 203.
48. Coats A.W. and Redfern J.P. (1964) *Nature* **201**, 68.

13 Physical Adsorption Measurement: Preliminaries

13.1 EXPERIMENTAL TECHNIQUES FOR PHYSICAL ADSORPTION MEASUREMENTS

The adsorbed amount as a function of pressure can be obtained by volumetric (manometric) and gravimetric methods, carrier gas and calorimetric techniques, nuclear resonance as well as by a combination of calorimetric and impedance spectroscopic measurements (for an overview see refs [1-3]). However, the most frequently used methods are the volumetric (manometric) and the gravimetric methods. The gravimetric method is based on a sensitive microbalance and a pressure gauge. The adsorbed amount can be measured directly, but a pressure dependent buoyancy correction is necessary. The gravimetric method is convenient to use for the study of adsorption not too far from room temperature. The adsorbent is not in direct contact with the thermostat and it is therefore more difficult to control and measure the exact temperature of the adsorbent at both high and cryogenic temperatures. Therefore, the volumetric method is recommended to measure the adsorption of nitrogen, argon and krypton at the temperatures of liquid nitrogen (77.35 K) and argon (87.27 K) [4].

The volumetric method is based on calibrated volumes and pressure measurements by applying the general gas equation. The adsorbed amount is calculated by determining the difference of the total amount of gas admitted to the sample cell with the adsorbent and the amount of gas in the free space. The void volume needs to be known very accurately. We discuss details with regard to this important matter in chapter 14.

Both volumetric and gravimetric methods allow adsorption to be measured under either static and quasi-equilibrium conditions. In quasi-equilibrium methods the adsorptive is continuously admitted to the sample at a certain, low rate. To obtain a scan of the desorption isotherm the pressure is continuously decreased. The most difficult point associated with the quasi-equilibrium procedure is that one needs to reach at any time of the experiment satisfactory equilibrium conditions. To check that equilibrium has been established the analysis should be repeated using slower gas rates (gas bleed rate). The validity of the analysis is strengthened if identical data are obtained at two different gas flows. If one can reach true equilibrium conditions, the main advantage of this method is that it provides isotherms of unsurpassed resolution. A detailed description of quasi-equilibrium methods is given in refs. [5-7].

In contrast to this quasi-equilibrium method, the continuous flow method proposed by Nelson and Eggertson [8], gives rise to a discontinuous, point-by-point adsorption, as is the case for the volumetric static method. This flow method is based on a continuous flow of a mixture of a carrier gas (helium) and adsorptive (e.g., nitrogen) through the powder bed. The change in gas composition due to the adsorption of nitrogen is monitored by a thermal conductivity detector. The method is still frequently used for single point surface area measurements. Hence, we will discuss both the static volumetric- and the dynamic flow method in more detail later in this book (chapters 14 and 15), because these two methods are the most frequently used for the surface area and pore size characterization of porous solids.

13.2 REFERENCE STANDARDS

Adsorbents are usually characterized using parameters such as specific surface area, pore volume and the pore size distribution. These quantities can be derived by analysis of gas sorption isotherms by applying an appropriate theory used to treat the adsorption and/or desorption data.

However, the results obtained for the surface area, pore size etc. are dependent on the applied theoretical method for data analysis and, to some extent, on the chosen experimental method. In order to overcome these problems, the use of certified reference materials and standardized measurement procedures allow one to check and calibrate the performance of sorption analyzers and to compare results from different laboratories. More than twenty certified reference materials for surface area and pore size analysis are now available from the four internationally recognized standard authorities including BAM (Germany), IRRM (European Community), LGC (UK), and NIST (USA) [9]. These reference materials generally consist of powders of inert materials such as alumina, titania, silica/quartz, carbons and silicon nitrides.

Much work was done recently in standardizing measuring methods for the surface area and pore size characterization of porous solids. A comprehensive survey of standards on surface structure characterization can be found in the review of Robens *et al* [10]. Such standards are available from the national standardization organizations, but as a result of the globalization of research and industries, standardization is shifted more to the international organizations, e.g., ISO (International Organization for Standardization, ASTM International (American Society for Testing and Materials).

13.3 REPRESENTATIVE SAMPLES

Often, for the purposes of laboratory analysis, it is necessary to obtain a small quantity of powder from a larger batch. For maximum accuracy and reproducibility, it is necessary that the sample chosen be representative of the larger initial quantity. Here, the term representative means that the sample must possess the same particle and pore size distributions and specific surface area as the larger quantity from which it was obtained.

To some extent, under even slight agitation, particles tend to segregate with the finer ones settling toward the bottom of the container. When poured from a container into a conical pile, the smaller particles will collect towards the center. This behavior is caused by large particles rolling over the smaller ones and the small particles settling through the voids between the larger ones.

It is generally impossible to make a segregated sample completely homogeneous by shaking, tumbling or any other technique. Often these attempts only further enhance the segregation process. Devices such as the spinning or rotary riffler can be used to obtain representative samples. Rifflers operate on the principle that a sample need not be homogeneous in order to be representative.

Such a riffler (shown in Fig. 13.1) operates by loading the powder sample into a vibrating hopper, which delivers the sample down a chute into eight rotating collectors. Both the delivery and rotational rates can be controlled.

The sample, when loaded in the hopper, will be segregated. Therefore, at any depth ℓ there will exist a particle diameter gradient $\Delta D/\Delta \ell$. The powder settles as it is delivered to the collectors at the rate $\Delta \ell/\Delta t$. Then,

$$\frac{\Delta D}{\Delta \ell} \cdot \frac{\Delta \ell}{\Delta t} = \frac{\Delta D}{\Delta t} \tag{13.1}$$

which is the rate of change of particle diameters leaving the hopper. If ψ is defined as the change in particle diameter entering each collector per revolution, then,

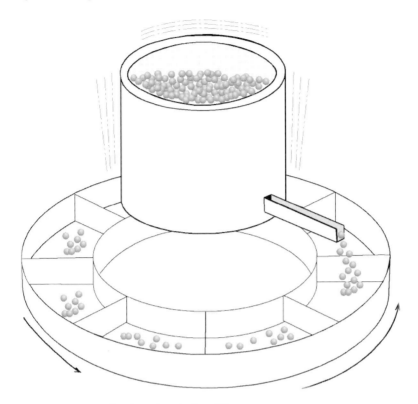

Figure. 13.1 Schematic drawing of a spinning riffler.

$$\psi = \frac{1}{8} \frac{\Delta D}{\Delta t} \tau \qquad (13.2)$$

where τ is the time per revolution. Substituting equation (13.1) into (13.2) gives

$$\psi = \frac{1}{8} \frac{\Delta D}{\Delta \ell} \cdot \frac{\Delta \ell}{\Delta t} \tau \qquad (13.3)$$

Recognizing that $\Delta \ell / \Delta t$ is proportional to the feed rate, F, yields

$$\psi = K \frac{\Delta D}{\Delta \ell} F \tau \qquad (13.4)$$

Equation (13.4) asserts that the change in particle diameter entering each collector can be made as small as necessary by decreasing the feed rate or increasing the collector's rotational rate. However, Hatton [11] argues that it is preferable to slow the feed rate rather than increase the rotational rate to provide better representation.

When the entire sample has been delivered, each collector will contain powder exactly representative of the initial batch. Each collector will also contain a size gradient from top to bottom. Therefore, if the quantity required for analysis is less than the amount in any single collector, the process must be repeated. This is achieved by placing the contents of one collector back into the hopper and re-riffling. There are rifflers available that can riffle samples of less than one gram to accommodate the need for small final samples.

Rifflers can be equipped with an optional sieve that can be placed on top of the hopper to exclude particles above a required size. The advantage of this arrangement lies in the fact that a single particle of 100 μm radius has the same weight as one million 1μm particles with only one-hundredth the surface area. If only a few of the large particles are present, they may not be properly represented in the final sample.

Often it is thought that the effectiveness of a riffler can be demonstrated by the uniformity of weight accumulated in each collector. This reasoning is incorrect if one considers that each collector will necessarily acquire a slightly different amount of sample if the collector diameters vary slightly. The only correct test for the effective performance of a riffler is to compare the contents of each collector in terms of particle size distribution or specific surface area.

Experiments with silica powder chosen from five depths in a two-pound container gave surface areas from top to bottom of 9.8, 10.2, 10.4, 10.5, and 10.7 m^2/g. When the same sample was poured into a conical pile, five random samples produced surface areas of 10.3, 11.0, 10.4, 10.0, and 10.6 m^2/g. However, when the sample was riffled in a spinning riffler with three size reductions, the subsequent analysis of the contents of five collectors gave 10.2, 10.1, 10.2, 10.2, and 10.1 m^2/g as the specific surface area.

13.4 SAMPLE CONDITIONING: OUTGASSING OF THE ADSORBENT

In order to obtain correct data it is required to remove all physically adsorbed material from the adsorbent surface to ensure a reproducible initial state of the adsorbent surface, especially one in which pores are obstructed by foreign species. This can be accomplished by vacuum pumping or purging with an inert gas at elevated temperatures. Vacuum is attractive, because it prepares the surface under the same conditions that are required to start a static volumetric adsorption experiment (i.e., to start a such an adsorption experiment the sample cell with adsorbent has to be evacuated). In addition, it also allows outgassing at lower temperatures than one would need if flow outgassing (purging) under atmospheric conditions were applied. A drawback of the vacuum method is certainly the problem of powder elutriation (see §13.5), which does not occur as readily in the flow method. An additional advantage of flow is that its setup is very easy (i.e., no expensive vacuum system is required).

The sample should be outgassed at the highest temperature that will not cause a structural change to the sample. In general, too low an outgassing temperature will cause lengthy preparation, and may result in lower than expected surface areas and pore volumes. In general, outgassing organics must be performed with care since most have quite low softening or glass transition points (e.g., magnesium stearate). In contrast, most carbon samples for instance can be outgassed quite safely at 573 K. Physisorbed water in nonporous or mesoporous materials will be lost at relatively low temperatures (< 473 K) under the influence of vacuum, but if adsorbed in narrow micropores, as they are present in some zeolites, high temperatures (up to 573 K) and long outgassing periods (often no less than 8 hours) are required. A special heating program is often needed, one which allows for a slow removal of most of the preadsorbed water at temperatures below 373 K accompanied by a stepwise increase in temperature until the final outgassing temperature is reached. This is done to avoid potential structural damage of the sample due to surface tension effects and so-called "steaming", i.e., hydrothermal alteration. In particular zeolites are sensitive to steaming, where the possibility of vaporization and re-condensation inside the pores can lead to structural changes.

In those instances where samples cannot be heated, the method of repetitive cycling [12] investigated by Lopez-Gonzales et al [13] can be utilized. They found that by repetitive adsorption and desorption the surface can be adequately cleaned to allow reproducible measurements. Usually three to six cycles are sufficient to produce a decontaminated surface. Presumably the process of desorption, as the sample temperature is raised, results in momentum exchange between the highly dense adsorbate leaving the surface and the contaminants. As the impurities are removed from the surface they will be carried out of the sample cell by the flowing gas. Thus the technique of repetitive cycling is an efficient means for removal of contaminants from the surface of a solid.

If the vacuum method is performed outgassing of the sample to a residual pressure of about 1 Pa (7.5×10^{-3} torr) to 0.01 Pa is considered to be satisfactory for most nonporous and mesoporous materials. This can be readily achieved by a combination of a rotary and diffusion pump in connection with a liquid nitrogen trap. The usual precautions, including a cold trap, should be taken to insure no contamination by the pump oil. However, as already mentioned, microporous materials such as zeolites require an outgassing at much lower pressures, i.e., below 0.01 Pa. Nitrogen adsorption occurs here at relative pressures P/P_0 even below 10^{-7} for pores of diameter below ca. 6 Å. Hence, the sample should also be outgassed at these very low pressures. This can be achieved by using a turbomolecular pump which, if coupled with a diaphragm roughing pump, allows the sample to be outgassed in a completely oil free system.

13.5 ELUTRIATION AND ITS PREVENTION

Elutriation, or loss of powder out of the sample cell, is caused by gas flowing too rapidly out of the cell, and is in particular problematic for vacuum outgassing of materials such as the ones prepared by the sol-gel method.

Wider stems and sample cells with larger bulbs can be beneficial in reducing elutriation. Wider stems reduce the velocity of the gas leaving the cell when evacuation begins and thus it is less likely to entrain powder particles and transport them upwards and out of the cell. The presence of a filler rod significantly increases gas velocity because of the internal dimensions. Hence, outgassing should always be performed *without* a filler rod inserted into the sample cell. In problematic cases the analysis needs to be performed without a filler rod, but some loss of resolution and/or sensitivity may result (for more information about the proper choice of sample cells and filler rods etc. see also chapter 14.6).

In certain cases it might be required to pump down the sample very slowly by controlling (manually or in an automated way) the opening of the valve, which connects the sample cell to the vacuum line. Elutriation problems are also often encountered during degassing of damp, "light" powders. This condition can be reduced or eliminated by pre-drying the samples in a conventional drying oven and outgassing under vacuum at room temperature for some time before heating to the final temperature, where the outgassing of the sample should be ultimately performed.

In the most difficult cases, it might be necessary to insert a small glass wool plug (or a glass frit) into the cell stem, however this has the disadvantage that the quality of the vacuum surrounding the sample is no longer known.

13.6 REFERENCES

1. Keller J.U., Robens E. and du Fresne von Hohenesche C. (2002) *Stud. Surf. Sci. Catal.* **144**, 387.
2. Mikhail R.Sh. and Robens E. (1983) *Microstructure and Thermal Analysis of Solid Surfaces*, Wiley, Chichester.
3. Kaneko K., Ohba T., Hattori Y., Sunaga M., Tanaka H. and Kanoh H. (2002) *Stud. Surf. Sci. Catal.* **144**, 11.

4. Sing K.S.W., Everett D.H., Haul R.A.W., Mouscou L., Pierotti R.A, Rouquerol J and Siemieniewska T. (1985) *Pure Appl. Chem.* **57,** 603.
5. Rouquerol J., Rouquerol F., Grillet Y. and Ward R.J (1988) *Stud. Surf. Sci. Catal.* **39,** 67.
6. Rouquerol F., Rouquerol J. and Sing K.S.W. (1999) *Adsorption by Powders & Porous Solids*, Academic Press, London.
7. Rouquerol J. (1997) In *Physical Adsorption: Experiment, Theory and Applications*, (Fraissard J. and Conner C.W, eds.) Kluwer, Dordrecht, p17.
8. Nelson F.M. and Eggertsen F.T. (1958) *Anal. Chem.* **30**, 1387.
9. [a]BAM: Bundesanstalt für Materialforschung und –prüfung, Richard-Willstätter-Str. 11, D-12489 Berlin, Germany, www.bam.de.; [b]NIST: National Institute of Standards and Technology, Gaithersburg, MD, USA, http://ts.nist.gov/ts/htdocs/230/232/232.htm; [c]LGC: LGC Promochem, Queens Rd, Teddington, Middlesex TW11 0LY, UK, www.lgcpromochem.com; [d]IRMM: Institute for Reference Materials and Measurements, Reference Materials Unit, attn BCR Sales, Retieseweg, B-2440 Geel, Belgium, www.irmm.jrc.be.
10. Robens E., Krebs K-F., Meyer K., Unger K.K. and Dabrowski A.(2002) *Colloids Surf. A*: *Physicochemical and Engineering Aspects,* 253.
11. Hatton T.A. (1978) *Powder Technol.* **19**, 227.
12. Harned H.S. (1920) *J. Am. Chem. Soc.* **42**, 372.
13. Lopez-Gonzales J. de D., Carpenter F.G. and Deitz V.R. (1955) *J. Res. Nat. Bur. Stand.* **55**, 11.

14 Vacuum Volumetric Measurement (Manometry)

14.1 BASICS OF VOLUMETRIC ADSORPTION MEASUREMENT

Many types of static volumetric vacuum adsorption apparatus have been developed [e.g., 1-7] and no doubt every laboratory where serious adsorption measurements are made has equipment with certain unique features. The number of variations is limited only by the need and ingenuity of the users. However, all volumetric adsorption systems have certain essential features, including a vacuum pump, one or more gas supplies, a sample container, a calibrated manometer, and a coolant. Fig.14.1a describes a historical set-up using an Hg-volume manometer instead of a pressure transducer; the volumes V_a, V_b, and V_c correspond to the calibrated reference volume in Fig. 14.1b, which refers to a simplified, modern static volumetric sorption apparatus.

   Fully automated and highly precise sorption analyzers are available which are designed in a way that they fulfill the needs of researchers in academia and industry, who are interested in measuring high resolution sorption isotherms with the highest possible accuracy. These instruments are also compatible with the requirements of product control, where fast and highly reproducible surface area - and pore size measurements are needed.

   The general procedure of volumetric gas adsorption measurements will be described on the basis of Fig. 14.1b. In order to measure an adsorption isotherm, known amounts of gas are admitted stepwise through the manifold (shaded area) into the sample cell. At each step, adsorption of the gas by the sample occurs and the pressure in the confined volume falls until the adsorbate and the remaining gas in the sample container are in equilibrium. The amount of gas adsorbed is the difference between the amount of gas admitted and the quantity of gas filling the void volume. The adsorbed amount (or volume) is computed by applying the general gas equation. The determination of the amount adsorbed requires that the manifold volume (V_m) as well as the void volume (V_v) is accurately known. The void volume is the volume of the sample cell, which is not occupied by the adsorbent. V_m calibration can be performed by expanding helium into a calibrated reference volume (see Fig. 14.1b) and by applying the general gas equation. Another possibility is to displace free volume by a reference volume made from material of stable dimensions (e.g., a stainless steel sphere of defined volume), or attaching a calibrated volume in place of the sample tube and expanding helium into this from the manifold.

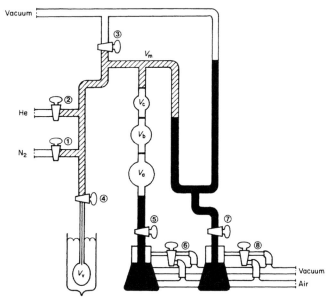

Figure 14.1a. Historical static volumetric apparatus. Solid dark areas represent mercury. Horizontal lines between V_a, V_b and V_c are fiducial marks.

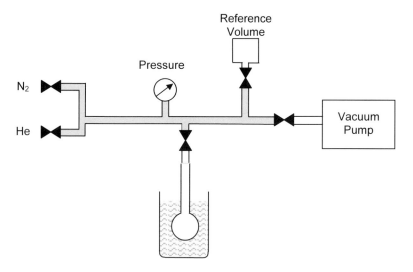

Figure 14.1b Simplified, modern static volumetric apparatus.

V_v, the void volume of the sample cell, can be determined by using helium before the sample cell is immersed into liquid nitrogen according to:

$$\frac{P_1 V_m}{T_m} = \frac{P_2 (V_m + V_v)}{T_m} \qquad (14.1)$$

where T_m is the manifold temperature, P_1 is the manifold pressure before helium is transferred into the sample cell, P_2 is the pressure in manifold and sample cell after the valve between manifold and sample cell was opened and helium transferred into the sample cell.

By convention, all gas volumes are converted to standard conditions, (T_{Std} = 273.15 K, P_{Std}= 760 torr), in which gas quantities are given in terms of standard volumes (i.e., at STP). A volume V can be converted to STP conditions according to: $V_{STP} = V(P/P_{Std})(T_{Std}/T)$. When V_{STP} is given in cm^3, dividing by 22414 cm^3 converts it into the number of adsorbed moles of adsorptive (one mole of ideal gas at standard temperature and pressure occupies a volume of 22414 cm^3).

Hence, if one expresses the volumes in equation (14.1) in STP units, one obtains

$$\frac{P_1 V_m}{T_m} = P_2 \left(\frac{V_m}{T_m} + \frac{V_v}{T_{Std}} \right) \qquad (14.2)$$

The apparent free space volume V_{yf} when the sample cell is immersed into the coolant bath (e.g., liquid nitrogen) can be determined again by using helium, because helium is essentially not adsorbed at 77.35 K and does not exhibit measurable non-ideal behavior at this temperature

$$\frac{P_1 V_m}{T_m} = P_3 \left(\frac{V_m}{T_m} + \frac{V_{vf}}{T_{Std}} \right) \qquad (14.3)$$

where P_1 is the pressure in the manifold (before helium is transferred into the sample cell), whereas P_3 is the pressure in manifold and sample cell after helium is transferred into the sample cell which is partially immersed in liquid nitrogen.

In the next step, the adsorptive (e.g., nitrogen) instead of helium is used and therefore adsorption occurs. The volume of adsorptive dosed into the sample tube is given by:

$$V_d = \left(\frac{P_m V_m}{T_m} - \frac{P V_m}{T_{me}} \right) \times \left(\frac{T_{Std}}{P_{Std}} \right) \qquad (14.4)$$

V_d = total volume of adsorptive dosed
P_m = the manifold pressure before the dose
T_m = manifold temperature before the dose
P = the equilibrium pressure after the dose
T_{me} = the manifold temperature at the time equilibrium is attained
(If the manifold temperature is kept constant T_{me} is equal to T_m)

The volume sorbed, V_s, after the first dose is given by

$$V_S = V_d - \left(\frac{P V_{vf}}{P_{Std}} \right) \qquad (14.5)$$

where V_{vf} is again the effective void calculated with the sample cell immersed into liquid nitrogen. The volume $V_{s,i}$, adsorbed on the ith dose is given by $V_{s,i} = V_{d,i} - (P_i V_{vf} / P_{Std})$, where $V_{d,i}$ is the volume dosed after the ith dose.

The errors in the vacuum volumetric method arise principally from two sources. The error in measured doses is cumulative, requiring very accurate calibration. In addition, the void volume correction becomes more significant at higher pressures, when measuring low surface areas and when more molecules are retained in the void volume than are adsorbed.

14.2 DEVIATIONS FROM IDEALITY

At the cryogenic temperatures used for physical adsorption characterization the correction for the non-ideality of real gases cannot be neglected and can be as large as ca. 5 % (e.g., for argon at 77 K). Hence, the gas contained in that volume of the sample cell immersed in a bath of liquid nitrogen, for example, needs to be corrected for non-ideality. The total void volume V_v of the sample cell partially immersed in liquid nitrogen consists of a warm-zone volume, $V_{vw,}$ (at ambient temperature) and a cold-zone volume, $V_{vc,,}$ (at the temperature of the coolant), viz

$$V_v = V_{vw} + V_{vc} \qquad (14.6)$$

and

$$V_{vc} = \frac{\left(V_{vf} - V_v\right)}{\left(1 - \dfrac{T_{cold}}{T_{amb}}\right)} \tag{14.7}$$

where V_{vf} is the free space volume determined with the sample cell immersed in the coolant, see equation (14.3), V_v is the free space of the sample cell determined at ambient temperature i.e., the geometric volume; see equation (14.1), V_{vc} is the volume of adsorptive within that volume of the sample cell immersed in the liquid nitrogen bath, T_{cold} is the temperature of the coolant (e.g., liquid nitrogen temperature), T_{amb} is ambient temperature. Hence, the adsorbed volume V_s including the correction for non-ideality is given by

$$V_S = V_d - \left(\frac{PV_{vf}}{P_{Std}} + \frac{P^2 V_{vc}\,\alpha}{P_{Std}}\right) \tag{14.8}$$

where α is the so-called "non-ideality factor", which is 6.6×10^{-5} (for nitrogen) if the pressure P is expressed in torr.

Emmett and Brunauer [5] derived an appropriate correction, which is linear with pressure, from the van der Waals equation (see table 14.1).

14.3 VOID VOLUME DETERMINATION

To generate accurate gas sorption data using the conventional volumetric static methods it is necessary (as mentioned before) to introduce a non-adsorbing gas such as helium prior (or after) every analysis in order to measure the void (free space) volumes at room temperature and at the temperature at which the adsorption experiment is performed.

The helium void volume measurement procedure is based on various assumptions: (i) helium is not adsorbed/absorbed on/into the adsorbent, (ii) helium does not penetrate into regions which are inaccessible for the adsorptive (e.g., nitrogen). However, these pre-requisites are not always fulfilled – in particular in case of microporous adsorbents (see discussion in §14.8). The use of helium can be avoided if the measurement of the void volume can be separated from the adsorption measurement by applying the so-called NOVA concept, which is described in §14.9.2. The determination of the void volume can be avoided totally by using difference measurements, i.e., an apparatus consisting of identical reference and sample cells, and the pressure difference being monitored by a differential pressure transducer.

Table 14.1 Some correction factors for non-ideality.

Gas	T, °C	α (torr^{-1}) from Emmett and Brunauer [5]
N_2	-195.8	6.58×10^{-5}
	-183	3.78×10^{-5}
O_2	-183	4.17×10^{-5}
Ar	-195.8	1.14×10^{-4}
	-183	3.94×10^{-5}
CO	-183	3.42×10^{-5}
CH_4	-140	7.79×10^{-5}
CO_2	-78	2.75×10^{-5}
	-25	1.56×10^{-5}
n-C_4H_{10}	0	1.42×10^{-4}
	25	4.21×10^{-5}

14.4 COOLANT LEVEL AND TEMPERATURE CONTROL

In an open dewar the cryogenic coolant such as liquid nitrogen and/or argon will evaporate, and will therefore change the level of cryogen around the sample cell stem and consequently the cold zone and warm zone volumes. Therefore, it is crucial that the specific position of the cryogen level on the sample cell stem is kept constant during the measurement. It should be maintained - unless otherwise compensated - at least 20 mm above the sample and constant to within at least 1-2 mm. This can be achieved by moving the dewar up, using a thermistor (sensor) coolant level control, a porous sleeve that surrounds the sample tube stem (to maintain the cryogen level by capillary action), or by periodic replenishing of the cryogen.

For maximum accuracy, the calibrated volumes and the manifold should be maintained at constant temperature, or alternatively, the temperature may be closely monitored, i.e., the actual manifold temperature need to be taken into account in the calculations of the adsorbed amount (see equation 14.4).

14.5 SATURATION VAPOR PRESSURE, P_0 AND TEMPERATURE OF THE SAMPLE CELL

The sorption isotherm is measured as a function of pressure until the saturation pressure P_0 of the bulk fluid is achieved. The term P_0 is defined as the saturated equilibrium vapor pressure exhibited by the pure adsorptive contained in the sample cell when immersed in the coolant (e.g., liquid nitrogen or argon).

As discussed in chapter 4, the thickness of an adsorbed (liquid-like) film, as well the pressure where pore filling and pore condensation occurs in a pore of given width, is related to the difference in chemical potential of the adsorbate (μ_a) and the chemical potential of the bulk liquid (μ_o) at the same temperature (i.e., the temperature at which the adsorption experiment is performed). This chemical potential difference can be related to the pressures P and P_0 of the vapor in equilibrium with the adsorbed film and the saturated liquid, respectively, by $\Delta\mu = (\mu_a - \mu_o) = RT\ln P/P_0$, where R is the universal gas constant and T is the temperature. Hence, the adsorbed amount is measured as a function of the ratio P/P_0 and the accurate monitoring of the saturation pressure is crucial in order to ensure the highest accuracy and precision for pore size and surface area analysis.

The saturation vapor pressure depends on temperature. This is illustrated in the schematic phase diagram in Fig. 14.2. The vapor pressure line, which defines the temperatures and pressures where vapor and liquid are in coexistence, terminates in the critical point. The relationship between saturation pressure and temperature is given by the Clausius-Clapeyron equation: $(dP_0/dT)_{coex} = \Delta H/T\Delta V$, where ΔH is the heat associated with the gas-liquid phase transition (heat of evaporation) and ΔV is the difference of molar volumes between coexisting vapor and liquid. For temperatures far below the critical temperature, ΔV corresponds to the molar volume of the vapor and, for relatively small temperature intervals, one can consider ΔH to be constant. Based on these assumptions the Clausius-Clapeyron equation can then be written as $\ln(P_{0,T2}/P_{0,T1}) = \Delta H/R(1/T_1 - 1/T_2)$. Thus, the saturation pressure increases exponentially with temperature. For instance, an increase of ca. 0.2 K results in a saturation pressure increase of ca. 20 torr for nitrogen at a temperature around 77 K.

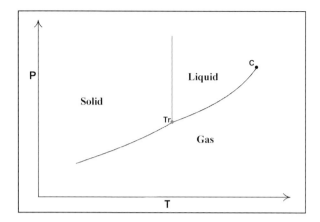

Figure 14.2 Schematic phase diagram of a fluid. The vapor pressure line, which defines the temperatures and pressures where gas and liquid are in coexistence, terminates at C, the critical point. Tr is the triple point.

When adsorption isotherms are measured at the liquid nitrogen (77.35 K at 760 torr) or liquid argon (87.27 K at 760 torr), the coolant is usually held in a dewar flask. The system is open to atmosphere and the temperature of the liquid therefore depends on both the ambient pressure and the presence of impurities in the liquid, which tends to elevate the boiling point. Depending upon the amount of dissolved impurities, such as water vapor, oxygen and other atmospheric gases, the pure liquid nitrogen contained in the P_0 cell therefore exists at a slightly elevated temperature (ca. 0.1 - 0.2 K), which results in a saturation pressure increase of ca. 10- 20 torr. Hence, in order to perform an accurate pore size analysis, P_0 should be measured with the highest resolution possible. In fact an error in P_0 of ca. 5 torr at a relative pressure of 0.95 will lead to an error of 10% in the calculated pore size. Preferably, P_0 should be directly measured by condensing nitrogen in a dedicated saturation pressure cell contained in the coolant and connected directly to a dedicated, high precision pressure transducer. This is illustrated in the schematic Fig. 14.3 (§14.8) This preferred configuration allows the saturation pressure to be updated for *every* datum point. However, the effect of using an incorrect saturation pressure for a surface area calculation is reduced by the nature of the BET plot, since both the ordinate and abscissa will deviate in the same direction, leaving the slope nearly constant. For example, in the case of a BET C constant of 100, a slope of 1000 and an intercept of 10, an error of 15 torr in a total of 760 torr will produce less than 1% error in surface area. Hence, in this case a direct measurement of the saturation pressure is not necessarily required; very often a value for the saturation pressure is estimated since, in the case of nitrogen adsorption at liquid nitrogen temperature, the real saturation pressure corresponds closely with ambient pressure. The effect of impurities in the liquid nitrogen bath on

P_0 are often taken into account by adding 10 torr to ambient pressure, which will give the corrected saturated pressure to within 5 torr.

14.6 SAMPLE CELLS

In general, sample cells should be designed to minimize the void volume (exceptions are tolerable for sorption measurements at very low pressures as described in §14.7 and §14.8). Samples with sufficiently low surface area may adsorb less volume than is required to fill the void volume! In such instances large errors can be generated unless the void volume is accurately measured and reduced to as small a value as possible. Ideally, the sample should be placed in an open-ended glass cell with a small internal volume (bulb) and narrow stem; the latter to minimize that cell void volume which emerges from the coolant bath and is not at a uniform temperature. However, if larger amounts of sample need to be measured a large bulb should be used, and in addition a wide bore stem if there are difficulties with regard to adding and removing sample. In order to minimize the void volume arising from the stem volume, filler rods should be used. However, the use of a filler rod is not recommended in the low-pressure range where thermal transpiration effects play an important role (see §14.8.3). In this relative pressure range one should avoid unnecessary restrictions to gas flow in the thermal-gradient zone.

14.7 LOW SURFACE AREA

The most advanced volumetric sorption analyzers allow surface areas as low as approximately $0.5 - 1 \text{ m}^2$ to be measured using nitrogen as the adsorptive. With small surface areas, the quantity of adsorptive remaining in the void volume is large compared to the amount adsorbed and indeed, the void volume error can be larger than the volume adsorbed. As already discussed in chapter 5.9, the number of molecules left in the void volume can be reduced by using adsorptives with low vapor pressures. Beebe, Beckwith, and Honig [8] were the first to use krypton for this purpose. Litvan [9] reported the vapor pressure of krypton at liquid nitrogen temperature as 2.63 torr, i.e., the saturation pressure of undercooled liquid krypton at this temperature (at 77.35 K krypton is ca. 38.5 K below its triple point and solidifies at a pressure of ca. 1.6 torr). Because of this much lower saturation pressure compared to that of nitrogen, the amount of krypton remaining in the void volume, at any given relative pressure, will be much less than that for nitrogen, whereas the amount of adsorption will be only slightly less (by approximately the ratio of cross-sectional areas of nitrogen and krypton, or about 16.2/20.2, see chapter 5). Another advantage of using adsorbents with low vapor pressure is that corrections for non-ideality are often not necessary.

Because of the very low saturation pressure of krypton at ~77K (i.e., 2.63 torr) the relative pressures in the classical BET range (i.e. 0.05 –0.3) correspond to absolute pressures below 1 Torr. Hence, in contrast to the experimental setup for regular nitrogen BET surface area analysis, a much more sophisticated experimental set-up is needed for krypton surface area analysis. Such an apparatus is shown in Fig. 14.3 and is described in more detail in chapter 14.6. In addition, at the very low pressures required for krypton analysis it may be necessary to correct for the so-called thermal transpiration phenomenon, which is described in §14.8.3.

14.8 MICRO- AND MESOPORE ANALYSIS

14.8.1 Experimental Requirements

As discussed in chapters 8 and 9, physical adsorption in micropores occurs at relative pressures substantially lower (very often down to a relative pressures of 10^{-7}) than in case of sorption phenomena in mesopores. Physical adsorption in microporous adsorbents can span a broad spectrum of pressures (up to seven orders of magnitude) hence special care is necessary for the pressure measurements. Consequently, more than one pressure transducer is necessary to measure all equilibrium pressures with sufficient accuracy.

In order to study the adsorption of gases like nitrogen and argon (at their boiling temperatures) within a relative pressure range of $10^{-7} \leq P/P_0 \leq 1$ with sufficiently high accuracy, it is desirable to use a combination of at least three transducers with maximum ranges of 1 torr, 10 torr and 1000 torr. In addition, one has to assure that the sample cell and the manifold can be evacuated to pressures as low as possible, which requires a suitable high vacuum pumping system. The achievable pressure over the sample should be lower than the pressure of the first experimental point and preferably the pumping system should be able to evacuate the manifold and sample cell to less than a relative pressure of 10^{-7}. The desired low pressure can be achieved by using a turbomolecular pump.

One such experimental setup, which fulfills these requirements in order to assure accurate and precise pressure measurements and a high vacuum environment, is shown schematically in Fig. 14.3. In addition to the obligatory pressure transducers in the dosing volume (manifold) of the apparatus, the analysis station of the volumetric sorption apparatus is also equipped with high precision pressure transducers dedicated to read the pressure just in the sample cell. Hence, the sample cell is isolated during equilibration, which ensures a very small effective void volume and therefore a highly accurate determination of the adsorbed amount. The saturation pressure P_0 is measured throughout the entire analysis by means of a dedicated saturation pressure transducer, which allows the vapor pressure to be monitored for every datum point. This leads to high accuracy and precision in the determination of P/P_0 and thus in the determination of the pore size distribution. It is advantageous to use a diaphragm pump to back the turbomolecular pump in order to guarantee a completely oil-free environment for (i) the adsorption measurement and (ii) the outgassing of the sample prior to the analysis.

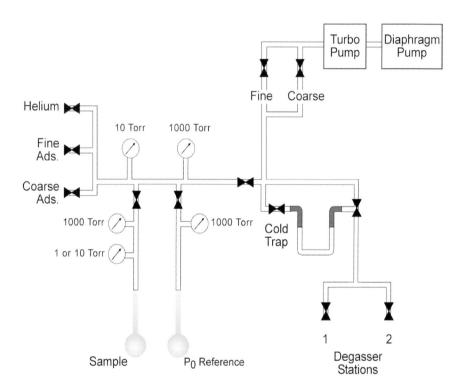

Figure. 14.3 Schematic representation of a high precision volumetric apparatus suitable for pore size analysis of micro- and mesoporous materials as well as for samples with low surface areas.

14.8.2 Micropore Analysis and Void Volume Determination

At the low pressures, where micropore analysis is performed, the void volume correction is relatively small, but because adsorption isotherms are usually performed over the complete relative pressure range up to 1, the void volume needs to be determined carefully. As discussed above (§14. 3), a non-adsorbing gas such as helium is used prior (or in principle after every analysis) to measure the void (free space) volumes at room temperature and at the temperature of the coolant. However, as already indicated in §14.3, the use of helium for the void volume calibrations may be problematic, and recent investigations have confirmed that some microporous solids may adsorb significant amounts of helium at liquid nitrogen temperature (so-called helium entrapment) [4, 10]. Therefore, after exposure of the sample to helium during free space measurements, it is highly recommended to evacuate the sample cell and repeat degassing of the sample at elevated temperatures (this so-called He-removal procedure should be performed at least at room temperature) before continuing with analysis. Another possibility is to determine the void volume with helium after completion of sorption measurement.

Another potential problem is that at these low temperatures (e.g., at the temperatures of liquid nitrogen, 77.35 K, and liquid argon, 87.27 K) and low pressures nitrogen and argon molecules can (in contrast to helium) be restricted from entering the narrowest micropores of some adsorbent due to diffusion limitations. As a consequence, nitrogen and argon sorption data obtained on such adsorbents can be affected by a small systematic void volume error.

14.8.3 Thermal Transpiration Correction

Another complication is that for gas pressures below ca. 80-100 millitorr (i.e., $P/P_0 < 10^{-4}$ for nitrogen and argon adsorption at 77 K and 87 K, respectively) pressure differences along the capillary of the sample bulb on account of the Knudsen effect have to be taken into account (i.e., thermal transpiration correction) [11]. Thermal transpiration results in a pressure gradient between the sample, which is kept at the (cryogenic) measurement temperature, and the pressure transducer, which is at room temperature, if the inner diameter of the tubing between the two parts of the system is very small compared with the mean free path of the gas. Knudsen [12] postulated that in case the mean free path exceeds several times the tube diameter, the pressure ratio between the part of the sample cell immersed into e.g. liquid nitrogen (P_{cold}) and the part of the system that is kept at room temperature (P_{warm}) is given by the simple relationship

$$\frac{P_{cold}}{\sqrt{T_{cold}}} = \frac{P_{warm}}{\sqrt{T_{warm}}} \qquad (14.9)$$

Then

$$P_{warm} = \sqrt{\frac{T_{warm}}{T_{cold}}} P_{cold} \qquad (14.10)$$

Equation (14.10) holds for the situation where the mean free path exceeds the cell stem dimension several times. If the sample is at 77 K and the pressure transducer used to measure the pressure in the sample cell is at 293K, the pressure measured at 293 K is nearly twice the true pressure at the sample. In practice the effect is much weaker and depends on the relative size of the mean free path, which depends on both the pressure and on the cell stem diameter. In order to determine the pressure ratio P_{cold}/P_{warm}, empirical approaches have been published [13-16]. One example is the empirical model of Liang et al [15], which is often employed to calculate thermal transpiration corrections for measured pressures.

$$\frac{P_2}{P_1} = \frac{\alpha \phi^2 x^2 + \beta \phi x + \sqrt{T_1/T_2}}{\alpha \phi^2 x^2 + \beta \phi x + 1} \tag{14.11}$$

where

$x = 0.133 P_2 d$
P_1, P_2 are in pascal
d = diameter of connecting tube (in m)
$\alpha(He) = 2.52$
$\beta(He) = 7.68[1-(T_2/T_1)^{0.5}]$

and ϕ is the pressure shift factor that varies for gases relative to the value 1.00 for helium and can be calculated by

$$0.27 \log\phi = \log D + 9.59 \tag{14.12}$$

where D is the molecular diameter of the gas, in meters. Since the effect of thermal transpiration is observed when the mean free molecular path becomes greater than the tube diameter, using a sample cell with a wide cell stem can minimize the effect. In addition any type of obstruction or restriction in the thermal gradient zone should be avoided (in this case it is not recommended to use a filler rod).

    The thermal transpiration effect does not occur at higher pressures, i.e., the mean free paths are smaller and the corresponding molecular collisions destroy the effect. Usually, correction is required only for pressures below 80-100 millitorr. Above this pressure the correction is negligible.

14.8.4 Adsorptives other than Nitrogen for Micro-and Mesopore Analysis- Experimental Aspects

The pore size- and volume analysis of microporous materials such as zeolites, carbon molecular sieves etc. is difficult, because the filling of pores of dimension 0.5 - 1 nm occurs at relative pressures of 10^{-7} to 10^{-5}, where the rates of diffusion and adsorption equilibration are very slow. At these low temperatures and pressures reduced heat transfer is also a problem: it takes some time to reach and maintain thermal equilibrium.

Argon fills micropores of dimensions $0.5 - 1nm$ at much higher relative pressures (i.e., $10^{-5} < P/P_0 < 10^{-3}$) compared to nitrogen, which leads to accelerated diffusion and equilibration processes and thus to a reduction in analysis time. Hence, it is of advantage to analyze microporous materials consisting of such narrow micropores by using argon as adsorptive at liquid argon temperature (87.27 K), see Fig. 9.8. Please note, however, that a combined and complete micro- and mesopore size analysis with argon is not possible at liquid nitrogen temperature (which is ca. 6.5 K below the triple point temperature of bulk argon). The pore size analysis of porous silica by argon adsorption at 77 K is limited to pore diameters smaller than ca. 15 nm [18] (see chapter 8 for more details). Of course, such a limitation does not exist for argon sorption at the liquid argon temperature (87.27 K); here pore filling and pore condensation can be observed over the complete micro-, meso-, and macropore size range. Hence, argon adsorption at liquid argon temperature offers a convenient way to obtain an accurate pore size analysis over the complete micro- and mesopore size range.

But, for argon at ~87K, as in the case of nitrogen adsorption at ~77K, a turbomolecular pump vacuum is still needed in order to achieve the very low relative pressures necessary to monitor the pore filling of the narrowest micropores. Associated with the low pressures is (as indicated above) the well-known problem of restricted diffusion, which prevents nitrogen molecules and also argon molecules from entering the narrowest micropores (i.e. ultramicropore which are present in activated carbon fibers, carbon molecular sieves, some zeolites etc.). This may lead to erroneous sorption isotherms, underestimated pore volumes etc. A possibility to overcome these problems exists: the use of CO_2 as adsorptive at close to room temperature. For example, the saturation pressure at 273K is ca. 26400 torr, i.e. in order to achieve the low relative pressures (down to $P/P_0 = 10^{-7}$) a turbomolecular pump vacuum is not necessary. With CO_2 adsorption up to 1 atm one can evaluate pores from the narrowest micropores up to ca. 1.5 nm. At these relatively "high" temperatures and pressures, significant diffusion limitations no longer exist, which leads to the observation that equilibration is achieved much faster compared to nitrogen and argon adsorption at ~77 K and ~87 K, respectively. Typically, a micropore analysis with nitrogen as adsorptive requires ca. 24 h or more, whereas in contrast an adequate micropore analysis using CO_2 can typically be completed in less ca. 5 h. The experimental setup needed to perform micropore analysis with CO_2 is much simpler than the corresponding experimental setup for micropore analysis by argon or nitrogen adsorption, because a turbomolecular pump, and low-pressure transducers are not needed.

In order to assess the micropore volume of pores that are not accessible even to CO_2 at room temperature, helium adsorption below the critical temperature of helium (i.e., below ~5 K) offers a viable alternative. Here, a special cryostat and a turbomolecular pump vacuum as well as highly accurate low pressure transducers *are* needed. Such an experimental set-up is described in ref. [19].

14.9 AUTOMATED INSTRUMENTATION

14.9.1 Multistation Sorption Analyzer
Highly precise sorption analyzers are available which allow high resolution sorption isotherms to be obtained in both micro-and mesopore ranges (see experimental setup shown in Fig. 14.3). Some modern, state of the art automatic sorption analyzers allow up to six sorption isotherms to be measured independently [20]. Such a system is depicted schematically in Fig. 14.4, which shows six sample cells and six P_0 stations connected to a common manifold by valves. A pressure transducer is connected to each cell and in this example, two are attached to the manifold (10 torr and 1000 torr full scale; which allow measurements at low relative pressures, i.e. micropore analysis). Each analysis station is also equipped with a P_0 cell; opening the appropriate valve to the manifold allows the saturation pressure for each station to be measured periodically.

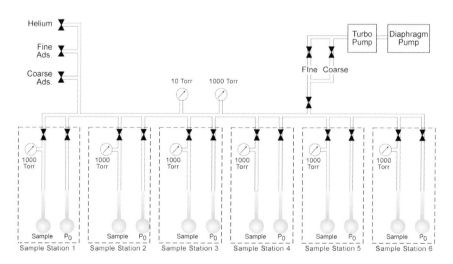

Figure 14.4 Schematic representation of an automated sorption system with six independent stations.

Prior to each analysis, during the initialization, the computer-controlled instrument admits adsorptive to the manifold with the valves to the sample cells open. The computer checks the output of each cell transducer to confirm that they agree with the primary pressure transducer mounted on the manifold by monitoring the pressure at several different values. Any transducer found to be misaligned is automatically recalibrated. After the void volume in each cell is determined by dosing with helium, the first cell is evacuated and then dosed with the adsorptive, e.g. nitrogen gas. The dosing algorithm requires that the manifold be brought to a pressure such that, when the valve to the cell is opened the cell pressure will reach a pre-set target pressure, assuming no adsorption occurs. Once the target pressure is achieved the cell is isolated from the manifold and the cell transducer monitors the approach to equilibrium. In this manner, the manifold is now free to dose another cell, while the first one is reaching its equilibrium pressure. If a cell does not reach its target equilibrium pressure after it is dosed, as sensed by the cell transducer, that cell will be redosed automatically as often as necessary until it reaches the target pressure. While waiting for equilibrium, after dosing a cell, the manifold is employed to redose other cells whose target pressure was not reached. In this manner, six adsorption and desorption isotherms can be obtained with significant timesavings while maintaining the ability to run mixed analysis types (any combination of surface area and/or pore size). The instrumentation allows the target pressure "tolerance" and the time allowed to reach the target pressure for each of the six sample cells to be set separately. Thus, with a tight tolerance and long equilibration times, the instrument will produce data at almost exactly the requested relative pressures. With looser tolerances and shorter equilibrium times, the analysis can proceed more rapidly and produce data points with less precision, but no less accuracy.

14.9.2 The NOVA Concept

A novel, automated volumetric technique called NOVA (*NO V*oid *A*nalysis) was recently introduced [21]. As discussed in §14.3, to generate accurate gas sorption data using the conventional volumetric static methods it is necessary to introduce a non-adsorbing gas such as helium prior to every analysis. The helium is used to measure void (free space) volumes, and define void volume zones (at room temperature and at some cryogenic temperature, e.g. that of liquid nitrogen, at which the analysis will be done).

However, the helium void volume measurement procedure is based on various assumptions: (i) helium is not adsorbed/absorbed on/into the adsorbent; (ii) helium does not penetrate into regions which are inaccessible for the adsorptive (typically nitrogen and argon). However, as also already mentioned, these requirements are not always fulfilled and the use of helium for dead-space calibrations has been, therefore, under discussion for many years.

The NOVA technique overcomes the aforementioned problems by separating the free space determination from the adsorption experiment itself. In this particular case the use of helium is no longer required. The NOVA technique is based on a separate calibration of sample cells prior to analysis. Sample cell calibrations are simply *blank* analyses of empty sample cells performed under the same experimental conditions (of temperature and relative pressure range) as the sorption measurements. The calibration/blank curve for each cell type is stored in the memory of the automated instrument and is recalled every time such a cell is used for an analysis with the NOVA apparatus. This calibration/blank curve is then automatically subtracted from the sample analysis curve and accurate adsorption/desorption data are obtained after minor corrections are made to account for both the relatively small sample volume and that volume's gas non-ideality contribution to the empty cell profile. The sample volume is obtained by automated pycnometric measurement with adsorptive at room temperature at the start of the analysis (or, for further time savings, by entering a known sample density).

In addition to these advantages, the principle of separating the void volume determination from the adsorption experiment (and consequently the elimination of the use of helium), shortens significantly the analysis time (especially in the case of a simple BET analysis). This is because the NOVA principle allows -in contrast to the situation of conventional volumetric static methods - the sorption experiment to be done without simultaneous void volume measurements. In addition, uncertainties related to gas non-ideality corrections are virtually eliminated because their influence is restricted to the small volume occupied by the sample. Because cell calibration and sample analysis are performed with the same gas, any contribution by adsorption on the sample cell walls is also subtracted automatically when generating isotherm data.

14.10 REFERENCES

1. Faeth P.A. (1962) *Adsorption and Vacuum Technique*, Institute of Science and Technology, University of Michigan, Ann Arbor.
2. Emmett P.H. (1941) *A.S.T.M. Symposium on New Methods for Particle Size Determination in the Sub-Sieve Range*, p95.
3. Harris M.R. and Sing K.S.W. (1955) *J. Appl. Chem.* **5**, 223.
4. Rouquerol F., Rouquerol J. and Sing K.S.W. (2000) *Adsorption by Powders & Porous Solids,* Academic Press, London.
5. Emmett P.H. and Brunauer S. (1937) *J. Am. Chem Soc.* **59**, 1553.
6. Conner W.C. (1997) In *Physical Adsorption: Experiment, Theory and Applications* (Fraissard J.and Conner W.C., eds) Kluwer, Dordrecht, p33.
7. Mikhail R.Sh. and Robens E. (1983) *Microstructure and Thermal Analysis of Solid Surfaces*, Wiley, Chichester.
8. Beebe R.A., Beckwith J.B. and Honig J.M. (1945) *J. Am. Chem. Soc.* **67**, 1554.

9.   Litvan G.G. (1972) *J. Phys. Chem*. **76**, 2584.
10. Staudt R., Bohn S., Dreisbach F. and Keller F. (1997) In *Characterization of Porous Solids IV* (B. McEnaney et al., eds.) *Spec. Publ.* **213**, Royal Society of Chemistry, Cambridge.
11. Edmonds T. and Hobson J. P. (1965) *J. Vac. Sci. Technol.* **2**, 182.
12. Knudsen M. (1910) *Ann. Physik* **31**, 210.
13. Takaishi T. and Sensui Y.(1963) *Trans. Faraday Soc.* **59**, 2503.
14. Miller G.A. (1963) *J. Phys. Chem.* **6**, 1359.
15. Liang S.C. (1953) *J. Phys, Chem.* **57**, 910.
16. Poulter K.F., Rodgers M.J., Nash P.J., Thompson T.J. and Perkin M.P. (1983) *Vacuum*, **33**, 311.
17. [a]Thommes M., Koehn R. and Froeba M. (2000) *J. Phys. Chem. B* **104**, 7932; [b]Thommes M., Koehn R. and Froeba M. (2002) *Applied Surface Science* **196**, 239.
18. Kuwabara H., Suzuki T. and Kaneko K. (1991) *J. Chem. Soc. Faraday Trans.* **87**, 19.
19. Kaneko K., Setoyama N., Suzuki T. and Kuwabara H. (1992) In *Proc. IVth Int. Conf. on Fundamentals of Adsorption*, Kyoto.
20. Shields J.E. and Lowell S. (1984) *Amer. Lab.*, November.
21. [a]Lowell S. (1994) U.S. Pat. No. 5,360,743; [b]Thomas M.A., Novella N.N. and Lowell S. (2002) U.S. Pat. No. 6,387,704 B1.

15 Dynamic Flow Method

15.1 NELSON AND EGGERTSEN CONTINUOUS FLOW METHOD

In 1951, Loebenstein and Deitz [1] described an innovative gas adsorption technique that did not require the use of a vacuum. They adsorbed nitrogen out of a mixture of nitrogen and helium that was passed back and forth over the sample between two burettes by raising and lowering attached mercury columns. Equilibrium was established by noting no further change in pressure with additional cycles. The quantity adsorbed was determined by the pressure decrease at constant volume. Successive data points were acquired by adding more nitrogen at the system. The results obtained by Loebenstein and Deitz agreed with vacuum volumetric measurements on a large variety of samples with a wide range of surface areas. They were also able to establish that the quantities of nitrogen adsorbed were independent of the presence of helium.

Nelson and Eggertsen [2], in 1958, extended the Loebenstein and Dietz technique by continuously flowing a mixture of helium and nitrogen through the powder bed. They used a hot wire thermal conductivity detector to sense the change in effluent gas composition during adsorption and desorption, when the sample cell was immersed into and removed from the bath, respectively. Fig. 15.1 illustrates a simplified continuous flow apparatus. Fig. 15.2 is a schematic of the flow path arrangement using a four-filament thermal conductivity bridge.

In Fig. 15.1, a mixture of adsorptive and carrier gas of known concentration is admitted into the apparatus at 'a'. Valve V_1 is used to control the flow rate. The analytical pressure is the partial pressure of the adsorptive component of the mixture. When the system has been purged, the detectors are zeroed by balancing the bridge (see Fig. 15.8). When the sample cell 'b' is immersed in the coolant, adsorption commences and detector D_B senses the decreased nitrogen concentration. Upon completion of adsorption, D_B again detects the same concentration as D_A and the signal returns to zero. When the coolant is removed, desorption occurs as the sample warms and detector D_B senses the increased nitrogen concentration. Upon completion of desorption, the detectors again sense the same concentration and the signal returns to its initial zero value. Wide tubes 'c' act as ballasts to (i) decrease the linear flow velocity of the gas ensuring its return to ambient temperature prior to entering D_B and (ii) to prevent air being drawn over D_B when the cell is cooled and the gas contracts.

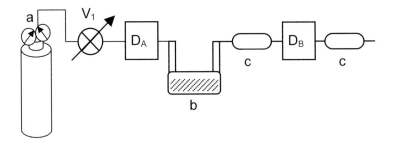

Figure 15.1 Simplified continuous flow apparatus

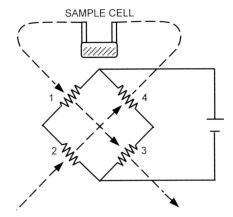

Figure 15.2 Gas flow path (dashed line) using a four-filament detector. D_A is formed by filaments 2 and 4, $D_B = 1$ and 3. This type of circuit is known as a Wheatstone bridge.

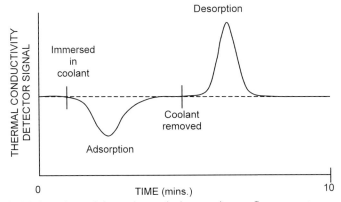

Figure 15.3 Adsorption and desorption peaks in a continuous flow apparatus.

Fig. 15.3 illustrates the detector signals due to adsorption and subsequent desorption. Figs. 15.4 and 15.5 illustrate a parallel flow arrangement which has the advantage of requiring shorter purge times when changing gas composition but is somewhat more wasteful of the mixed gases. The symbols shown in Fig. 15.4 have the same meaning as those used in Fig. 15.1.

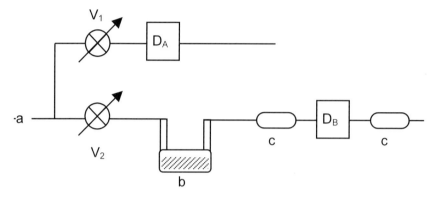

Figure 15.4 Parallel flow circuit.

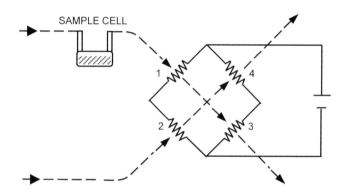

Figure 15.5 Parallel flow path using a four-filament bridge. D_A is formed by filaments 2 and 4 are, 1 and 3 comprise D_B. Dashed lines are gas flow paths.

15.2 CARRIER GAS (HELIUM) AND DETECTOR SENSITIVITY

To be effective, the carrier gas must fulfill two requirements. First, it cannot be adsorbed at the coolant temperature; second, it must possess a thermal conductivity sufficiently different from that of the adsorptive that small concentration changes can be detected. Usually helium is used as a carrier gas; hence, some discussion regarding the possible influence of helium is necessary.

The forces leading to liquefaction and adsorption are the same in origin and magnitude. Gases substantially above their critical temperature, T_c, cannot be liquefied because their thermal energy is sufficient to overcome their intermolecular potential. Although the adsorption potential of a gas can be greater than the intermolecular potential, helium is, nevertheless, not adsorbed at liquid nitrogen temperature (~77 K) because this temperature is still more than 14 times the critical temperature of helium (~5.3 K).

Furthermore, in order to be considered adsorbed, a molecule must reside on the surface for a time τ at least as long as one vibrational cycle of the adsorbate normal to the surface. The time for one vibration is usually of the order of 10^{-13} seconds which, by equation (15.1), makes τ about 2×10^{-13} seconds at 77 K.

$$\tau = 10^{-13} \exp(100/RT) = 1.91 \times 10^{-13} \sec \qquad (15.1)$$

The value of 100 cal/mol chosen for the adsorption energy of helium is consistent with the fact that helium has no dipole or quadrupole and is only slightly polarizable. Thus, it will minimally interact with any surface. Based upon reflections of a helium beam from LiF and NaCl cleaved surfaces, de Boer [3] estimated the adsorption energy to be less than 100 cal/mol. At 77K the velocity of a helium atom is 638 m/sec, so that in 1.91×10^{13} sec it will travel $1.91 \times 10^{13} \times 638 \times 10^{10} = 1.2$ Å. Thus, the condition that the adsorbate molecules reside near the surface for one vibrational cycle is fulfilled by the normal velocity of helium and not by virtue of its being adsorbed. Stated in alternate terms, the density of helium near a solid surface at 77 K is independent of the surface and is the same as the density remote from the surface. Molecular collisions with the adsorbed film by helium will certainly be no more destructive than collisions made by the adsorbate. In fact, helium collisions will be less disruptive of the adsorbed film structure since the velocity of helium is, on the average, 2.6 times greater than that of nitrogen at the same temperature, while a nitrogen molecule is 7 times heavier. Thus, the momentum exchange due to nitrogen collisions will be the more disruptive. The thermal energy of helium at 77 K is about 220 cal/mol. The heat of vaporization of nitrogen at 77 K is 1.335 kcal/mol, which may be taken as the minimum heat of adsorption. A complete exchange of thermal energy during collisions between a helium atom and an adsorbed nitrogen molecule would not be sufficient to cause desorption of the nitrogen.

To understand the effect of the carrier gas on the response of the thermal conductivity detector, consider the steady state condition that prevails when the resistive heat generated in the hot wire filament is exactly balanced by the heat conducted away by the gas. This condition is described by equation (15.2)

$$i^2R = ck(t_f - t_w)$$ (15.2)

where i is the filament current, R is the filament resistance, k is the thermal conductivity of the gas mixture, and t_f and t_w are the filament and the wall temperatures, respectively. The constant c is a cell constant that reflects the cell geometry and the separation of the filament from the wall, which acts as the heat sink.

When the gas composition is altered due to adsorption or desorption, the value of k changes by Δk, which in turn alters the filament temperature by Δt_f. Under the new conditions, equation (15.2) can be rewritten as

$$i^2R = ck(k + \Delta k)(t_f + \Delta t_f - t_w)$$ (15.3)

Equating the right hand sight of equations (15.2) and (15.3) gives

$$k(t_f - t_w) = (k + \Delta k)(t_f + \Delta t_f - t_w)$$ (15.4)

By neglecting the term $\Delta k \Delta t_f$, a second order effect, equation (15.4) rearranges to

$$\Delta t_f = \frac{\Delta k}{k}(t_w - t_f)$$ (15.5)

The change in filament resistance, ΔR, is directly proportional to the small temperature change Δt_f and is given by

$$\Delta R = \alpha R \Delta t_f$$ (15.6)

where α is the temperature coefficient of the filament, dependent on its composition, and R is the filament resistance at temperature t_f; thus

$$\Delta R = \alpha R \frac{\Delta k}{k}(t_w - t_f)$$ (15.7)

Equation (15.7) requires that Δk be as large as possible for maximum response under a fixed set of operating conditions.

A fortunate set of circumstances leads to a situation in which the same molecular properties that impart minimal interactions or adsorption potentials also lead to the highest thermal conductivities. Molecules of large mass and many degrees of vibrational and rotational freedom tend to be more polarizable and possess dipoles and quadrupoles which give them higher boiling points and stronger interactions with surfaces. These same properties tend to reduce their effectiveness as thermal conductors.

Helium possesses only three degrees of translational freedom and hydrogen the same, plus two rotational and one vibrational degree. However, because of hydrogen's low weight, it has the highest thermal conductivity of all gases, followed by helium. Either of these two gases fulfills the requirement for adequately high thermal conductivities so that Δk in equation (15.7) will be sufficiently large to give good sensitivity with any adsorbate. Helium, however, is usually used in continuous flow analysis because of the hazards associated with hydrogen.

Fig.15.6 is a plot of the thermal conductivity of mixtures of helium and nitrogen obtained on an apparatus similar to that described in the next section. Characteristically, the thermal conductivity of most mixtures does not vary linearly with concentration. The slope of the curve at any point determines the value of Δk and, therefore, the detector response. Fig. 15.6 also illustrates that the greater the difference between thermal conductivities of the adsorbate and carrier gas, the higher will be the slope and therefore the detector response.

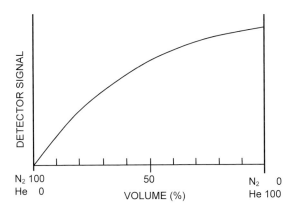

Figure 15.6 Thermal conductivity bridge response.

The shape of the curve shown in Fig. 15.6 is fortuitous in as far as the continuous flow method is concerned. For reasons to be discussed later, the desorption signal (see Fig. 15.3) is generally used to calculate the adsorbed volume. When, for example, 1.0 cm^3 of nitrogen is desorbed into

9.0 cm^3 of helium, the concentration change is 10%. However, when 1.0 cm^3 of nitrogen is desorbed into 9.0 cm^3 of a 90% nitrogen-in-helium mixture, the absolute change is only 1%. Therefore, the increase in slope at high nitrogen concentrations enables smaller concentration changes to be detected when data at high relative pressures are required.

Fig. 15.6 was prepared by flowing helium through one detector while varying the helium to nitrogen concentration ratio through the second detector.

15.3 DESIGN PARAMETERS FOR CONTINUOUS FLOW APPARATUS

The thermal conductivity (T.C.) detector consists of four filaments embedded in a stainless steel or brass block that acts as a heat sink. The T.C. detector is extremely sensitive to temperature changes and should be insulated to prevent temperature excursions during the time in which it takes to complete an adsorption or desorption measurement. Long-term thermal drift is not significant because of the calibration procedure discussed in the next section and, therefore, thermostating is not required. Fig. 15.7 shows a cross-sectional view of a T.C. block and the arrangement of the filaments relative to the flow path. The filaments shown are electrically connected, external to the block, and constitute one of the two detectors.

The filaments must be removed from the flow path, unlike the conventional 'flow over' type used in gas chromatography, because of the extreme flow variations encountered when the sample cell is cooled and subsequently warmed. Flow variations alter the steady state transport from the filaments, leaving them inadequate time to recover before the concentration change from adsorption to desorption is swept into the detector. When this occurs, the baseline from which the signal is measured will be unstable. By removing the filaments from the flow path and allowing diffusion to produce the signal, the problem of perturbing the filaments is completely solved. However, the tradeoff is nonlinear response characteristics. Since the thermal conductivity of the gas mixture is already non-linear with concentration, this additional nonlinearity poses no further problems, and is accommodated by calibration of signals (see §15.4).

A suitable electronic circuit provides power to the filaments and a means of zeroing or balancing the T.C. bridge, adjusting the filament current, attenuating the signal, and adjusting the polarity is shown in Fig. 15.8. Signals produced by adsorption or desorption can be fed to a data acquisition recorder for a continuous trace of the process, and/or to a digital integrator for summing the area under the adsorption and desorption curves.

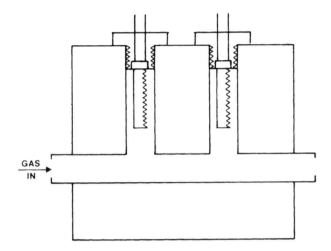

Figure 15.7 Thermal conductivity block with filaments located out of the flow path.

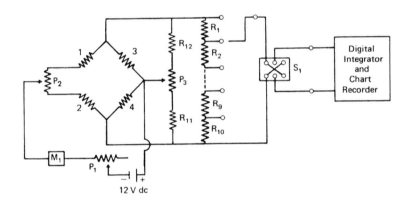

Figure 15.8 Thermal conductivity bridge electronic circuit. 12V dc power supply, stable to 1mV (ripple is not significant due to thermal lag of filaments); P_1, 100 ohms for filament current control; M_1 milliammeter, 0-250 mA; P_2, 2 ohms for coarse zero. Filaments 1 and 4 are detector 1, 2 and 3 are detector 2. P_3, 1 ohm for fine zero; R_{11}, R_{12}, padding resistors ~64ohms; R_1 –R_{10}, attenuation resistors 1, 2, 4, …512 ohms; S_1 (D.P.D.T.) switch for polarity. Attenuator resistors are ¼ % ww, lowest temperature coefficient; all others are 1%.

Fig. 15.9 is the flow schematic for a commercial continuous flow apparatus. Any one of up to four gas concentrations can be selected from from premixed tanks. Alternatively, adsorptive and carrier can be blended internally by controlling their flows with individual needle valves. A third choice is to feed mixtures to the apparatus from two linear mass flow controllers. Flow meters (operating under pressure to extend their range) indicate input flow rates. A flow meter at the very end of the flow path is used to calibrate the flow meters. Pressure gauges indicate the input pressure under which the flow meters operate. An optional cold trap removes contaminants from the blended gas stream. After leaving the cold trap, but before flowing into detector D_A, the gas passes through a thermal equilibration tube that reduces the linear flow velocity and thereby provides time for warming back to ambient temperature.

After leaving D_A the flow splits, the larger flow goes to the sample cell at the analysis station, then to a second thermal equilibration tube and a flow meter used to indicate the flow through the sample cell. The smaller flow, controlled by its own needle valve merges with the sample cell effluent before entering detector D_B. The equilibration tubing downstream of D_B serves as a ballast to prevent air from entering D_B when the sample is immersed in the coolant. For high surface areas, the large quantities of desorbed gas can be diverted to a long path (far right). This prevents the gas from reaching the detector before the flow has returned to its original rate.

Splitting the flow as described above serves as a means of diluting the adsorption and desorption peaks in order provide infinitely variable signal height adjustment, in addition to using the stepwise electronic attenuator shown in Fig. 15.8.

A slow flow of adsorptive is directed to the 'out' septum and then to a sample cell positioned at the degas station. The gas flowing through this part of the circuit provides a both source of adsorptive for calibrating detector signals and as a purge for the degas station. Known quantities of adsorptive are injected into the analysis flow through the septum labeled 'in' to simulate a desorption signal for calibration purposes.

To measure the saturated vapor pressure, pure adsorptive is admitted to the P_0 station, when immersed in liquid nitrogen, until it liquefies. The equilibrium pressure is measured on the adjacent transducer.

A diverter valve at the analysis station ensures continuity of flow through the system even when the sample cell is removed.

In order to avoid contamination of the degassed sample when transferring from the degas station to the analysis station, the cells are mounted in spring-loaded self-sealing holders that close when disconnected and open when placed in position.

Sample cells consist of a wide variety of designs for various applications. Fig. 15.10 illustrates seven cells used for various types of samples. Their specific applications and limitations are given in more detail in Section 15.7. Each of the cells shown is made of Pyrex® glass. They are easily filled and cleaned. The cells range from four to five inches in length with stem inside and outside diameters of 0.15 and 0.24 inches, respectively.

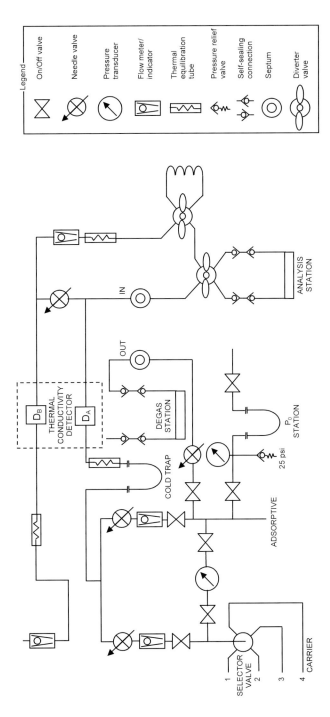

Figure 15.9 Flow diagram for a continuous flow system for adsorption measurements according to the Nelson-Eggertsen method. Not all commercial instruments are equipped with all the features shown.

15.4 SIGNALS AND SIGNAL CALIBRATION

The signal intensity created by an adsorption or desorption peak passing through the detector is dependent upon the attenuator setting, the filament current, and the design of the T.C. detector. Also, as stated previously, the detector response is nonlinear. These circumstances require that the adsorption or desorption signals be calibrated by introducing a volume of carrier or adsorbate gas into the flow stream. An expeditious and accurate method of calibration is the withdrawal of a sample of adsorbate from the 'out' septum (see Fig. 15.9) with a precision gas syringe and the injection of a known volume into the flow stream through the 'in' septum.

Usually the desorption peak is calibrated because it is free of tailing. By immersing the cell in a beaker of water immediately after removal from the liquid nitrogen, the rate of desorption is hastened. Heat transfer from the water is more rapid than from the air; therefore, a sharp desorption peak is generated. The calibration signal should be within 20% of the desorption signal height in order to reduce detector nonlinearity to a negligible effect.

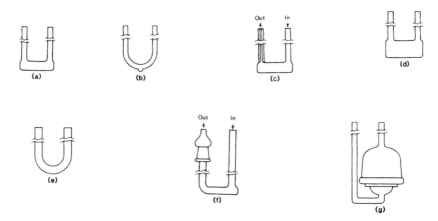

Figure 15.10 Sample cell designs (a) Conventional sample cell – for most powder samples with surface areas greater than 0.2 m^2 in the cell. For samples less than 0.2 m^2, this cell can be used with krypton as the adsorptive. (b) Micro cell –used for very high surface area samples or for low area samples that exhibit thermal diffusion signals. Because of the small capacity of the micro cell, low area samples must be run on high sensitivity settings. (c) Capillary cell – useful for minimizing thermal diffusion signals. Because of the small capacity of the micro cell, low area samples must be run on high sensitivity settings. (d) Macro cell – used with krypton when a large quantity of low area sample is required. Also used for chemisorption when total uptake is small. (e) Large U-tube cell – for larger particles or bulk samples of high area with nitrogen or low area with krypton. (f) Pellet cell – used for pellets or tablets. High surface area with nitrogen or low area samples with krypton. (g) Monolith catalyst cell-for monolithic catalysts and other samples of wide diameter that must be measured as one piece.

The area under the desorption peak, A_d, and the area under the calibration peak, A_c, are used to calculate the volume, V_d, desorbed from the sample according to equation (15.8)

$$V_d = \frac{A_d}{A_c} V_c$$
(15.8)

where V_c is the volume of adsorptive injected. Equation (15.8) requires no correction for gas nonideality since the volume desorbed is measured at ambient temperature and pressure. Because desorption occurs at room temperature, it is complete and represents exactly the quantity adsorbed. For vapors adsorbed near room temperature, the sample can be heated to ensure complete desorption.

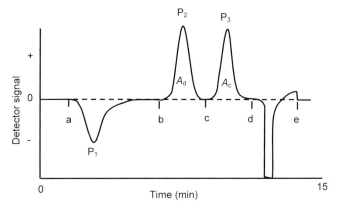

Figure 15.11 Complete cycle for one datum point.

A detailed analysis of the signal record, shown in Fig. 15.11, corresponding to a complete adsorption, desorption, calibration and concentration change cycle discloses that at point a, the sample cell is immersed in the coolant; this action produces the adsorption peak P_1. Point b represents the removal of the coolant bath which leads to the desorption peak P_2. The calibration peak P_3 results from the calibration injection made at point c. At d, a new gas concentration is admitted into the apparatus, which produces a steady base line at e where the detector is re-zeroed and the cycle repeated. The total time for a cycle is usually 15 minutes. Some timesavings can be achieved by combining the purge step (d-e) with the adsorption step (a-b). Familiarity with the apparatus usually allows the operator to choose the correct volume for calibration at particular attenuator and filament current settings or, alternatively, a calibration table can be prepared. If speed is essential, the flow rate can be increased to hasten the cycle. However, this is at the risk of warming the adsorbent if excessive flow rates are used. Flow rates of 12-15 cm^3/min allow ample time for the gas to equilibrate thermally before reaching the sample powder with all of the cells shown in Fig. 15.10. The presence of helium, with its high thermal conductivity, ensures rapid thermal equilibrium. Therefore, immersion depth of the cell is not critical provided that about 6-7cm (2½ inches) is in the coolant at the flow rates given previously.

Sample cells are not completely filled with powder; room is left above the surface for the unimpeded flow of gas. Although the gas flows over the powder bed and not through it, lower flow rates aid in ensuring against elutriation.

The areas under the adsorption and desorption peaks are usually not exactly the same. This observation is related to the changing slope of Fig. 15.6. Adsorption produces concentration changes to the right, in the direction of decreased sensitivity, while desorption produces signals in the direction of increased sensitivity.

If calibration of the adsorption signal is desired, it is necessary to inject a known quantity of helium. The amount of helium used to calibrate the adsorption signal will usually vary considerably from the amount of nitrogen required for the desorption calibration. This situation arises because, for example, if 1 cm^3 of nitrogen is adsorbed out of a 10% flowing mixture, it will produce the equivalent of 9.0 cm^3 of helium. Therefore, calibration of the adsorption signal will require nine times more helium than the corresponding volume of nitrogen needed to calibrate the desorption signal. If C_{N_2} and C_{He} are the concentrations of nitrogen and helium in the flow stream and if V_{He} is the volume of helium used for calibration, then the volume of nitrogen adsorbed, V_{ads}, is given by

$$V_{ads} = V_{He}\left(\frac{C_{N_2}}{C_{He}}\right)\left(\frac{A_{ads}}{A_{cal}}\right) \qquad (15.9)$$

where A_{ads} and A_{cal} are the areas under the adsorption and calibration signals, respectively.

When small signals are generated, it is difficult to make accurate injections of the required small amounts of gas. Karp and Lowell [4] have offered a solution to this problem that involves the injection of larger volumes of adsorbate diluted with the carrier gas. When a volume containing a mixture of nitrogen and helium, V_{mix}, is injected into the flow stream, the equivalent volume of pure nitrogen, V_{N_2}, is given by

$$V_{N_2} = V_{mix} \left(\frac{X_{N_2} - X'_{N_2}}{X'_{He}} \right) \tag{15.10}$$

where X_{N_2} is the mole fraction of nitrogen in the calibration mixture and X'_{N_2} and X'_{He} are the mole fractions of nitrogen and helium in the flow stream.

The signals must propagate through the system at identical flow rates. Calibration at a flow rate other than the flow rate associated with the adsorption or desorption peaks can lead to serious errors because the width of the peaks and therefore the peak areas are directly proportional to the flow rate. A good two-stage pressure regulator and needle valve provide adequately constant flow rates over the short time required for desorption and calibration.

Precision gas calibrating syringes can be obtained in various size ranges with no more than 1% volumetric error. Constant stroke adapters provide a high degree of reproducibility. Often, in the BET range of relative pressures, the calibration volumes remain nearly constant because the increased volume adsorbed at higher relative pressures tends to be offset by the decrease in the detector sensitivity. Thus, the same syringe may be used for a wide range of calibrations, which results in the syringe error not effecting the BET slope and only slightly altering the intercept, which usually makes a small contribution to the surface area. A syringe error of 1% will produce an error in surface area far less than 1% for those BET plots with a slope greater than the intercept or for high C values.

15.5 ADSORPTION AND DESORPTION ISOTHERMS BY CONTINUOUS FLOW

To construct the adsorption isotherm, the adsorption, desorption, and calibration cycle shown in Fig. 15.11 is repeated for each datum point required. Errors are not cumulative since each point is independently determined. Relative pressures corresponding to each data point are established by measuring the saturated vapor pressure using any of the preceding methods or by adding 15 torr to ambient pressure. Thus, if X is the mole fraction of adsorbent in the flow stream, the relative pressure is given by

$$\frac{P}{P_0} = \frac{XP_a}{P_a + 15} \tag{15.11}$$

where P_a is ambient pressure in torr. At the recommended flow rates of 12-15 cm^3/min, the flow impedance of the tubing does not raise the pressure in the sample cell.

The method used to construct the adsorption isotherm cannot be used to build the desorption isotherm. This is true because each data point on the adsorption curve reflects the amount adsorbed by a surface initially free of adsorbate. The desorption isotherm, however, must consist of data points indicating the amount desorbed from a surface that was previously saturated with adsorbate and subsequently equilibrated with adsorbate of the desired relative pressure. Karp *et al* [5] demonstrated that the desorption isotherm and hysteresis loop scans can be made in the following manner. First, the sample is exposed, while immersed in the coolant, to a flow of pure adsorbate. The flow is then changed to the desired concentration, leading to some desorption until the surface again equilibrates with the new concentration. The coolant is then removed and the resulting desorption signal is calibrated to give the volume adsorbed on the desorption isotherm. The above procedure is repeated for each datum point required, always starting with a surface first saturated with pure adsorbate.

To scan the hysteresis loop from the adsorption to the desorption isotherm, the sample, immersed in the coolant, is equilibrated with a gas mixture with a relative pressure corresponding to the start of the scan on the desorption isotherm. The adsorbate concentration is then reduced to a value corresponding to a relative pressure between the adsorption and desorption isotherms. When equilibrium is reached, as indicated by a constant detector signal, the coolant is removed and the resulting desorption signal is calibrated. Repetition of this procedure, each time using a slightly different relative pressure between the adsorption and desorption isotherms, yields a hysteresis scan from the adsorption to the desorption isotherm.

To scan from the desorption to the adsorption branch, pure adsorbate is first adsorbed, then the adsorbate concentration is reduced to a value giving a relative pressure corresponding to the start of the scan on the desorption isotherm. When equilibrium is established, as indicated by a constant base line, the adsorbate concentration is increased to give a relative pressure between the desorption and adsorption isotherms. After equilibrium is again established, coolant is removed and the resulting signal is calibrated to yield a data point between the desorption and adsorption isotherms. This procedure repeated, each time using a different final relative pressure, will yield a hysteresis loop scan from the desorption to the adsorption isotherm.

Figs. 15.12 and 15.13 illustrate the results obtained using the above method on a porous amorphous alumina sample. A distinct advantage of the flow system for these measurements is that data points can be obtained where they are desired and not where they happen to occur after dosing, as in the vacuum volumetric method. In addition, desorption isotherms and hysteresis scans are generated with no error accumulation, void volume measurements, or nonideality corrections.

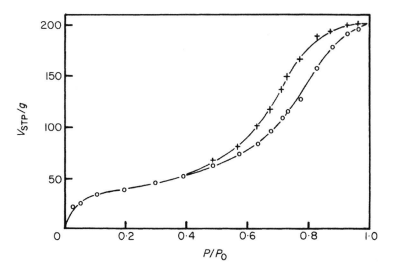

Figure 15.12 Adsorption and desorption isotherms of N_2 for 0.106 g sample of alumina Adsorption o, Desorption +

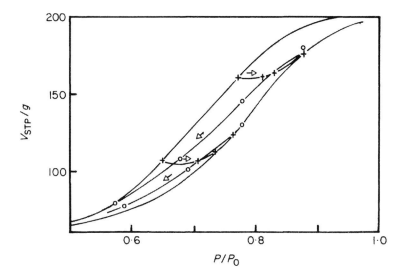

Figure 15.13 Hysteresis loop scan for same sample as Fig. 15.12. Adsorption o, Desorption +

15.6 LOW SURFACE AREA MEASUREMENTS

The thermal conductivity bridge and flow circuits shown in Figs. 15.8 and 15.9 are capable of producing a full-scale signal (1.0 mV) when 0.01 cm^3 of nitrogen are desorbed into a 30% nitrogen and helium mixture. To achieve stable operating conditions at this sensitivity, the thermal conductivity block requires some time to equilibrate thermally and the system must be purged of any contaminants.

A desorbed volume of 0.001 cm^3, using nitrogen as the adsorbate, will correspond to about 0.0028 m^2 (28 cm^2) of surface area if a single adsorbed layer were formed. An equivalent statement is that 0.0028 m^2 is the surface area, measured by the single point method, on a sample which gives a high C value, if 0.001 cm^3 were desorbed. Assuming that a signal 20% of full scale is sufficient to give reasonable accuracy for integration, then the lower limit for surface area measurement using hot wire detection is about 0.0006 m^2 or 6 cm^2. With the use of thermistor detectors, the lower limit would be still smaller.

Long before these extremely small areas can be measured with nitrogen, the phenomenon of thermal diffusion obscures the signals and imposes a higher lower limit [7]. Thermal diffusion results from the tendency of a gas mixture to separate when exposed to a changing temperature gradient. The sample cell is immersed partially into liquid nitrogen. Hence, when the gas mixture enters and leaves the sample cell it encounters a very sharp thermal gradient. This gradient exists along the arm of the tube for *ca.* 2 cm above and below the liquid level, and consequently gases will tend to separate. The extent of separation is proportional to the temperature gradient, the difference in the molecular weights of the two gases and their relative concentrations. The heavier gas tends to settle to the bottom of the cell and as its concentration builds up, a steady state is soon achieved and then the concentration of gas entering and leaving the cell is the same as it was initially. The build up of the heavier gas is only a fraction of a percent and even as low as a few parts per million. Therefore, it does not affect the quantities adsorbed in any measurable way.

However, when the bath is removed and the cell warms up the steady state is disturbed and the slight excess of heavier gas generates a signal followed by a signal due to the excess of lighter gas held up in the "in" arm of the cell. These signals are observed as a negative signal before or after the desorption signal and can generate errors in the integration of the desorption signal. This effect begins to manifest itself with nitrogen and helium mixtures when the total area in the cell is approximately 0.1 - 0.3 m^2

In a static mixture of gases, the amount of thermal diffusion is a function of the time rate of change of the temperature gradient, the gas concentration, and the difference in masses of the molecules. In a flowing gas mixture, in the presence of adsorption, it is difficult to assess the exact amount of thermal diffusion. Lowell and Karp [8] measured the effect of thermal diffusion on surface areas using the continuous flow method. Fig. 15.14 illustrates a fully developed anomalous desorption signal caused by thermal diffusion. As a result of the positive and negative nature of the signal, accurate integration of the true desorption peak is not possible.

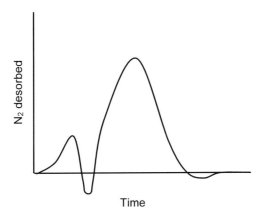

Figure 15.14 Signal shape from desorption of a small volume of nitrogen.

Table 15.1 shows the results of measuring the surface area of various quantities of zinc oxide using a conventional sample cell, Fig. 15.10a. When the same sample was analyzed using a micro cell, Fig. 15.10b, the results obtained were considerably improved, as shown in Table 15.2.

The onset of thermal diffusion depends on the gas concentrations, the sample surface area, the rate at which the sample cools to bath temperature, and the packing efficiency of the powder. In many instances, using a conventional sample cell, surface areas less than 0.1 m^2 can be accurately measured on well-packed samples that exhibit small interparticle void volume. The use of the micro cell (Fig. 15.10b) is predicated on the latter of these observations. Presumably, by decreasing the available volume into which the denser gas can settle, the effects of thermal diffusion can be minimized. Although small sample quantities are used with the micro cell, thermal conductivity detectors are sufficiently sensitive to give ample signal.

Another cell design that aids in minimizing the effects of thermal diffusion is the capillary cell, Fig. 15.10c. By using capillary tubing on the vent side of the cell, a sufficiently high linear flow velocity is maintained to prevent that arm from contributing to the problem. The large sample capacity of the capillary cell, compared to the micro cell, produces sufficient desorption signal to often make the thermal diffusion effect negligibly small.

Table 15.1 Data obtained using conventional cell (measured using the single-point BET method with 20% N_2 in He).

Weight	Actual area (m^2)	Measured area (m^2)	Deviation (%)	Signal shape at start of desorption peak
1.305	5.07	5.07	0	
0.739	2.87	2.87	0	
0.378	1.47	1.45	1.4	
0.177	0.678	0.686	−1.2	
0.089	0.345	0.327	5.2	
0.049	0.190	0.166	12.6	
0.0190	0.0730	0.0481	34.1	
0.0101	0.0394	0.0192	51.3	

Lowell [9] published a method to circumvent the problem of thermal diffusion by using an adsorbate with a low vapor pressure, such as krypton, at liquid nitrogen temperature. The coefficient of thermal diffusion $D(t)$ is given by [10]

$$D(t) = \frac{N_1/N_{tot} - N_1'/N_{tot}}{\ln T_1/T_2} \qquad (15.12)$$

where N_1 and N_1' are the adsorbate concentrations at the absolute temperatures T_2 and T_1, respectively with $T_2 > T_1$. The term N_{tot} is the total molecular concentration of adsorbate and carrier gas. Because of krypton's low vapor pressure, its mole fraction in the BET range of relative pressures is of the order of 10^{-4}. This small value causes the difference between N_1/N_{tot} and N_1'/N_{tot} nearly to vanish, with the consequence that no obscuring thermal diffusion signals are generated.

Attempts to increase the size of nitrogen adsorption or desorption signals, by using larger sample cells, results in enhanced thermal diffusion signals due to the increased void volume into which the helium can settle. However, when krypton is used, no thermal diffusion effect is detectable in any of the sample cells shown in Fig. 15.10.

Table 15.2 Data obtained using U-tube cell.

Weight ZnO (g)	Actual Area (m²)	Measured Area (m²)*
0.0590	0.2280	0.2300
0.0270	0.1050	0.1020
0.0045	0.0175	0.0161

*This value is corrected by 15 cm² for the cell wall area, as estimated from the cell dimensions. Desorption peaks from an empty U-tube cell gave areas of 12- 17 cm².

The adsorption signals using krypton-helium mixtures are broad and shallow because the adsorption rate is limited by the low vapor pressure of krypton. The desorption signals are sharp and comparable to those obtained with nitrogen, since the rate of desorption is governed by the rate of heat transfer into the powder bed.

With krypton, the ability to use larger samples of low area powders facilitates measuring low surface areas because larger signals are generated in the absence of thermal diffusion. Also, as is true for nitrogen, krypton measurements do not require void volume evaluations or nonideality corrections, nor is thermal transpiration a factor as in the volumetric measurements.

15.7 DATA REDUCTION—CONTINUOUS FLOW

Table 15.3 can be used as a work sheet for calculating specific surface areas from continuous flow data. The data in the lower left corner are entered first and are used to calculate the other entries. In the example shown, nitrogen is the adsorbate.

Column 1 is the mole fraction of adsorbate in the flow stream. Column 2 is obtained as the product of P_a and column 1. Column 2, when divided by column 3, gives the relative pressure, which is entered in column 4 and from which columns 5 and 6 are calculated. Column 7 is the volume required to calibrate the desorption signal and column 8 is the corresponding weight of the calibration injection, calculated from the equation in the lower left side of the work sheet. The terms A_s and A_c are the areas under the signal and calibration peaks, respectively. Columns 11-13 are calculated from the data in the previous columns. The data in column 13 are then plotted versus the corresponding relative pressures in column 5. The slope s and intercept i are calculated and the value of W_m is found as the reciprocal of their sum. Equation (4.13) is used to obtain the total sample surface area, S_t, and dividing by the sample weight yields the specific surface area, S.

15.8 SINGLE POINT METHOD

The assumption of a zero intercept reduces the BET equation to equation (5.3). This assumption is, of course, not realizable since it would require a BET C constant of infinity. Nevertheless, many samples possess sufficiently high C values to make the error associated with the single-point method acceptably small (see Chapter 5 and Table 5.2).

Using the zero intercept assumption, the BET equation can be written as

$$W_m = W\left(1 - \frac{P}{P_0}\right) \qquad \text{(cf. 5.11)}$$

from which the total surface area can be calculated by

$$S_t = W\left(1 - \frac{P}{P_0}\right)\frac{\overline{N}}{\overline{M}}A_x \qquad \text{(cf. 5.12)}$$

From the ideal gas equation of state

$$W = \frac{P_a V \overline{M}}{RT} \qquad (15.13)$$

so that

$$S_t = \left(1 - \frac{P}{P_0}\right)\frac{P_a V \overline{N}}{RT}A_x \qquad (15.14)$$

where P_a and T are the ambient pressure and absolute temperature, respectively, $\overline{N}$ is Avogadro's number, A_x is the adsorbate cross-sectional area, and V is the volume adsorbed.

Table 15.3 Multipoint BET surface area data sheet.

Date: _____ Sample: Titania Total Weight: 12.0434g

Operator: _____ Outgas Procedure: He purge, 1 hr., 150°C Tare: 11.3430g

Sample Weight: 0.7004g

1	2	3	4	5	6	7	8	9	10	11	12	13
X_{N_2}	P (torr)	P_0 (torr)	P/P_0	P_0/P	$(P_0/P) - 1$	V_c (cm³)	W_c (g)	A_s	A_c	$W = \frac{A_s}{A_c} W_c$	$W[(P_0/P)-1]$	$\frac{1}{W[(P_0/P)-1]}$
0.050	37.68	768.6	0.0490	20.398	19.398	2.00	0.00230	1066	1083	0.00226	0.0438	22.81
0.100	75.36	768.6	0.0980	10.199	9.199	2.20	0.00252	1078	1050	0.00259	0.0238	41.97
0.200	150.7	768.6	0.1961	5.096	4.096	2.50	0.00287	1168	1098	0.00305	0.0125	80.05
0.2995	225.7	768.6	0.2937	3.405	2.405	2.60	0.00298	1031	879	0.00350	0.0084	118.80

Ambient pressure, $P_a = 753.6$ mm Hg

Vapor pressure, $P_0 = 768.6$ mm Hg

Ambient temperature, T = 295 K

Adsorbate molecular weight, $\overline{M} = 28.01$

Adsorbate area, $A_x = 16.2 \times 10^{-20}$ m²

Calibration gas weight, $W_c = \dfrac{P_a \overline{M} V_c}{(6.235 \times 10^4) T}$

Plot $\dfrac{1}{W[(P_0/P)-1]}$ vs. P/P_0

Slope, $s = 391.9$

Intercept, $i = 3.52$

$W_m = \dfrac{1}{s+i}$

$W_m = 0.00253$ g

Total surface area,

$S_t = \dfrac{W_m \left(6.023 \times 10^{23}\right) A_x}{\overline{M}}$

$S_t = 8.81$ m²

$S = S_t/W$

$S = 12.58$ m²/g

Using nitrogen as the adsorptive at a concentration of 0.3 mole fraction and assuming P_0 is 15 torr above ambient pressure, equation (15.14) can be expressed as

$$S_t = W\left(1 - \frac{0.3P_a}{P_a + 15}\right)\frac{P_a V \overline{N}}{RT} A_x \tag{15.15}$$

Assuming the ambient pressure P_a is 760 Torr and ambient temperature T is 295 K, equation (15.15) reduces to

$$S_t = 2.84V \ (\text{m}^2) \tag{15.16}$$

Thus, the total surface area contained in the sample cell is given by the simple linear relationship above when V is in cubic centimeters. By calibrating the desorption signal, A_{des}, with a known volume of nitrogen, V_{cal}, equation (15.16) can be rewritten as

$$S_t = 2.84\frac{A_{des}}{A_{cal}}V_{cal} \tag{15.17}$$

where A_{des} and A_{cal} are the integrated areas under the desorption and calibration signals, respectively.

Modern commercial single point instruments contain a linearization network that corrects for the hot wire nonlinearity. This procedure allows a built-in digital integrator to integrate the signals linearly so that the surface area is given directly on a digital display. An advantage is that the analysis time for a BET surface area determination is extremely short, usually less than ten minutes [10].

15.9 REFERENCES

1. Loebenstein W.V. and Deitz V.R. (1951) *J. Res. Nat. Bur. Stand.* **46**, 51.
2. Nelson F.M. and Eggertsen F.T. (1958) *Anal. Chem.* **30**, 1387.
3. de Boer J.H. (1953) *The Dynamical Character of Adsorption*, Oxford University Press, London, p33.
4. Karp S. and Lowell S. (1971) *Anal. Chem.* **43**, 1910.
5. Karp S., Lowell S. and Mustaccuiolo A. (1972) *Anal. Chem.* **44**, 2395.
6. Kourilova D. and Krevel M. (1972) *J. Chromat.* **65**, 71.
7. Lowell S. and Karp S. (1972) *Anal. Chem.* **44**, 1706.
8. Lowell S. (1973) *Anal. Chem.* **45**, 8.
9. Benson S.W. (1960) *The Foundation of Chemical Kinetics*, McGraw-Hill, New York, p188.
10. Quantachrome Instruments, Boynton Beach, FL.

16 Volumetric Chemisorption: Catalyst Characterization by Static Methods

16.1 APPLICATIONS

The vacuum volumetric, or static, method is used to determine the monolayer capacity of a catalyst sample from which certain important characteristics such as active metal area, dispersion, crystallite size, etc., may be derived by the acquisition of adsorption isotherms.

The volume of a suitably reactive gas adsorbed by a suitably prepared catalyst sample is measured as a function of gas pressure at a fixed (isothermal) sample temperature. Repeated measurements at different temperatures can be used to calculate heats of adsorption. Most, if not all, industrial catalysts are amenable to vacuum volumetric measurements when both appropriate preparation conditions and adsorptive are employed.

16.2 SAMPLE REQUIREMENTS

The sample to be analyzed should have sufficient metal area to yield measurable quantities of gas uptake. However, this is not merely a function of total metal content, but rather how well it is distributed or dispersed. The exposed area associated with a low, but well-dispersed, metal content (0.1% platinum on alumina for example) is easier to measure than even a pure metal (not supported) that has relatively large particle size (e.g. platinum gauze as might be used for ammonia oxidation). In many cases, previous experience from an earlier analysis of a similar material will guide the analyst in selecting the appropriate sample size. However, since prior knowledge cannot be guaranteed, the general rule is that "more is better than less" within the limitations set by the size of the sample cell size to be used. Nevertheless, a well-designed, modern apparatus should have the ability to detect as little as 1 μmole of adsorption.

It is quite possible there arises a need to analyze a used catalyst by static chemisorption methods. However, there exists the real possibility that such a sample is contaminated, probably heavily, with products and by-products of the reaction process in which it has been used. Such contamination has the potential to volatilize into the measuring system, causing real leaks across valve seats and/or unstable pressures due to its own vapor pressure. This obviously undesirable condition should be avoided, and care taken to clean used catalysts before loading them on the chemisorption analyzer. One very suitable method employs soxhlet extraction. The sample is suspended in a porous (often paper/fabric) basket and "bathed" in a continuous stream of condensed vapors from a refluxing flask of solvent. The sample is gradually cleansed of soluble contaminants, which concentrate in the liquid solvent.

16.3 GENERAL DESCRIPTION OF EQUIPMENT

The vacuum volumetric analyzer consists of at least the following: a vacuum pump or pumps, a flow-through sample cell, a fixed volume (manifold) from which adsorbate is dosed to the sample cell, a system of valves to allow evacuation and isolation of the sample cell, a system of valves to admit helium gas (for void volume determination) and active gases (adsorptives), a means for accurately measuring gas pressure (transducer), a means for preparing the sample *in-situ* (heating device and controller) and a data collection/reduction system (microcomputer/PC). A suitable, automated apparatus is represented in Fig 16.1. In certain application areas, primarily studies of metal hydrides, a volumetric apparatus is also known as a Sievert's apparatus.

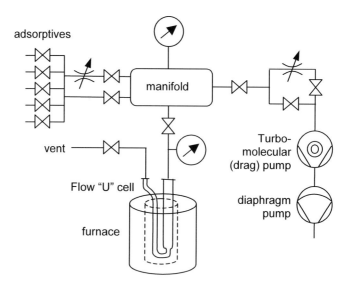

Figure 16.1 Diagram of a modern, automated vacuum-volumetric chemisorption analyzer

16.4 MEASURING SYSTEM

Quantitative measurements are made by first admitting a known pressure of adsorptive into a dosing manifold, whose volume is known. If the temperature of the manifold is constantly monitored, there is no need for thermostatting. The transducer mounted on the manifold indicates a pressure drop due to the transfer of gas into the sample cell when the valve leading to the sample cell is opened. The moles of gas dosed, n_{dose}, into the sample cell is calculated thusly:

$$n_{dose} = \frac{\Delta P_{man} V_{man}}{R T_{man}}$$ (16.1)

where ΔP_{man} is the change in pressure in the manifold, V_{man} is the internal geometric volume of the manifold, R is the gas constant and T_{man} is the temperature of the manifold (in kelvin).

To enhance sensitivity towards the chemisorption process that, in poorly dispersed and/or low-loading samples, may involve extremely small amounts of adsorption, it is preferable to limit the volume in which the adsorption process is monitored. This is done in order to (i) minimize the pressure drop due to merely filling the void volume and (ii) maximize the resulting pressure changes due to sorption. It is desirable therefore to isolate the sample cell after dosing and use a transducer affixed to the sample cell to follow the equilibration process and thereby to determine the actual degree of adsorption. The amount of gas remaining unadsorbed in the void volume of the cell, n_{void} is calculated as

$$n_{void} = \frac{P_{cell} V_{void}}{R T_{void}}$$ (16.2)

where P_{cell} is the equilibrated pressure just in the sample cell, V_{void} is the effective internal volume of the sample cell, transducer, connections and accounting for the presence of sample, and T_{void} is the effective temperature of the void volume. It is usual that V_{void} is determined as a preliminary step in the analysis by expanding helium from the manifold into the cell void assuming the temperature of the void to be the nominal analysis temperature at which the sample itself is maintained:

$$V_{void} = V_{man} \left(\frac{T_{void}}{T_{man}} \right) \left(\frac{P_1}{P_2} - 1 \right)$$ (16.3)

where P_1 and P_2 are initial and final pressures respectively. *Note: it is not required that the __entire__ cell void be kept at a single temperature!* However, since the thermal profile of the cell might vary with internal pressure (due to the increased conduction and convection with increasing quantity of gas), it is appropriate to determine the effective void volume as a function of pressure. Commonly just two values are used for this step, for example approximately one-tenth and one atmosphere.

The amount of gas adsorbed, n_{ads}, is computed as the difference between the amount dosed (from the manifold) and the amount remaining in the void volume. Combining (16.1) and (16.2) yields

$$n_{ads} = \left(\frac{\Delta P_{man} V_{man}}{RT_{man}} \right) - \left(\frac{P_{cell} V_{void}}{RT_{void}} \right) \tag{16.4}$$

16.4.1 Pressure Measurement

All volumetric measurements are made from pressure changes by employing the ideal gas law, $PV = nRT$. Therefore, high quality pressure transducers should be employed combined with high-resolution electronics. Ideally, the system should be leak checked using helium. Since the plateau in the chemisorption region typically extends over the pressure range 13.3kPa (100 mmHg) to 100kPa (one atmosphere), a transducer with approximately 1000 mmHg full-scale is commonly employed. Should much lower pressure data be required, though not essential for the measurements described in the theory section, a lower range transducer or transducers is/are required, say 10 and/or 1 torr full scale. Since a high vacuum system will be employed, the low range transducers play a significant role in determining that a high vacuum has indeed been achieved, one that might not be confirmed by using the high-range transducer alone.

16.4.2 Valves

The valves that admit gas into the measuring system should be constructed of materials compatible with the gases employed. Stainless steel and chemically resistant elastomer seals (such as EPDM) should be used. See §16.4.6 for detailed discussion regarding chemical compatibility below. Furthermore, they should consume little or no power in the fully open or fully closed state so as not to introduce a local heating effect that could cause a pressure rise not due to the sorption process, and thereby lead to erroneous data.

16.4.3 Vacuum

It is important that all gas, which is not to be used in the measurement, be removed from the system as completely as possible before commencing the analysis. It is especially important to ensure that no air or other oxidizing gases be present that might react and de-activate a freshly reduced catalyst sample. Therefore, it is usual not to rely on the vacuum level afforded by rotary oil pumps, but to demand the much-enhanced vacuum level that is possible by using a high-speed turbomolecular pump or hybrid turbo-drag pump. Preferably the high vacuum system shall benefit from an oil-free fore pump such as is afforded by a dry diaphragm pump. This eliminates oil backstreaming and the necessity to maintain a cold trap between the pumping system and sample.

16.4.4 Sample Cell

It is normal that the catalyst sample be prepared under some sort of heating regimen prior to analysis. Therefore, the sample cell is most commonly constructed from quartz glass that can withstand the very high temperatures sometimes used. The cell is usually fabricated in the form of a "U" that permits both inert and reactive gases to be flowed over and through the sample bed (see 16.6 Pretreatment, below). The exit side of the cell shall have a means by which it can be closed (e.g. automatic solenoid valve) to allow the cell to be evacuated, at least at the onset of the analysis phase. The use of a quantity of loosely packed quartz wool both above and below the sample bed is recommended to prevent powder being elutriated, or lost, from the cell. A volume-filling quartz rod is beneficial in reducing the void-volume in the sample cell thereby (i) magnifying pressure changes due to the sorption process and increasing instrument sensitivity and (ii) minimizing the effects due to errors in void volume determination.

16.4.5 Heating System

Whilst the temperature of analysis is typically in the range 300K to 400K, a high temperature mantle or furnace is required to properly condition the sample prior to analysis. According to the sample type, heat may only be required to dry, or outgas the sample in which case a mantle capable of no more than 350 °C is sufficient. For higher temperatures, up to 1,100 °C if so desired, a tube furnace is usual. So as not to jeopardize the vacuum integrity of the sample cell, the controlling thermocouple is placed close to the sample region between the inner face of the furnace and the sample cell. Nowadays, both the temperature programming and positioning of the furnace around the sample cell can be fully automated, as can the transition between pretreatment and analysis temperatures.

16.4.6 Gases and Chemical Compatibilities

High purity gases should always be used. Helium, employed in the determination of void volume in the sample cell, should be at least 99.999% by volume. Hydrogen used for both reduction (pretreatment) and analysis should also be at least 99.999% by volume. Carbon monoxide is generally available in at least 99.95% by volume, though 99.99% (or greater) is preferred. Other gases should be of a comparable quality. If high purity gases are not immediately available, then dry "scrubbers" or gas purification systems commonly associated with gas chromatographs may be used to purify the gases before admitting them into the chemisorption analyzer. If such devices are used, they must be fitted with particulate filters to prevent contamination of the instrument valve seals!

Due to the necessary reactive nature of the gases used in chemisorption studies, it is important to anticipate the possibility of undesirable reactions between the gases used and the materials of construction used in the instrument. Generally, corrosion resistant stainless steel is used for internal plumbing, but care should still be taken to ensure complete compatibility between all gases and the exact type of steel. Elastomeric seals used in valves and connections in general offer the greatest challenge. There is no universal elastomer that is 100% compatible with all gases and vapors. It is possible that the seals inside the instrument will have to be changed between runs with different gases. Always consult with the instrument manufacturer before using a gas that is not explicitly authorized. Some commonly used gases, metal and elastomer pairings are listed below:

Table 16.1 Chemical compatibility chart

Gas	Steel		Elastomer		
	304	316	Buna	EPDM	Viton
CO	1[a]	1[a]	1[a]	1[a]	1[a]
H_2	1[a]	1[a]	1[a]	1[a]	1[a]
O_2	1[b]	1[c]	2[d]	1[e]	1[e]
NH_3 (dry)	1[a]	1[a]	2[a]	1[a]	3[a]
SO_2 (dry)	3[a]	1[a]	3[a]	1[a]	1[a]

1= recommended; 2 = acceptable; 3 =incompatible. Sources: a) Cole Parmer Instrument Company, Vernon Hills, CA, USA; b) The BOC Group, Murray Hill, NJ, USA; c) Matheson Tri-Gas, Montgomeryville, PA, USA; d) eFunda, Inc., Sunnyvale CA, USA; e) Parker Seal Group, Irvine, CA, USA. This is not an exhaustive list and is for illustrative purposes only.

16.5 PRETREATMENT

Many, if not most, catalyst samples that are received into the laboratory for analysis by chemisorption are not in a chemical form suitable for the immediate determination of metal area, dispersion, etc. This is so because such samples are some compound, be it the oxide, carbonate or nitrate, etc., whereas it is the zero-valent metal that most often must be analyzed. Therefore, those samples must undergo various procedures to reduce the compounded metal into its elemental form. This combination of procedures, or *pretreatment*, usually takes the form of a series of heat treatments combined with exposure to reactive gases.

16.5.1 Heating

The sample may first be heated to a temperature sufficient to desorb loosely bound gases and vapors, principally water. At first, it might not be apparent why moisture need be desorbed from an oxide sample if it is to be subsequently reduced with the concomitant production of water! However, the quantity of physisorbed water in the pores of the catalyst support can greatly exceed the water of reduction, and severely restrict access of reducing gas to metal oxide particles. Therefore, the reduction may not proceed to completion with the rapidity one might initially anticipate... unless the sample is dried prior to initiating the reduction step [1].

Too high a temperature may cause decomposition reactions, of carbonates for example, which may or may not be desirable. A detailed thermal analysis, by thermogravimetry (TGA) or differential scanning calorimetry (DSC) is often very useful in determining appropriate temperatures during pretreatment. Extremely rapid changes in temperature are to be avoided, to prevent the elutriation of powder by the flash generation of steam for example, or to prevent a runaway exotherm during exposure to a reactive gas.

The raising of sample temperature should be performed in a linear fashion, albeit at different rates. Modern PID (*proportional-derivative-integral*) controllers, tuned to the particular heating device employed, are capable of increasing temperature by as little as one degree per minute to as much as fifty degrees per minute. Note that the highest heating rates may not be possible at the higher temperatures due to heat losses. There is no "golden rule" regarding ramp-rates so the effect of heating rate on the final results should be determined by experimentation. However, ten to twenty degrees per minute has been found to be suitable for many materials. Once a suitable sample temperature has been attained, to promote chemical reduction for example at a reasonable rate, it is usual to hold the temperature for a fixed period rather than to continue increasing the temperature to the furnace maximum. This is to prevent sintering the material, which results in a rapid and undesirable loss of surface area. A TPR analysis (see Chapter 17, §17.7) is an appropriate exploratory step to determine appropriate temperature for reduction. However, it is unlikely that a sample which has undergone a complete TPR profiling, i.e. one well beyond the temperature of maximum reduction rate, is suitable for subsequent isothermal analysis due to sintering. Use a fresh sample for the measurement of the isotherm.

16.5.2 Atmosphere

The gas or gases used in pretreatment determine the form in which the sample will ultimately be analyzed. Whether the gases are flowed over the sample, held static or the sample be exposed to vacuum also affects the final state of the surface.

Most samples require some form of outgassing or degassing, that is, the removal of adventitiously adsorbed gases and vapors (see §16.5.1). The use of a dry, inert gas, such as helium (or argon or nitrogen depending on the nature of the sample and the temperature used) is normal. This inert purge gas will be required at least one more time, to purge the reactive gas(es) used in the chemical pretreatment.

If the sample submitted for analysis is an oxide (or carbonate), then it is normally reduced to the metal. Hydrogen is commonly used since it is readily available in high purity, and though flammable, easy to handle. Carbon monoxide can be used, though its purity cannot always be guaranteed. It is more toxic though, but can be handled safely if proper precautions are taken, which may include installing a carbon-monoxide detector (readily available from any reputable general laboratory equipment catalog). Diluted hydrogen (typically 5% hydrogen, 95% nitrogen) is also widely used, (i) to virtually eliminate the flammability hazard and (ii) limit the amount of hydrogen available to the sample to better control or limit the reaction rate.

Occasionally, it may be necessary to oxidize the sample to ensure a homogeneous starting point for subsequent reduction. Or oxidation may be required to bring the sample to a higher oxidation state or valency for that is the state in which the sample (depending on the application) will be analyzed. In either case, air or depleted air (i.e. relatively low oxygen content) can be employed. Depleted air may be preferred to prevent a runaway exotherm. Carbon dioxide is mildly oxidizing and can, for example, be used to stabilize the surface of freshly reduced nickel oxide without incurring bulk oxidation, which might otherwise occur if oxygen is used. Never allow oxidizing and reducing gases to mix inadvertently, especially over a hot catalyst!

A static gas atmosphere is rarely used, except in those cases where a self-generated steam atmosphere (from the reduction of an oxide with hydrogen) is used to promote steam sintering actually during the pretreatment phase of sample analysis. Of course, other static atmospheres can be used where their effects on the final analytical outcome need to be studied in their own right.

A flowing stream of gas preferably through (rather than over) the sample bed is more usual since the flow carries away the products of degassing (thermally induced desorption) and reaction. The flow rate can be important, especially where the heating rate is relatively high and evolution of reaction products is extremely rapid. The absolute flow rate is important (ten to thirty milliliters per minute is typical) but so is the space velocity, SV. That is, how many bed volumes of reactive gas flow through the bed per hour? For example, given a bed volume of two milliliters and five-percent hydrogen flowing at twenty milliliters per minute, the space velocity would be thirty. This could be achieved using pure hydrogen and a flow rate of just one-milliliter per minute, but that most likely would be insufficient to adequately purge the sample bed of reaction products such as water. More usually, one would use pure hydrogen at the higher volumetric flow rate to increase the space velocity to 600 for thorough reduction.

$$SV = 60\frac{F}{V_{bed}} \times \frac{C\%}{100} \qquad (16.5)$$

Where F is the volumetric flow rate in the usual cm^3min^{-1}, V_{bed} is the geometric volume of the powder bed (in cm^3) and $C\%$ is the concentration of reducing gas in the total gas stream (in percent). Using the conversion factor 60 (mins h^{-1}) gives SV in units of reciprocal time, i.e. h^{-1}.

Removal of reducing gas(es) at reduction temperature, is a vital step in the preparation of the sample, otherwise they would be adsorbed at the metal surface upon cooling thus negating the subsequent adsorption isotherm. A simple purge is effective at removing large volumes of reducing gas(es) quickly, especially from the voids within the powder bed. However, it is somewhat less effective at removing the same gases from *within* pores of the support, there being no bulk flow inside the pore network. Rather, diffusion of remaining reactive gas(es) is slowed by the presence of essentially static purge gas molecules in the pores. Evacuation of the sample is an alternative to a flowing gas purge, but is limited by diffusion kinetics, that is, desorbed molecules must find their own way out of the sample cell (slowly) rather than being swiftly displaced by a steady stream of inert gas. Final traces of reducing gas(es) do however diffuse more rapidly from the pores in the absence of a purge gas, i.e. under vacuum. Therefore, the authors recommend for post-reduction clean-up a combination of flow and evacuation to ensure complete removal of reducing gas(es) and products.

It is important, however, to include at least one period of vacuum at the end of the entire pretreatment process if only to establish that purge gases have themselves been removed, that they were indeed successful in removing volatile reactants and gaseous products, and to establish that a leak-free atmosphere exists in the sample cell before proceeding to the measurement phase.

Table 16.2 Summary of pretreatment protocol recommended for sample type: 1% Pt on Al_2O_3.

Step	Action	Purpose
1	Purge with He, 20 ml min^{-1} at 140°C, 0.5h	Remove moisture.
2	Purge with H_2, 20 ml min^{-1} at 10° min^{-1}.	Reduce surface oxides and remove H_2O produced.
3	Maintain H_2 flow at 400°C for 2h.	Ensure complete reduction and remove H_2O produced..
4	Purge with He, 20 ml min^{-1} at 400°C, 2h.	Remove H_2.
5	Cool to 40°C under vacuum.	Remove He.

16.6 ISOTHERMS

The intention of measuring the adsorption isotherm is to determine the volume of gas adsorbed by the sample as a function of pressure and thereby calculate important, quantitative characteristics such as active metal area. Therefore, it is important to select not only the correct adsorptive (reactive analysis gas), but also an appropriate pressure range and to obtain any repeat isotherms which might be necessary.

16.6.1 Reactive Gas

A specific gas might be chosen simply to investigate the exact nature of the interaction of that gas with a given solid. In this case any gas, which is compatible with the materials of construction of the apparatus or instrument being used, may be employed. However, predictable behavior, at least with respect to stoichiometry, dictates that either hydrogen or carbon monoxide be the gas of choice, according to the metal being analyzed (see Table 16.2). Carbon monoxide does offer increased sensitivity for many analyses compared to hydrogen in that the stoichiometry is often one, rather than two (as is the case with hydrogen.) Twice as many CO molecules than hydrogen are therefore required to saturate a metal surface. This may be an advantage when either the loading or total amount of sample is low. Oxygen, though reactive with most metal surfaces, has few applications – silver probably being the most important.

16.6.2 The Combined Isotherm

The first isotherm measured represents the total adsorption on both strong and weak sites. It includes therefore some physisorption and spillover that may be undesirable in the final calculations. Some workers make use of this data without further recourse. Nevertheless, one cannot measure strong chemisorption alone without having first measured combined adsorption! Typically, the data extends over the pressure range from one-tenth to one atmosphere. Previous experience may indicate that a shorter pressure range, up to one third or one half atmosphere for example, may be sufficient. The temperature can be any temperature achievable with the apparatus used. However, a constant temperature in the range twenty to fifty degrees Celsius is usual. Use a dewar of water at room temperature, or a mantle or furnace at the higher temperatures, to thermostat the sample cell.

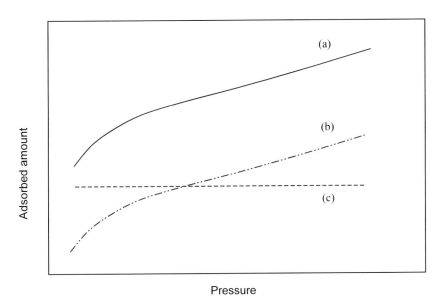

Figure 16.2 Representation of chemisorption isotherms: (a) combined strong plus weak adsorption, (b) weak adsorption, (c) strong adsorption (by difference).

Some typical examples of analysis conditions are given in table 16.3. It is not meant to be exhaustive, but to illustrate the variety of methods employed by different workers, though the general similarities should be evident.

Table 16.3 Some published conditions for catalyst characterization by chemisorption isotherm(s).

Metal	Support	Redn Temp (°C)	Redn Time (h)	Anal Temp (°C)	Gas	Stoich	Ref
Pt	C	300	2°/min	RT[1]	H_2	2	1
Pt	Zeolite KL[2]	300	2	23	H_2	1.9+/-0.2	2
Pt	Hexagonal faujisite	300	2	23	H_2	1.5	3
Pt	Al_2O_3	300-825	1	25	H_2	<1	6
Pt	LTL zeolite[3]	400	1	RT	H_2	2	7
Pt	SiO_2	400	1	RT	H_2	2	7
Pt	WOx/ZrO_2	200, 400	1	40	H_2	2	13
Pt	WOx/ZrO_2	200, 400	1	40	O_2	-	13
Pt	WOx/ZrO_2	200, 400	1	40	CO	-	13
Pt	-	300	2	35	CO	1	14
Pd	LTL zeolite	400	1	RT	CO	1	7
Pd	SiO_2	300	1	RT	CO	1.4	8
Ru	Al_2O_3, C, TiO_2, SiO_2	300	3	-	-	-	10
Ru	C	450	2	35	H_2	2	11
Ru	C	450	2	35	CO	uncertain	11
Ru	C	200	2	35	H_2	2	9
Ru	Al_2O_3, C, TiO_2, SiO_2	300	3	-	-	-	10
Ru/Ir	NaY	400	2	23	H_2	~2 (Ir)[4] ~4 (Ru)[4]	4
Rh	Al_2O_3	320	-	25.	CO	0.5	5
Rh	Al_2O_3	200-500	1	25	H_2	1 to 1.66	6
Ir	Al_2O_3	500	1	25	H_2	<0.25	6
Ni	Al_2O_3, C, TiO_2, SiO_2	500	2	?	H_2	?	10
Fe	SiO_2	N/a[5]		-78	CO	?	12

Notes

[1] Room temperature. Ambient temperature +/- 15° is typical of many isotherm measurements.

[2] KL zeolite is a potassium rich zeolite with a pore structure of straight, non-intersecting channels of linked cages.

[3] Zeolite LTL is a crystalline aluminosilicate molecular sieve with a microporous channel formed by linear interconnection of lobe-shaped cages (1.1 nm in diameter, 0.6 nm in length) through apertures of 0.71 nm in diameter [3].

[4] Strong chemisorption only after subtracting reversible weakly adsorbed.

[5] Direct preparation by thermal decomposition of iron carbonyl.

16.6.3 The Weak Isotherm

Before a second *reversible* isotherm is measured, the sample must be evacuated fully to remove weakly bound gas molecules. This is done at the first analysis temperature without removing the sample from the apparatus. A second isotherm is measured, again at the same temperature and using the same adsorptive as the first isotherm, which represents the re-adsorption of weakly adsorbed molecules.

16.6.4 The Strong Isotherm

In the vacuum-volumetric technique, the strong *irreversible* chemisorption isotherm is not actually measured but calculated as the difference between the combined isotherm and the weak isotherm. Where data do not occur at exactly the same pressure in both isotherms, it is sufficient to linearly interpolate between points. It is the extrapolation of these data to zero pressure that most often yields the most realistic estimation of metal nanoparticle, or *crystallite*, size.

16.6.5 Multiple Isotherms

These are required when calculating heats of adsorption and monolayer uptake by the Temkin and Freundlich methods (see Chapter 12). Two or more isotherms (combined plus weak if preferred) are measured at two distinct temperatures that typically differ by at least ten degrees, but could vary by 100 degrees or more. These may be acquired using the same or fresh sample of catalyst, with appropriate re- or pretreatment between isotherms. It is possible using modern computer controlled equipment to measure multiple isotherms on the same sample completely automatically without any operator intervention after the initial loading of the sample onto the analyzer.

16.7 REFERENCES

1.   Burwell R.L. and Taylor H.S. (1936) *J. Am. Chem. Soc*. **58**, 1753.
2.   Joo S.H., Choi S.J., Oh I., Kwak J., Liu Z., Terasaki O. and Ryoo R. (2001) *Nature* **412**, 169.
3.   Cho S.J., Ahn W-S., Hong S.B and Ryoo R. (1996) *J. Phys. Chem.* **100**, 4996.
4.   Ihee H., Bécue T., Ryoo R., Potvin C., Manoli J-M. and Djéga-Mariadassou G. (1994) *Stud. Surf. Sci. Catal.* **84**, 765.
5.   Kip B.J., Duivenvoorden F.B.M., Koningsberger D.C. and Prins R. (1987) *J. Catal.* **105**, 26.
6.   Mojet B.L., Miller J.T., Ramaker D.E. and Koningsberger D.C. (1999) *J. Catal.* **186**, 373.
7.   Barton D.G., Soled S.L., Meitzner G.D., Fuentes G.A. and Iglesia E. (1999) *J. Catal.* **181**, 57.
8.   Vergunst T., Kapteijn F. and Moulijn J.A. (1998) In *Preparation of Catalysts IV* (Delmon B. *et al*, eds.) Elsevier, p175.

9. Shin E.W., Kang J.H., Kim W.J., Park J.D and Moon S.H. (2002) *Appl. Catal., A* **223**, 161.
10. Kusserow B., Schimpf S. and Claus P. (2003) *Adv. Synth. Catal.* **345**, 289.
11. Liang C., Wei Z., Xin Q. and Li C. (2001) *Appl. Catal., A.* **208**, 193.
12. Toebes M.L., Prinsloo F.F., Bitter J.H., van Dillen A.J. and de Jong K.P. (2003) *J. Catal.* **214**, 78.
13. Cho S.J. and Ryoo R. (2001) *J. Phys. Chem., B* **105**, 1293.
14. van't Blik H.F.J., van Zon J.B.A.D., Huizinga T., Vis J.C., Koningsberger D.C. and Prins R. (1985) *J. Am. Chem. Soc.* **107**, 3139.
15. Suslik K.S., Hyeon T. and Fang M. (1996) *Chem. Mater.* **8**, 2172.

17 Dynamic Chemisorption: Catalyst Characterization by Flow Techniques

17.1 APPLICATIONS

Under conditions of dynamic flow, controlled heating rates can be used to acquire characteristic reaction rate curves that can be used to classify, or fingerprint, different catalysts.

The dynamic flow method is also used to determine the monolayer capacity of a catalyst sample from which the usual important characteristics such as active metal area, dispersion crystallite size, etc., are determined by titrating, or saturating, the surface with a reactive gas.

17.2 SAMPLE REQUIREMENTS

A sample to be titrated should have, just as for vacuum volumetric measurements, sufficient active metal to cause enough gas adsorption that can be detected by the system used. Too much active metal, however, can lead to excessively long analysis times because of the large volume of gas that would have to be introduced incrementally to saturate the active surface. Previous experience will generally dictate the correct sample size to be used. Typically, no more than 1g of sample is required even though the active metal content might be quite low. Nevertheless, a sensitive flow instrument should be able to detect as little as 2μL of adsorption.

Those temperature-initiated changes that consume species from the gas phase could, if the sample amount were too great, entirely deplete a flowing gas mixture of the reactive component. Care should be taken to recognize the potential for this condition (by estimating the concentration of reactants in the sample) and making suitable adjustments as to quantity and/or gas composition.

Details regarding equipment and conditions for the different types of analyses are given in the following sections.

17.3 GENERAL DESCRIPTION OF EQUIPMENT

The flow analyzer consists of a means of regulating gas flow rates, a sample cell station, valves for directing gas to various parts of the system, a detector capable of responding to changes in gas composition and a means of controlling sample temperature. The various functions and design considerations of the equipment's discrete components are outlined below.

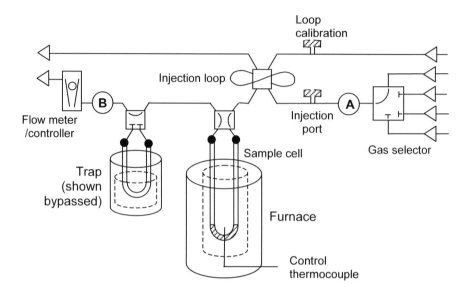

Figure 17.1 Simplified schematic showing essential components of a flow chemisorption analyzer, where **A** and **B** comprise a TCD. (*In-situ* sample cell thermocouple not shown).

17.3.1 Flow Path

A typical design is shown in Fig. 17.1. A gas-switching (selector) valve permits the use of a variety of gases that can be permanently connected to the instrument for convenience and to prevent admission of atmospheric gases when switching between gases. A bypass valve allows gas flow to continue uninterrupted while a sample cell is removed.

A diversion valve is useful to cause gas exiting the sample to flow through a trap to remove unwanted reaction products, such as water vapor, which might otherwise interfere with the signal detection. The trap may be surrounded by a cryogenic liquid to freeze out condensable vapors, or can be filled with a dry adsorbent such as 3Å zeolite. A gas sampling and injection system, such as a syringe and septum or loop, is necessary to titrate the sample with active gas in a pulse fashion.

17.3.2 Sample Cell

Because of the high temperatures that may be necessary in order to bring about reaction between sample and gas (e.g. reduction of a stable oxide), the sample cell is constructed, in the form of a U-tube, from quartz. Borosilicate glass is sufficient if the temperature will not exceed 350°C. The sample should be crushed to a particle size of approximately 1mm and added to the bottom portion of the cell such that it does not extend beyond the hot zone of the heating device employed. Small quartz wool plugs are added carefully at each end of the sample bed to prevent elutriation or loss of fine particles.

If the sample is a powder, then care must be taken not to over-compact the bed so as to cause excessive pressure drop and consequential low flow rates. If that did happen, the gas might "channel" through the bed through cracks that develop through the bed. Channeling is undesirable in that it does not allow proper contact between the gas and solid. To prevent this, it is beneficial to mix a powder sample with coarse (e.g. 1mm) quartz granules prior to adding the mix to the sample cell.

17.3.3 Gases

The same consideration regarding the purity of gases should be given as for vacuum volumetric measurements (see Chapter 16, §16.5.2). The carrier gas is of particular importance since it is constantly flowing over the sample. Even very small amounts of undesirable reactant can very quickly completely de-activate the sample!

Table 17.1 Recommended gas mixtures according to analysis type.

Analysis type	Carrier gas	Reactive gas
H_2 pulse titration	100% N_2 or Ar	100% H_2
CO pulse titration	100% He	100% CO
TPR	95% N_2 or Ar	5% H_2
TPO	98% He	2% O_2
TPD (H_2)	100% N_2 or Ar	n/a
TPD (CO)	100% He	n/a
TPD (NH_3)	100% He	n/a
TPD (CO_2)	100% He	n/a

For pulse titration, the carrier gas must have a significantly different thermal conductivity from the reactive titration gas. If hydrogen is the gas to be injected, then nitrogen or argon is commonly used. If carbon monoxide is the adsorbate, then helium is the appropriate choice of carrier gas. Similarly, during desorption studies helium is the preferred carrier gas.

Temperature programmed reactions require the use of *mixtures* of pure gases. An inert gas is used to dilute the reactive gas so that the TCD will respond to a change in the reactive gas concentration. Flowing only reactive gas would not afford any means by which the detector could respond, except to the products of reaction, which is better performed by a mass spectrometer (see below). See table 17.1 for guidelines on suitable combinations of gases.

17.3.4 Heating

Whilst the temperature of a titration analysis is typically in the range zero to 100°C, a high temperature mantle or furnace is required to properly condition the sample prior to analysis. According to the sample type, heat may only be required to dry, or outgas the sample in which case a mantle capable of no more than 350°C is sufficient. For higher temperatures, up to 1,100°C if so desired, a tube furnace is usual. For temperature programmed studies (TPR, TPD, etc.) the temperature must be increased in a controlled linear fashion. A so-called PID (proportional, integral, derivative) controller is used to ensure accurate and repeatable heating rates, which typically range from five degrees to fifty degrees per minute. The controlling thermocouple should be close to the sample zone, between the sample cell and the inner wall of the furnace or mantle. However, since the conditions during the heating program are not isothermal, the *sample temperature* is measured by a thermocouple (type K or similar, suitably sheathed to improve chemical resistance) inserted into the sample cell and, if possible, in direct contact with the sample bed.

17.3.5 Pulse Injection

Precision gas-sampling syringes offer the greatest flexibility in terms of the range of injection volumes, from just a few microliters to many milliliters. Reactive titration gas is withdrawn from a supply line via a septum and injected upstream (before) the sample cell through a second septum.

Injection loops offer fixed volume injections but still require that they be calibrated using syringes, if only to correct for the internal volume of the valve via which they are operated.

17.3.6 Detector

A thermal conductivity detector (TCD), or katharometer, is adequate for quantitative measurements and routine TPR/TPD studies. This detector responds to the change in the thermal conductivity of the gas. More specifically, in this application, the relative thermal conductivities before and after the sample cell are compared. Thus, the change in gas composition due to reaction with the sample, or desorption from it, can be continuously monitored. The filament should be of material that is resistant to the gases encountered. Rhenium/tungsten filaments are oxidation resistant and perform well with all commonly encountered dry gases. Gold-plated filaments do not fair well with ammonia.

A mass spectrometer can be a valuable addition to the flow technique because of its ability to differentiate multiple gases. For example, oxidation of carbonaceous species might yield both carbon dioxide and monoxide. Similarly the carbon dioxide hydrogenation reaction can yield a number of hydrocarbons and, if only partially reduced, carbon monoxide. A non-specific detector such as a TCD could not show whether two species were in fact present. Of course, it is possible to use a mass-spectrometer as the sole means of detection. See §17.9 for details and further examples.

17.4 PRETREATMENT

Many catalyst samples that are received into the laboratory for analysis by chemisorption are not in a chemical form suitable for the determination of metal area, dispersion, etc. This is so because such samples are actually the oxide, carbonate or nitrate, etc., whereas it is the zero-valent metal that must be analyzed. Therefore, those samples must undergo various procedures to reduce the compounded metal into its elemental form. This combination of procedures, or pretreatment, usually takes the form of a series of heat treatments combined with exposure to reactive gases.

Typical conditions of temperature, heating rates and gases to be used are discussed in chapter 16. If chemical modification of the sample is neither necessary nor desired, at least a drying stage should be incorporated into the overall analysis. It is important to remove adventitiously adsorbed vapors, especially water, which would otherwise interfere with the actual analysis. The flow-technique affords the opportunity to combine pretreatment with a TPD and/or TPR characterization of the sample.

17.5 PULSE TITRATION

This method requires the flowing of an inert gas over the sample and then injecting a quantity (pulse) of adsorbate into the flowing gas upstream of the sample cell. If the appropriate signal range setting is not known, then test injections should be made with the sample cell bypassed. The signal response from such an injection represents the maximum that will be achieved.

As the first pulse, or slug, of adsorbate is carried over the sample, some adsorption will normally occur. The detector will respond to the quantity of adsorbate that is not chemisorbed. If all the gas in the pulse is reacted with the sample, then no signal will be produced! It is important to record every injection made. Subsequent adsorbate injections will produce successively larger peaks as the surface approached saturation, and more of what is injected each time will be in excess. The surface is saturated when two or more successive peaks exhibit the same (within reason, say 5%) signal area. The total volume adsorbed, V_{ads}, is given by

$$V_{ads} = V_1^{inj} - \left(\frac{A_1^{inj}}{A_1^{sat}} V_1^{inj} \right) + V_2^{inj} - \left(\frac{A_2^{inj}}{A_2^{sat}} V_2^{inj} \right) \ldots + V_i^{inj} - \left(\frac{A_i^{inj}}{A_i^{sat}} V_i^{inj} \right) \quad (17.1)$$

where V_i^{inj} is the volume of gas injected for each pulse, A_i^{inj} is the signal area resulting from each corresponding injection and A_i^{sat} is the signal area resulting from the same volume injected at saturation (i.e. when no further adsorption occurs). The terms in parentheses represent the volumes of gas which are not adsorbed and which must be subtracted from the volumes injected to give the total volume adsorbed.

Whilst not absolutely necessary, it is normal to inject the same volume of gas every time. Note, those pulses which are not completely adsorbed by the sample give rise to signal areas less than A_i^{sat} and so for maximum accuracy a calibration curve should be prepared to eliminate detector non-linearity as a source of error. After saturation, pulses of varying volumes are injected so as to match the signal areas from partially adsorbed injections. Thus, the actual volumes adsorbed can be corrected. Ideally, the amount of sample and injection volumes are adjusted such that one, perhaps two, pulses are entirely adsorbed, two or three partially adsorbed, and two or three that represent wholly unadsorbed pulses after the sample has reached saturation. In this way, the analysis is optimized for no more than eight injections.

V_{ads} represents the nominal volume adsorbed calculated from the displacement of the syringe (or swept loop volume). This volume must be corrected to standard conditions, V_{STP}, to account for ambient temperature and pressure in order to properly calculate the amount of gas adsorbed.

$$V_{STP} = V_{ads} \times \left(\frac{P_{STP}}{P_{amb}} \right) \times \left(\frac{T_{amb}}{T_{STP}} \right) \quad (17.2)$$

where P_{amb} and T_{amb} are the ambient atmospheric pressure and temperature (in kelvin) of the gas syringe respectively and P_{STP} and T_{STP} are standard pressure (760 mm Hg) and temperature (273.15 K) conditions respectively. Note, the sample may or may not be at ambient temperature, but it is the temperature of the injected gas that must be used for the calculation.

The continual flow of carrier gas over the sample removes weakly adsorbed gas such that only strongly adsorbed gas remains, at least at the temperature of the test. Hence, in the flow method

$$V_{STP} = V_m \qquad (17.3)$$

where V_m is the monolayer volume. Thus, the number of gas molecules adsorbed by surface metal atoms, and the resulting metal area, dispersion and crystallite size values, are easily computed (see Chapter 12, §12.5 – 12.7).

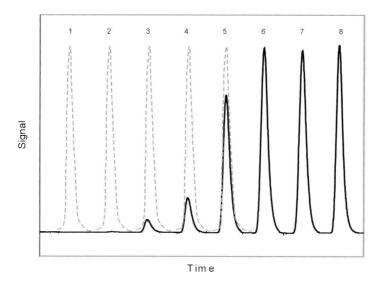

Figure 17.2 Chemisorption titration. Solid line represents detector signal after the sample. Dotted line represents volume injected. Pulses 1 and 2 are essentially entirely adsorbed, peaks 3, 4 and 5 represent increasingly less adsorption, peaks 6, 7 and 8 indicate saturation of sample surface.

17.6 ADDITIONAL REQUIREMENTS FOR TEMPERATURE PROGRAMMED METHODS

17.6.1 Programmed Heating

To ensure that the signal resulting from changes in gas concentration can be properly ascribed to sample characteristics, the heating rate must be linear. Otherwise the detector signal will have superimposed upon it changes due to oscillating temperatures or varying heating rates, since the observed reaction rate is dependent on the heating rate. The PID controller should be "tuned" to the particular heating device being used such that the heating rate is linear over the temperature range of interest. It may be necessary to retune or input different PID parameters for different ramp rates and maximum temperatures.

It is preferable to control the heating rate from the furnace or mantle thermocouple, not the sample temperature, since the effects of any local exotherm (or endotherm) in the sample are strongly ameliorated and will not upset the heating program.

17.6.2 Sample Temperature

Since these measurements require that the conditions be non-isothermal, there is necessarily a difference between temperature of the heating device and the sample temperature. The sample temperature lags that of the heating device, less so if the heating rate is slow, more so if the heating rate is fast. Therefore, the position of the sample thermocouple is more important during temperature-programmed measurements than it is, for example, during pretreatment prior to pulse titration. Furthermore, if the sample temperature is recorded from inside the sample cell then local variations in temperature due to exothermic or endothermic reactions can be observed.

17.7 TEMPERATURE PROGRAMMED REDUCTION

Temperature programmed reduction (TPR) is most commonly used to determine the "ease of reducibility" of a metal oxide or oxygen-containing metal compound. Therefore, hydrogen diluted with an inert gas (nitrogen - or argon if the undesirable formation of nitrides is anticipated) is used. Five percent (by volume) is a useful concentration since this affords not only a great deal of instrument sensitivity, but also a high safety factor. Water is the main product of reaction and should be removed (e.g. cold trap) from the gas flow prior to the detector to afford a "clean" signal. Carbon monoxide diluted in helium is a milder alternative, but in this case carbon dioxide is the main reaction product that must be removed (trapped) using e.g. dry ice/solvent as coolant.

The sample typically undergoes minimal preparation, save the removal of water by purging with pure inert gas, albeit at an elevated temperature and one that does not otherwise materially alter the sample. Resident moisture can markedly affect the reducibility of oxides at lower temperatures [1]. Allow the sample to cool before admitting reactive gas.

The reducing gas mixture should be allowed to purge the sample cell and detector thoroughly before heating begins. After the heating ramp is started, the ramp rate should not be altered because this will introduce spurious peaks in the TPR signal. Both detector signal and sample temperature should be recorded as a function of time. If the reduced sample will subsequently be analyzed by pulse titration for metal area etc., it is important to avoid unnecessarily high temperatures, which normally result in sintering and a significant loss of active metal area - unless, of course, the effect of high temperatures on sintering is actually being studied!

The results of a TPR analysis normally consist of a graphical plot of detector signal versus sample temperature. It is important to record the instrument settings (flow rate, detector current, signal range setting, etc.) so that direct comparisons can be made between different samples and/or heating rates. The peak reduction temperature of NiO has been shown to shift to lower temperatures by incorporation of alkali metal ions [2]. Conversely, barium increases the reduction temperature of ruthenium [3]. Interestingly, gold was shown by TPR to exist partially oxidized on a ferric oxide support [4]. Ceria, an important component of three-way automotive exhaust catalysts because of its redox behavior

$$CeO_2 \leftrightarrows CeO_{2-x} + {}^{x}/_{2}O_2 \tag{17.4}$$

is commonly characterized by TPR, according to the following scheme

$$2CeO_2 + H_2 \rightarrow Ce_2O_3 + H_2O \tag{17.5}$$

Of particular interest is the shift in its reduction profile when the ceria is supported and/or promoted, for example in a mixture with ZrO_2 [5, 6], copper oxide [7], nickel (oxide) [8]

The amount of reactive gas "consumed" by the sample can be quantified by making injections of carrier gas. The ratio of peak areas gives the nominal volume of gas adsorbed when normalized according to the active gas concentration. Alternatively, the signals corresponding to reducing gas mixture (normally set to zero) and to pure, inert diluent gas (maximum signal) are established. Integrating the peak area between these limits as a function of time, knowing the volumetric flow rate and active gas concentration, yields the total volume adsorbed. This method can only be used for detectors that are linear over the signal range used, unless gas mixtures of intermediate concentration are available in order to calibrate the detector response. A reference standard such as WO_3 has also been used for TCD calibration [9].

17.8 TEMPERATURE PROGRAMMED OXIDATION

Temperature programmed oxidation (TPO) is performed in an entirely analogous manner to TPR wherein diluted oxygen (strong oxidizing agent) /

carbon dioxide (mild oxidizing agent) is substituted for the reducing gas. TPO can be particularly useful for investigating different forms of carbons, such as graphite, amorphous carbon and lately of particular interest, carbon nanotubes. It has been reported that TPO is sufficiently sensitive to differentiate between single-walled nanotubes (SWNT) and multi-walled nanotubes (MWNT)[10, 11]. Carbon deposits from coking and laydown [12-17] and metal carbides, especially that of molybdenum [18-20] are equally amenable to differentiation by TPO.

17.9 TEMPERATURE PROGRAMMED DESORPTION

The only experimental difference between TPR/O and temperature programmed desorption (TPD) is that the change in gas composition monitored by the TCD is as a result of gas and vapors being desorbed from the sample rather than those being adsorbed by it. Therefore, a pure, single-component carrier gas is flowed over the sample. When desorbing previously adsorbed hydrogen (from a pulse titration experiment for example), nitrogen or argon should be used. If some other species is of interest, then helium is preferred. A mass spectrometer negates that particular requirement. Note however, nitrogen (mass28) should not be used as carrier when working with carbon monoxide (also mass 28)!

17.9.1 Some Specific Applications

Different adsorbed *probe* gases are used for different applications depending on the nature of the surface chemistry to be evaluated. The most commonly used methods are highlighted below.

17.9.1.1Acid/Base. Ammonia TPD is of particular interest since the results represent the relative acid site strength distribution on the solid surface. Zeolites, for example, typically exhibit both a low temperature peak corresponding to "weak" sites and a higher temperature peak corresponding to "strong" sites. The strength is not only determined by Si:Al ratio, but can also be modified by cation substitution [21-26] or otherwise modified[27]. Sulfated ZrO_2 catalysts as revealed by NH_3 TPD have more and stronger acid sites than their unmodified counterparts [28]. The acidity of various titania based catalysts were investigated by NH3 TPD [29] wherein zeolite H-ZSM-5 was used to calibrate the detector response. H-ZSM-5 has been extensively studied in its own right [21, 30-35]. Like most zeolites, H-ZSM-5 displays two peaks in the NH_3 TPD profile (see Fig. 17.3).

The low temperature peak (Lewis/weak Brønsted) is eliminated in the presence of moisture [37], which is equivalent to saturating with NH_4OH rather than gaseous NH_3. NH_4^+ binds only to strong Brønsted sites, not Lewis acid sites [38] so it is reasonable to ascribe the high temperature peak to the strong Brønsted acid sites [21]. Conversely, the high temperature peak

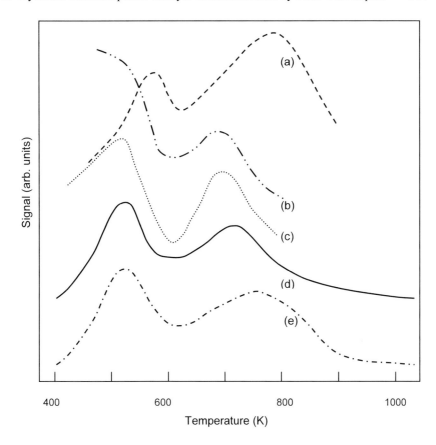

Fig 17.3 Ammonia TPD profiles from H-ZSM-5 zeolites: (a) adapted from Ref. [31], (b) adapted from Ref. [21], (c) adapted from Ref. [27], (d) Ref [36], (e) adapted from Ref. [32].

can be completely suppressed by pretreatment with formaldehyde or Na_2CO_3 [31]

Basic sites can be investigated by using carbon dioxide as the sorbed gas. Di Cosimo et al [39] found three CO_2 TPD peaks from MgO to be in agreement with infrared spectra of adsorbed CO_2 They ascribed the lowest temperature peak to bicarbonate species on weakly basic hydroxyl groups; intermediate basic sites being revealed by the second peak which represents bidentate carbonate species on Mg-O sites and lastly, strongly basic surface O^{2-} anions form unidentate carbonates. Multiple CO_2 desorption peaks have also been reported from a mixed metal oxide (fuel cell anode material)[40] and molybdenum nitride [41].

17.9.1.2 Oxidizers. O_2 desorption from gold [42] silver [43], bimetallic Pd-Rh [44] cerium oxide [45] and ruthenium dioxide [46, 47] and CO desorption from platinum [48] and copper [49] are examples of TPD studies of oxidation catalysts.

17.9.1.3 Reducers. H_2 TPD has been used to characterize hydrogenation catalysts such as ruthenium [50] and to investigate the behavior of hydrogen spillover on supported platinum hydrogenation catalysts [51]. It has also been used to determine the kinetics of hydrogen desorption from copper [52, 53] and the activation energy for desorption from silica-supported nickel [54].

Note that the evolved gas or vapor is not restricted to having been previously adsorbed, but can be the result of decomposition. For example, simple carbonates yield carbon dioxide when heated to a sufficiently high temperature.

17.10 MASS SPECTROMETRY

Some gases do not undergo simple adsorption onto or desorption from a catalyst surface. If other species are present at the surface of the appropriate catalyst type, then a reaction between different adsorbed gases may occur during a TPR analysis. For example, a nickel surface that has been stabilized by the adsorption of a surface layer of carbon dioxide will catalyze a reaction between the carbon dioxide and hydrogen to give some methane. This can occur at quite low temperatures. A simple TCD cannot realistically distinguish between the various components that make up the flowing gas downstream of (after) the sample.

However, a mass spectrometer (MS) can quickly obtain a profile of the gas composition according to the molecular weight of ionized (or fragmented) species. In the above example a mass spectrometer would clearly show the reduction in hydrogen concentration as the reaction with adsorbed carbon dioxide proceeded. In addition, a profile of methane generation (from the catalyzed reaction) would be just as easily generated. Simple thermal desorption of the carbon dioxide could be confirmed by the presence of mass 44 in the effluent gas. The dissociative adsorption of acetone on carbon nanotubes [55] was demonstrated by TPD-MS.

Gas should normally be taken from as close to the sample as possible. A capillary cell can be inserted directly into the cell if desired. This method will sample all gases present including water vapor, the signal from which may overwhelm any others present during TPR. Gas can be taken into the mass spectrometer after a trap so as to remove large amounts of gas or vapor that is not of direct interest. However, removing some gas before the TCD (if present) renders its signal somewhat inaccurate, at least quantitatively. If a complete TCD signal is required, i.e. one that represents *all* of the changes in gas composition, then gas must be taken into the mass spectrometer after the TCD. However, this may introduce a significant time-delay between what is actually happening in the cell and what the mass spectrometer "sees".

17.11 METAL PARAMETERS

Table 17.2 Some useful physical parameters of metals commonly encountered during chemisorption studies.

Metal	Atomic Mass	Density (g/ml)	Cross Sectional Area (nm^2/atom)
Cobalt, Co	51.933	8.90	0.0662
Copper, Cu	63.546	8.92	0.0680
Iridium, Ir	192.220	22.42	0.0862
Iron, Fe	55.847	7.86	0.0614
Molybdenum, Mo	95.940	10.20	0.0730
Nickel, Ni	58.690	8.90	0.0649
Palladium, Pd	106.420	12.02	0.0787
Platinum, Pt	195.080	21.45	0.0800
Rhenium, Re	186.207	20.53	0.0649
Rhodium, Rh	102.906	12.40	0.0752
Ruthenium, Ru	101.070	12.30	0.0614
Silver, Ag	107.868	10.50	0.0870

17.12 REFERENCES

1. Burwell R.L. and Taylor H.S. (1936) *J. Am. Chem. Soc.* **58**, 1753.
2. Moon H-D., Lim T-H. and Lee H-I. (1999) *Bull. Korean Chem. Soc*. **20**, 1413.
3. Liang C., Wei Z., Luo M., Ying P., Xin Q. and Li C. (2001) *Stud. Surf. Sci. Catal.* **138**, 283.
4. Hoa Z., An L. and Wang H. (2001) *Sci. China, Ser. B* **44**, 596.
5. Fernández-García M., Martínez-Arias A., Iglesias-Juez A., Belver C., Hungaria A.B., Conesa J.C. and Soria J. (2000) *J. Catal*. **194**, 385.
6. Kim D.H., Ahn B.J., Koslova A., Kung H.H. and Kung M.C. (2002) *ACS Petr. Chem. Div. Preprints* **47**, 380.
7. Adamopoulos O., Zhang Y., Croft M., Zakharchenko I., Tsakalakos T. and Muhammed M. (2001) *Mat. Res. Soc. Symp. Proc*. **676**, Y8.11.1.
8. Wang S. and Lu G.Q. (1998*) Appl. Catal. B* **19**, 267.
9. Biscardi J.A., Meitzner G.D. and Iglesia E. (1998) *J. Catal*. **179**, 192.
10. Kitiyanan B., Alvarez W.E., Harwell J.H. and Resasco D.E. (2000) *Chem. Phys. Lett*. **317**, 497.
11. Herrera J.E. and Resasco D.E. (2003) *Chem. Phys. Lett*. **376**, 302.
12. Wang S. and Lu G.Q. (2000) *J. Chem. Technol. Biotechnol*. **75**, 589.
13. Altin O., Bock G. and Eser S. (2002) *ACS Petr. Chem. Div. Preprints* **47**, 208.
14. Lee H.C., Woo H.C., Ryoo R., Lee K.H. and Lee J.S. (2000) *Appl. Catal. A* **196**, 135.
15. Łojewska J. (2001) *Stud. Surf. Sci. Catal*. **139**, 13.

16. Nagaoka K., Okamura M. and Aika K. (2001) *Catal. Commun.* **2**, 255.
17. McIntosh S., Vohs J.M. and Gorte R.J. (2003) *J. Electrochem. Soc.* **150**, A470.
18. Teixeira da Silva V.L.S., Schmal M., Schwartz V. and Oyama S.T. (1998) *J. Mat. Res.* **13**, 1977.
19. Kim Y.H., Borry R.W. and Iglesia E. (2000) *Micropor. Mesopor. Mater.* **35–36**, 495.
20. Ding W., Li S., Meitzner G.D. and Iglesia E. (2001) *J. Phys. Chem., B* **105**, 506.
21. Ahn B.J., Park J.R. and Chon H. (1989) *J. Korean Chem. Soc* (Korean) **33**, 177.
22. Wang. B., Lee C.W., Cai T-X. and Park S.E. (2001) *Bull. Korean Chem. Soc.* **22**, 1056.
23. Katada N., Kageyama Y. and Nawa M. (2000) *J. Phys. Chem., B* **104**, 7561.
24. Takami M., Yamazaki Y. and Hamada H. (2001) *Electrochemistry* **69**, 98.
25. Amin N.A.S. and Angorro D.D. (2003) *J. Nat. Gas Chem.* **12**, 123.
26. Pérez-Ramírez J., Mul G., Kapteijn F., Moulijn J.A., Overweg A.R., Doménech A., Ribera A. and Arends I.W.C.E. (2002) *J. Catal.* **207**, 113.
27. Lü R., Tangbo. H., Wang. Q. and Xiang S. (2003) *J. Nat. Gas Chem.* **12**, 56.
28. Bi Y. and Dalai A.K. (2003) *Can. J. Chem. Eng.* **81**, 230.
29. Sivalingam G., Nagaveni K., Madras G. and Hegde M.S. (2003) *Ind. Eng. Chem. Res.* **42**, 687.
30. Hunger B., Hoffmann J., Heitzsch O. and Hunger M. (1990) *J. Therm. Anal.* **36**, 1379.
31. Joo O-S., Jung K-D. and Han S-H. (2002) *Bull. Korean Chem. Soc.* **23**, 1103.
32. Lin H-e. and Ko A-N. (2000) *J. Chinese Chem Soc.* **47**, 509.
33. Wang. B., Lee C.W., Cai T-X. and Park S.E. (2001) *Catal. Lett.* **76**, 219.
34. Matsuura H., Katada N. and Niwa M. (2002) Presented at: *2nd International FEZA Conference*, Taormina, Italy, Sept. 1-5.
35. Singh A.P. and Venkatesan C. (2003) *Bull. Catal. Soc. India* **2**, 43.
36. Thomas M.A. *This work, previously unpublished.*
37. Igi. H., Katada N. and Niwa M. (1997) In *Proceedings of the International Symposium on Microporous Crystalline Materials, ZMPC '97*, Waseda University, Tokyo, Japan, August 24-27, p113.
38. Jentys A. and Lercher J.A. (2001) In *Introduction to Zeolite Science and Practice* 2nd Edn. (van Bekkum H. *et al*, eds.) Elsevier, Amsterdam, p345.
39. Di Cosimo J.I., Díez V.K., Xu M., Iglesia E. and Apesteguía C.R. (1998) *J. Catal.* **178**, 499.
40. Sauvet A-L., Fouletier J., Gaillard F. and Primet M. (2002) *J. Catal.* **209**, 25.
41. McGee R.C., Bej S.K. and Thompson L.T. (2003) "*Characterization of Base Sites on Molybdenum Nitride Catalysts*" presented at the 18th NAM (North American Catalysis Society) Cancun, Mexico.
42. Bondzie V.A., Parker S.C. and Campbell C.T. (1999) *Catal. Lett.* **63**, 143.
43. Huang W.X., Teng J.W. and Bao X.H. (2001) *Surf. Interface Anal.* **32**, 179.
44. Rassoul M., Gaillard F., Garbowski E. and Primet M. (2001) *J. Catal.* **203**, 232.
45. Yang L., Kresnawahjuesa O. and Gorte R.J. (2001) *Catal. Lett.* **72**, 33.
46. Kim Y.D. and Over H. (2001) *Top. Catal.* **14**, 95.
47. Over H. (2002) *Appl. Phys., A* **75**, 37.
48. Heiz U., Sanchez A., Abbet S. and Schneider W-D. (1999) *Eur. Phys. J. D* **9**, 1.
49. Martínez-Arias A., Fernández-García M., Gálvez O., Coronado J.M., Anderson J.A., Conesa J.C., Soria J. and Munuera G. (2000) *J. Catal.* **195**, 207.
50. Zhang Z., Jackson J.E. and Miller D.J. (2001) *Appl. Catal. A* **219**, 89.
51. Miller J.T., Meyers B.L., Barr M.K., Modica F.S. and Koningsberger D.C. (1996) *J. Catal.* **159**, 41.
52. Genger T., Hinrichsen O. and Muhler M. (1999) *Catal. Lett.* **59**, 137.
53. Tabatabaei J., Sakakini B.H., Watson M.J. and Waugh K.C. (1999) *Catal. Lett.* **59**, 143.
54. Arai M., Nishiyama Y., Masuda T. and Hashimoto K. (1995) *Appl. Surf. Sci.* **89**, 11.
55. Chakrapani N., Zhang Y.M., Nayak S.J., Moore J.A., Carroll D.L., Choi Y.Y. and Ajayan P.M. (2003) *J. Phys. Chem. B* **107**, 9308-9311.

18 Mercury Porosimetry: Intra and Inter-Particle Characterization

18.1 APPLICATIONS

The forced intrusion of liquid mercury between particles and into pores is routinely employed to characterize a wide range of particulate and solid materials. Most materials can be analyzed so long as the sample can be accommodated in the instrument, which typically restricts the sample dimensions to no more than 2.5cm. Those materials that amalgamate with mercury (zinc and gold for example) cannot be analyzed unless extreme steps are taken to passivate the surface. The exact pore size range that can be measured depends predominantly on the instrument pressure range but also on the contact angle employed in the Washburn equation. The largest pore size that can be determined is limited by the lowest filling pressure attainable and the smallest pore size by the highest pressure achievable.

Mercury intrusion is an extremely useful analytical tool in the investigation of compaction and sintering processes. The formation of true pores from the void spaces between loosely packed powder particles during pelletizing, tableting or granulation can be easily monitored. The coalescence of pores and subsequent elimination of pore volume, with the concomitant shrinkage of formed pieces, is similarly amenable to quantification by mercury intrusion. Therefore, this method is most appropriate in the fields of ceramics, catalyst manufacture, solid dosage forms (pharmaceuticals) and other similar processing of materials. Fibers, fibrous mats, fabrics and filter media in general can be analyzed for pore size, fiber diameter, permeability and tortuosity in a single measurement.

The particle size distribution of powders can theoretically be determined in the range from 2500 micrometers to 25 nm. Mercury intrusion is particularly useful in this regard for those materials that are so strongly agglomerated (magnetic ferrites for example) such that they can be neither easily, nor sufficiently, dispersed for particle size analysis by more commonly employed methods - such as laser diffraction or sedimentation.

18.2 WORKING WITH MERCURY

Mercury should be handled with the same care that any other laboratory chemical is given. The current Occupational Safety and Health Administration (OSHA) permissible exposure limit (PEL) for mercury vapor is 0.1 mg/m^3 of air [1]. The National Institute for Occupational Safety and Health (NIOSH) has (at time of going to press) established a recommended exposure limit (REL) for mercury vapor of 0.05 mg/m^3 as a time-weighted average (TWA) for up to a 10-hour workday and a 40-hour workweek [2].

Safe working conditions are quite easy to maintain if a few simple guidelines are adhered to. Always limit the amount of mercury being used at one time. Use appropriate personal protective clothing and equipment effective in preventing skin contact with mercury vapor. Ensure that containers of mercury are securely capped when not actually being poured from, or into. Handle containers of mercury, including sample cells, in a well-ventilated area. Use the mercury vapor traps supplied on the equipment and never override or disable any safety device.

> *Observe your local safety regulations regarding the use*
> *and disposal of mercury and mercury-contaminated waste.*

18.3 EXPERIMENTAL REQUIREMENTS

Although many mercury porosimeters have been commercialized, they all have several essential components, which are perhaps different in their detailed design but nevertheless are common to each apparatus. These components include the following:

- Cell (penetrometer) to hold the test sample.
- Vacuum filling apparatus to remove air from the cell and pores within the sample and for transferring mercury into the sample cell.
- Pressure generator.
- High-pressure vessel to contain the sample cell.
- Hydraulic fluid to transmit the pressure from the generator to the sample cell.
- Measuring circuits to monitor change in mercury volume as a function of applied pressure.

18.4 SAMPLE CELL

The sample cell or *penetrometer* (sometimes called a dilatometer) is used both to contain the sample and to facilitate the measurement of intrusion and extrusion volumes. In general, the penetrometer is a glass vessel plus some means of closing off one end. The sample is contained within a glass bulb of several ml capacity and a capillary bore glass tube or "stem" is attached to the bulb. If both bulb and stem are formed as a single piece of glass, a metal cap is placed over the sample bulb after the sample has been added to the bulb. Alternatively, a separate stem is attached to the bulb. A suitable means of clamping the two pieces together completes the penetrometer assembly.

Since the capillary stem forms the means by which the volume intruded and/or extruded will be recorded, the volume contained therein is crucial to the sensitivity and success of the measurement. The stem volume must be sufficiently large to accommodate the entire volume of both inter-particle voids and/or pores to be filled. However, it should not be so large such that the volume intruded constitutes an excessively small percentage of the stem volume. Experience and the anticipated pore volume will dictate the most appropriate stem diameter to be used. The amount of sample can also be adjusted up or down to make better use of the stem volume.

Cells to be used with powders should have means to reduce the likelihood of particles being elutriated from the sample bulb. See §18.13 for further considerations when working with powders.

18.5 VOLUME MEASUREMENT

Several methods have been employed to measure the change in mercury level within the penetrometer stem as intrusion and extrusion take place. The most common method today uses a metal sheath coaxially surrounding the stem that constitutes one plate of a capacitor, the mercury within the stem serving as the second plate. The capacitance changes with the mercury level due to the variation in the effective plate area (the overlap between mercury capillary and metal sheath).

A follower probe, once popular, could also monitor the change in mercury level. As the level decreased, electrical contact between the probe and the mercury level was broken. This actuated a drive motor that moved the probe, through a pressure seal, until contact was reestablished. The drive motor angular displacement (number of whole or partial rotations) was proportional to the change in mercury level. A third method (now obsolete) involved the use of a platinum or platinum iridium alloy resistance wire placed coaxially inside the penetrometer stem. As the mercury level decreased, the amount of resistance wire exposed increased, thereby providing a voltage that increased linearly with decreasing mercury level. Platinum or platinum-iridium alloy was used as the wire because neither amalgamates with mercury.

18.6 CONTACT ANGLE

The accuracy of the measurement of pore radii, into which intrusion occurs, is limited by the accuracy to which the contact angle between mercury and the sample surface is known. For comparative purposes, it is adequate to choose a reasonable value for the contact angle (typically 130° - 140°) in order to ascertain whether or not two samples have the same pore volume and pore-size distribution. However, if absolute data are required, it is necessary to measure the contact angle very accurately. While the great majority of materials exhibit intrusion contact angles near 140°, one must recognize that the nature of the cosine function is such that, at angles near 140°, the value of the cosine changes substantially with the angle. For example, an error of even 1° at 140° would introduce an error of slightly more than 1.4% in cosθ and thus an identical uncertainty in the pore radius. Not only is it necessary to measure the contact angle accurately but it should also be measured under the same conditions that prevail in mercury porosimetry. This requires that the contact angle be measured as the *advancing* angle [3,4] on a clean surface under vacuum and using the conditions that match those of the actual intrusion as closely as possible.

18.6.1 Dynamic Contact Angle

A mercury contact anglometer [5] that measures the contact angle on powders and meets the above requirements is illustrated in Fig. 18.1. First, a cylindrical hole of known radius is made in a compacted bed of powder around a precision diameter pin that, upon removal, produces a cylindrical open-ended (through) pore.

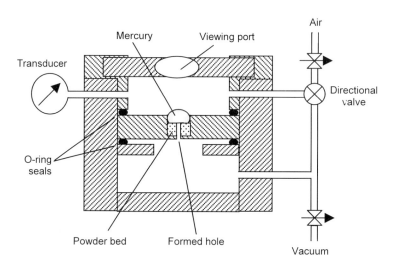

Figure 18.1 Incipient intrusion contact anglometer.

The prepared sample is then placed within the cavity of the anglometer. As shown in the figure, mercury of fixed, known volume is placed on the powder (to a known height and, therefore, known pressure head). O-ring seals prevent the flow of gas from the upper to the lower chamber. After placing the holder with powder sample and mercury into the anglometer housing, both upper and lower chambers are evacuated. Air is then allowed to bleed slowly into the upper chamber and a sensitive transducer monitors the pressure. A digital display, with an appropriate offset to allow for

Table 18.1. Contact angle measurements of mercury on various materials.

Material	Mean contact angle (θ°)	Standard deviation
Dimethylglyoxime	139.6	0.45
Galactose	140.3	0.43
Barium chromate	140.6	0.41
Titanium oxide	140.9	0.55
Zinc oxide	141.4	0.34
Dodecyl sodium sulfate	141.5	0.44
Antimony oxide	141.6	0.88
Fumaric acid	143.1	0.27
Starch	147.2	0.68
Carbon	154.9	1.20

the mercury pressure head, indicates the pressure or the corresponding value of cosθ, which is proportional to the pressure. When the breakthrough pressure is reached, mercury is forced through the precision-bore hole into the lower chamber. A fast electronic circuit responds to the rapid drop in pressure sensed by the transducer and locks the pressure (or cosθ) reading at the highest recorded value. Table 18.1 illustrates the results of six successive contact angle measurements on each of ten powders, along with the reproducibility, expressed as standard deviation using the anglometer described above.

Using low-pressure porosimetry, Winslow [6] measured contact angles by determining the breakthrough pressure that was required to force mercury into numerous holes drilled into the surface of solid discs.

18.6.2 Static Contact Angle

Another method [7] for measuring the contact angle utilizes a small drop of mercury placed on the surface of a smooth bed of powder, which assumes a near spherical shape. As additional mercury is added to the drop, the height increases until it reaches a maximum value. Further additions of mercury increase the drop diameter with no additional increase in the height. Fig. 18.2 illustrates the change of drop shape as mercury is added. The relationship between contact angle θ and the maximum drop height h_{max} is given by

$$\cos\theta = 1 - \frac{\rho g h_{max}^2}{2\gamma} \tag{18.1}$$

where ρ and γ are the density and surface tension of mercury respectively, and g is the acceleration due to gravity.

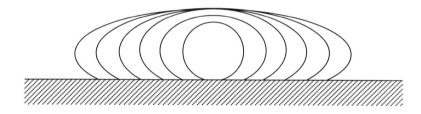

Figure 18.2 Shape of mercury drop with increasing volume on a surface, under the influence of gravity.

For a drop of given volume Ω, the diameter ϕ of the drop as viewed normal to the surface of the solid is given [8] by

$$\phi = 2\left(\frac{3}{\pi}\right)^{1/3} \frac{1}{(2+\cos\theta)^{1/3}(1-\cos\theta)^{2/3}} \Omega^{1/3} \tag{18.2}$$

The contact angle can also be evaluated from the maximum diameter of the drop, ϕ_{max}, and its diameter at its base across area of contact with the solid surface ϕ_{base} [9, 10]

$$\sin(180-\theta) = \frac{\phi_{base}}{\phi_{max}} \tag{18.3}$$

The so-called "sessile drop" methods measure a static contact angle and not the advancing angle. The measurement is difficult to perform under vacuum conditions and further difficulties arise because the powder tends to float on the mercury drop while the drop is prone to sinking into the powder bed, unless it is tightly compacted.

Optical devices are sometimes used for the measurement of contact angle wherein the operator must attempt to establish the tangent to the contact angle of a drop of mercury resting on a plane surface. This method has never proven reliably accurate because if its inherent subjectivity. Different experimenters will inevitably measure substantially different contact angles and even the same person will observe different angles on the same material on different occasions.

Table 18.2 Some other contact angle (θ) values.

Material	θ (°)	Reference
Mortar	130	[11]
NaY zeolite	132	[12]
Alumina	135	[13]
Steel	130	[14]
Silicon	137	[10]
Beryllium oxide	140	[15]
Magnetic tape	140	[16]
Quartz sand (Ottawa)	140	[17]
Titania	140	[12]
Portland cement	140.7	[18]
Silica Fume	142.5	[18]
Silica	145	[19]
Alumina	144.5	[20]
Borosilicate glass	153	[21]
Activated carbon	155	[12]
Titanium oxide	160	[21]
Norit Carbon	163	[12]

18.7 A MODERN POROSIMETER

The very wide pore size range measurable by mercury porosimetry is only made possible by virtue of a very wide *pressure* range – sub ambient to many hundreds of MPa. No one technology can reasonably achieve this and provide simultaneous analysis of more than one sample. Therefore, it is convenient to measure large pores (low pressure) by means of a pneumatic apparatus, and small pores (high pressure) by hydraulic means. Each has its own engineering constraints, but nowadays *both systems are incorporated into a single unit* for convenience, mercury containment and operator safety. The separate tasks entailed in low and high-pressure work are described individually below.

18.8 LOW PRESSURE MEASUREMENTS

To measure pores larger than 7 μm, corresponding to a pressure of approximately one atmosphere, it is necessary to monitor intrusion at pressures less than 1 atmosphere. Since the sample cell must be filled with mercury, and as much air removed from the sample's pores as is reasonably practicable, both functions (fill and measure) can be conveniently incorporated into a single apparatus (see Fig. 18.3)

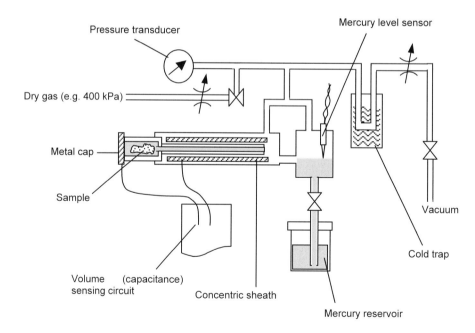

Figure 18.3 Low-pressure section of porosimeter.

18.8.1 Sample Cell Evacuation

The assembled penetrometer(s) containing the sample(s) is (are) installed into the sub-ambient pressure porosimeter in a horizontal or near-horizontal orientation via a leak-tight compression fitting. Air is pumped from the sample cell(s) at a sufficiently slow rate as to prevent elutriation of fine powder from the bulb into the cell stem. Evacuation is continued until the pressure falls to less than 50 μm Hg (50 millitorr). Dry samples evacuate more quickly and a cold-trap between the sample cell and vacuum system assists in attaining the best possible vacuum level. It also prevents mercury vapor from reaching the vacuum pump.

18.8.2 Filling with Mercury

The open end of the cell stem(s) is (are) submerged under mercury by one of two means - or even a combination: mercury can be drawn up to cover the end of the stem(s) by allowing ambient atmosphere or other dry gas to force the mercury up from a reservoir, and/or the penetrometer(s) can be rotated downward such that the open end of the stem enters a pool of mercury. Air or gas is slowly admitted above the liquid to force mercury into the stem(s) and to fill the void spaces around the sample(s) in the penetrometer bulb(s). When electrical contact is made between the mercury and an electrode cap, the cell has been filled with mercury. Excess mercury is allowed to drain from around the end of the stem(s), returning the penetrometer(s) to the horizontal if it were tipped in the filling process. If no low-pressure intrusion is anticipated, or is of no interest, the filled penetrometer(s) is (are) transferred directly to the high-pressure porosimeter chamber.

18.8.3 Low Pressure Intrusion-Extrusion

A slow, continuous bleed of gas into the measuring station causes the pressure to increase that forces mercury to intrude into inter-particle voids and large pores. The pressurization is continued until atmospheric pressure is attained, or up to 50 psia depending on whether the porosimeter is connected to a supply of pressurized gas or not. The change in the length of the mercury column in the cell stem is monitored as a function of applied pressure and converted to a volume reading by the associated electronics. If so desired, the pressure can be decreased from whatever maximum is achieved, thus permitting the acquisition of extrusion data.

This completes a full low-pressure intrusion-extrusion cycle, which can be repeated a number of times if so desired to investigate the repeatability of the process. When the sample cell is removed from the low-pressure apparatus, it can be inserted into the high-pressure cavity for high-pressure analysis (of smaller pores). This requires that the cell be moved from the horizontal to the vertical position with the open end of the cell stem uppermost. This creates an additional pressure head of 0.03 – 0.04 MPa above ambient on the sample. Since in most low-pressure analyses data has been acquired up to at least 0.17 MPa, this additional pressure head proves to be no inconvenience.

18.9 HIGH PRESSURE MEASUREMENTS

The pressure vessel, or high-pressure cavity, containing the penetrometer(s) consists of a thick-walled stainless-steel cylinder with removable caps on the top and bottom. The penetrometer(s) is (are) installed through the top opening. The top cap is equipped with a means of venting air and oil, and a check valve.

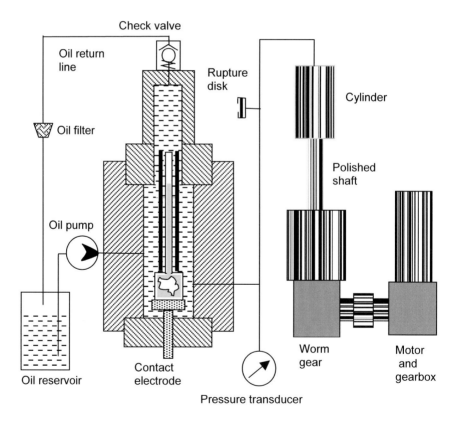

Figure 18.4 High-pressure section of porosimeter.

The removable bottom cap, which houses the contact electrode, allows for easy draining should a cell break or other foreign matter enter the high-pressure cavity This pressure vessel is connected to the pressure generator via high–pressure tubing. The pressure generator, driven by a powerful electric motor through a reducing gearbox, uses a threaded shaft, or worm gear, to drive a stainless-steel piston into a narrow cylinder filled with hydraulic oil. As the piston penetrates into the cylinder, the oil in the entire apparatus is compressed. The total volume displaced by the piston is approximately 20 ml, which is sufficient to produce pressure up to 414 MPa.

It is extremely convenient if the hydraulic system is provided with an oil reservoir, electric pump and solenoid vent valve to permit the automatic purging of the high-pressure cavity with oil. This serves the important purpose of expelling air from the cavity, which would otherwise limit the maximum pressure achievable due to its merely being compressed when the piston advances. The presence of air in the cavity can also have a deleterious effect on the accuracy of the intrusion volume measurement. Furthermore, the oil delivery system may be fitted with a filter and return line from the vent back to the oil reservoir. This allows clean oil to be automatically re-circulated, reducing operating costs and operator effort.

18.10 SCANNING METHOD

A mercury porosimeter capable of producing essentially continuous data of both intrusion and extrusion curves was developed by Quantachrome Corporation [22]. The rate of pressurization is controlled by the motor speed of the pressure generator system. The motor speed can be set to a fixed, constant value and permits detailed measurement of intrusion and extrusion in as little as six minutes. However, a microcomputer can adjust the pressurization and depressurization rate in inverse proportion to the rate of intrusion or extrusion respectively. Thus, the porosimeter provides maximum speed in the absence of intrusion or extrusion and maximum resolution when most required, that is, when intrusion or extrusion is occurring rapidly with changing pressure.

Some of the advantages of scanning porosimetry include the following: essentially continuous plots of both intrusion and extrusion curves are obtained. At no time along the intrusion curve does the pressure relax - therefore, no opportunity is provided for a small quantity of extrusion to occur in the region that should represent intrusion (i.e. increasing pressure) only. Unlike the incremental, or stepwise, method which can produces a relatively limited number of data points, the presentation of continuous scans eliminates the need to rerun an analysis to obtain information between data points for additional resolution in the regions of interest. Similarly, continuous scanning does not require any prior knowledge as to the likely intrusion/extrusion behavior of the sample in order to obtain high quality data.

The effects of mercury compression and the compressive heating of the hydraulic oil largely (and the commensurate expansion of mercury) compensate each other. Therefore, the need to make blank runs (no sample) is unnecessary for all but the most exacting analysis. The porosimeter shown in Fig. 18.4 further benefits from its having a film of hydraulic oil as part of the dielectric around the penetrometer stem. The oil's dielectric constant changes with pressure to produce almost zero volume signals from empty cells. Blank runs made on empty cells filled only with mercury show less than 1% of full-scale signal over the entire operating range from 0 to 414 MPa. Blank cell data is merely subtracted from a sample run to yield net intrusion/extrusion data. This already small correction can be refined by recording blank data using an incompressible volume slug that displaces the same volume of mercury as the sample under test.

18.11 STEPWISE METHOD

Alternatively, the pressure can be incremented or decremented by preprogrammed amounts between every datum point. This is the method of data generation that must be used by reciprocating pressure intensifiers. Some characteristics of this method may include:

- A hold or equilibration period must also be programmed for each point.
- The operator, prior to the measurement, must choose the target pressure values and this therefore assumes some prior knowledge of the sample under test.
- During the hold period, the pressure may fall and so during the intrusion curve elements of hysteresis may be generated when none should be encountered.

Nevertheless, the step-wise procedure facilitates the acquisition of data according to the Reverberi method [23, 24].

18.12 MERCURY ENTRAPMENT

It is generally observed that not the entire intruded volume of mercury is extruded from the sample during depressurization. This retained, or *entrapped* mercury results from the inability of mercury to flow back out of a pore network sufficiently quickly when the external pressure is dropped. Typically, the faster the depressurization the greater the quantity of entrapped mercury. It is the limited tensile strength of the thread of mercury running through the pore network that is largely responsible. See also §18.13 and 18.15. A used sample may appear to "weep" mercury on standing, especially if the extrusion phase was performed quickly.

An everyday example of the entrapment phenomenon occurs in a medical thermometer. When the expanded column of mercury cools, the mercury column breaks at the constriction in the fine capillary. A thread of mercury is separated from the remainder of the mercury that recedes into the bulb. This "entrapped" mercury allows the temperature to be recorded remotely from the point of measurement, even after the thermometer has cooled.

18.13 WORKING WITH POWDERS

Fine and especially free flowing powders can present an additional difficulty to the analyst. During the evacuation phase, particles might be entrained in the flow of air leaving the cell. Not only does this influence the actual quantity of material being analyzed, it can deposit said particles in the stem i.e. measuring portion of the cell! As the mercury meniscus passes such entities, sudden steps are created in the volume signal that does not represent intrusion (or extrusion). Particles may also be deposited on valve seats causing problematical leaks.

This loss of powder from the confinement of the sample bulb, known as elutriation, should be controlled and eliminated as far as is reasonably practicable. First, the powder should be dry since the evolution of water vapor under the influence of low pressure can exacerbate the problem. Next, the penetrometer chosen should cause the airflow to reverse direction when leaving the cell. This can be achieved, for example, by the stem portion of the cell being manufactured to form a U or hook inside the base of the cell bulb. A wad of filter material, e.g. glass wool, can be used inside the cell at the entrance to the stem, but in this case, it is essential that a blank run be performed with the identical quantity of filter material to account for the volume of mercury intruded between its fibers.

All powders can be compressed to some extent under the influence of the pressure exerted by the "envelope" of mercury around the powder mass. The resulting decrease in bed volume is recorded as apparent intrusion. It is important therefore to recognize the presence of such behavior so as not to incorrectly ascribe it to intrusion into large pores. If the apparent pore size exceeds that of the known particle size, then clearly powder compression is probable. Free flowing powders tend to compress via a simple logarithmic packing process [25], whereas cohesive powders "buckle" in a fashion more akin to that exhibited by highly porous solids [26,27] in a generally more linear fashion. Ultimately the powder reaches a point known as the jamming transition [28-30], at which particle rearrangement ceases. Mercury is then able to penetrate the remaining voids at pressures that are somewhat characteristic of the underlying particle size distribution. Powder compression in the porosimeter is irreversible and leads to significant entrapment on depressurization.

It is possible for the powder to compress against the metal electrode cap, essentially insulating it from the mercury. Thus, relying on automatic electrical sensing of complete cell filling might cause the analysis to start at an unnecessarily high pressure. This can be averted by incorporating a metal spring into the cell, such that it provides an electrical path through the powder.

18.14 INTER/INTRA PARTICLE POROSITY

The mercury porosimeter records the change in volume of mercury enveloping the sample as a function of applied pressure. The instrument does not differentiate between intrusion of mercury into voids *between* particles or into pores *within* particles. However, it is important to recognize when reporting mercury intrusion volumes that "*pore volume*" is normally taken to mean the intraparticle volume, and steps should be taken to eliminate or subtract volume due to interparticle voids.

Often, it is possible, with experience to make a successful subjective judgment as to when interparticle filling is complete. If not, some other data reduction technique might indicate a change in intrusion process, e.g. fractal analysis, since the fractal dimension is expected to differ according to filling of uncorrelated pore spaces.

18.15 ISOSTATIC CRUSH STRENGTH

Hollow spheres of silica (glass) ceramic or resin are used a variety of applications, generally to reduce the density of the medium in which they are borne. Cenospheres, hollow microspheres found in the by-product flyash, which is produced, in coal-burning power plants, are used as fillers in plastics and particularly in lightweight construction materials such as stucco, spackle, fire insulation and soundproofing. Specially manufactured glass bubbles can be mixed with resins to produce *syntactic* foam. Such materials are used, amongst other applications for deep-sea insulation and buoyancy because of their greater crush resistance than simple polymer foam.

The isostatic crush strength of both microspheres and the composite materials into which they are incorporated can be easily determined by pressurization in a mercury penetrometer. The volume change recorded as a function of increasing pressure reveals the crush strength distribution. It is usual to interpret the data in terms of "survival rate". That is, those pressures at which 10%, 50% and 90%, for example, apparent pore volume remains "intact". The crush strength depends on both sphere size and wall thickness. Apparent entrapment will be significant (up to 100%) since the crushing mechanism is irreversible.

18.16 REFERENCES

1.   U.S. Department of Labor Occupational Safety & Health Administration (OSHA), Safety and Health Topics: Health Guidelines.
2.   NIOSH (1992) Publication No. 92-100.
3.   Penn L.S. and Miller B. (1980) *J. Colloid Interface Sci.* **77**, 574.
4.   Adam N.K. (1948) *Trans. Faraday Soc.* **44**, 5.
5   Shields J.E. and Lowell S. (1982) *Powder Technol.* **31**, 227.
6.   Winslow D.N. (1978) *J. Colloid Interface Sci.* **67**, 42.
7.   Osipow L.I. (1964) In *Surface Chemistry*, Reinhold, New York, p233.
8.   Quéré D., Azzopardi M.-J. and Delattre L. (1998) *Langmuir* **14**, 2213.
9.   Smithwick R.W. (1988) *J. Coll. Interface Sci.* **123**, 482.
10. Latorre L., Kim J., Lee J., de Guzman P-P., Lee H.J., Nouet P. and Kim C.J. (2002) *J. Microelectromech. Systems* **11**, 302.
11. Cebeci O.Z., Al-Noury S.I. and Mirza W.H. (1988) *Stud. Surf. Sci. Catal.* **39**, 611.
12. Groen J.C., Peffer L.A.A. and Perez-Ramirez J. (2002) *Stud. Surf. Sci. Catal.* **144**, 91.
13. Oya M., Takahashi M., Iwata Y., Jono K., Hotta T., Yamamoto H., Washio K., Suda A., Matuo Y., Tanaka K. and Morimoto M. (2002) *Am. Ceram. Soc. Bull.* **81**, 52.
14. Hubert C. and Swanson D. (2001) *GSFC Flight Mechanics Symposium,* NASA.
15. Smithwick R.W. (1982) *Powder Technol.* **33**, 201.
16. Huisman H.F. and Rasenberg C.J.F.M. (1983/84) *Philips Tech. Rev.* **41**, 260.
17. Winslow D.N. (1984) *Surf. Colloid Sci.* **13**, 259.
18. Cohen M.D., Olek J. and Dolch W.L. (1990) *Cem. Concr. Res.* **20**, 103.
19. Simon J., Saffer S. and Kim C.J. (1997) *J Microelectromech. Systems* **6**, 208.
20. Lowell S. and Shields J.E. (1981) *J. Colloid Interface Sci.* **80**, 192.
21. *Determining Pore Volume Distribution of Catalysts by Mercury Intrusion Porosimetry*, D4284, ASTM International, West Conshohocken, PA, USA.
22. Lowell S. (1979) US Patent 4,170,129.
23. Svata M. (1971/72) *Powder Technol.* **5**, 345.
24. Reverberi. A. (1966) *Ann. Chim.* (Italy) **56**, 1552
25. Thomas M.A. and Coleman N.J. (2001) *Colloids Surf. A* **187-188**, 123.
26. Alié C., Pirard R. and Pirard J.P. (2001) *J. Non-Cryst. Solids* **292**, 138.
27. Pirard R., Sahouli B., Blacher S. and Pirard J.P. (1999) *J. Colloid Interface Sci.* **217**, 216.
28. Sellitto M. and Arenzon J.J. (2000) *Phys Rev E* **62**, 7793.
29. Edwards S.F. and Grinev D.V. (2001) In *Jamming & Rheology: Constrained Dynamics on Microscopic and Macroscopic Scales* (Liu A.J. and Nagel S.R., eds.) Taylor & Francis, New York.
30. Coniglio A. and Nicodemi M. (2000) *J. Phys.: Condens. Matter* **12**, 6601.

19 Density Measurement

19.1 INTRODUCTION

In many areas of powder technology, the need to measure the powder volume or density often arises. For example, powder bed porosities in permeametry, volume specific surface areas, sample cell void volume, and numerous other calculated volumes, all require accurately measured powder densities or specific volumes. It is appropriate, therefore, to introduce density measurements of solids.

19.2 TRUE DENSITY

The true or real density is defined as the ratio of the mass to the volume occupied by that mass. Therefore, the contribution to the volume made by pores or internal voids must be subtracted when measuring the true density. Regardless of whether pores and internal voids are present, the density of fine powders is often not the same as those of larger pieces of the same material because, in the process of preparing many powders, those atoms or molecules located near the surface are often forced out of their equilibrium positions within the solid structure. On large pieces of material, the percentage of atoms near the surface is negligibly small. As the particle size decreases, however, this percentage increases, with its resultant effect upon the density.

If the powder has no porosity, the true density can be measured by displacement of any fluid in which the solid remains inert. The accuracy of the method is limited by the accuracy with which the *fluid volume* can be determined. Usually, however, the solid particles contain pores, cracks, or crevices, which will not easily be completely penetrated by a displaced liquid. In these instances, the true density can be measured by using a gas as the displaced fluid. Apparatus used to measure solid volumes are often referred to as pyknometers or pycnometers after the Greek 'pyknos', meaning thick or dense. Once the sample volume and mass have been determined, the density is readily calculated.

There are many commercially available pycnometers for the determinations of the true density of powders, most of which operate on the principle of gas displacement, helium being the most frequently used gas because of its inertness and small size which enables it to penetrate even the smallest pores.

An automatic, manometric gas expansion pycnometer offers extreme simplicity of operation along with great speed and accuracy. A schematic diagram of such a pycnometer is shown in Fig. 19.1.

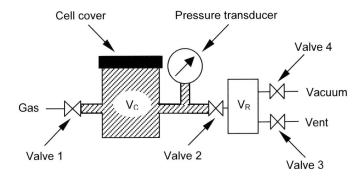

Figure 19.1 Gas expansion pycnometer for true density determination of powders, porous and irregular solids.

The volume V_C in the shaded area is considered the sample cell. After purging the system with helium through valves 1,2 and 3 or to vacuum through valves 2 and 4, all the volumes are brought to ambient pressure through valves 2 and 3. The sample cell is then pressurized to P_1, say 1 atm. above ambient. Valve 2 is then opened to connect the reference volume, V_R, to that of the cell. Consequently, the pressure drops to a lower value P_2 in the cell, while increasing from ambient to P_2 in the reference volume. Using P_1 and P_2, the powder volume, V_P, is calculated from

$$V_P = V_C + \frac{V_R}{1 - (P_1/P_2)} \qquad (19.1)$$

Equation (19.1) is derived as follows: at ambient pressure, P_a, the state of the system is described as

$$P_a V_C = nRT_a \qquad (19.2)$$

where n is the number of moles of gas occupying the volume V_C at P_a, R is the gas constant, and T_a is ambient temperature. When a sample of volume V_P is placed in the sample cell, equation (19.2) can be rewritten as

$$P_a(V_C - V_P) = n_a RT_a \qquad (19.3)$$

where n_a is the number of moles of gas in the sample cell at P_a and T_a. When the system is pressurized to P_1, it is described by

$$P_1(V_C - V_P) = n_1 R T_a \qquad (19.4)$$

where n_1 is the number of moles of gas in the sample cell at the new pressure, P_1. After volume V_R is added to that of the sample cell and the pressure falls to P_2, then

$$P_2(V_C - V_P + V_R) = n_1 R T_a + n_R R T_a \qquad (19.5)$$

where n_R is the number of moles of gas in the added volume at P_a. Substituting $P_a V_R$ for $n_R R T_a$ into equation (19.5) gives

$$P_2(V_C - V_P + V_R) = n_1 R T_a + P_a V_R \qquad (19.6)$$

Combining equations 19.4 and 19.6 results in

$$P_2(V_C - V_P + V_R) = P_1(V_C - V_P) + P_a V_R \qquad (19.7)$$

Solving equation (19.7) for the sample volume, V_P, and letting P_a be zero, since all measurements are made above ambient pressure, results in

$$V_P = V_C + \frac{V_R}{1 - (P_1 / P_2)} \qquad (cf.19.1)$$

The measurement of V_C and V_R is accomplished by two pressurizations, one with the sample cell empty ($V_P = 0$) and one with a calibrated *"blank"* of known volume in the sample cell. For an empty cell, equation 19.1 becomes

$$V_P = 0 = V_C + \frac{V_R}{1 - (P_1 / P_2)} \qquad (19.8)$$

When a calibrated *"blank"* of known volume is used, its volume, V_{cal}, can be expressed as

$$V_{cal} = V_P = V_C + \frac{V_R}{1-(P'_1/P'_2)} \quad (19.9)$$

Combining equations (19.8) and (19.9) and solving for V_R gives

$$V_R = \frac{V_{cal}}{\left[\dfrac{1}{(P_1/P_2)-1}\right]-\left[\dfrac{1}{(P'_1/P'_2)-1}\right]} \quad (19.10)$$

Simple substitution of the calculated value for V_R into equation (19.8) gives the cell volume, V_C, which can also be expressed as

$$V_C = \frac{V_{cal}}{\left[\dfrac{1-(P_1/P_2)}{1-(P'_1/P'_2)}\right]-1} \quad (19.11)$$

The calculated values for V_C and V_R are recorded for subsequent use in the working equation (19.1).

The above derivation assumes ideal gas behavior, which is closely obeyed at pressures near ambient room temperature by both pure helium and pure nitrogen. Helium is preferred, however, because of its smaller size. It is necessary to avoid using any gas that can be even slightly adsorbed. If so much as a thousandth of a monolayer of nitrogen were adsorbed, the equivalent volume of gas would be on the order of 0.001 cm^3 for each 2.84m^2 of area. Since the larger sample cells used in the type of apparatus described in Fig. 19.1 can hold 100 cm^3 or more, the total surface area of the sample can be hundreds or even thousands of square meters. Thus, errors of 0.1-1.0 cm^3 can be incurred due to very small amounts of adsorption. This is another reason why the use of helium is recommended in any gas pycnometer, except when it penetrates the solid matrix (e.g. cellulose and low density, especially cellular, polymers).

A second source of error, encountered with high area powders, is the annulus volume that exists between the powder surface and the closest approach distance of a gas molecule. If the closest approach of the helium atom to the powder surface is 0.5 Å, or 5×10^{-11} m, and the specific area of the powder is 1000 m^2/g, there will exist an annulus volume of 5×10^{-8} m^3/g or 5×10^{-2} cm^3/g. This represents a density error of 5% on materials with densities near unity. The use of gases with molecules larger than helium will exacerbate this error, again favoring helium as the preferred gas.

19.3 APPARENT DENSITY

When the fluid displaced by powder does not penetrate all the pores, the measured density will be less than the true density. When densities are determined by liquid displacement, an apparent density is obtained which often differs from the true density by gas displacement, and can differ according to the liquid used because of the different degrees to which they penetrate small pores.

The presence or absence of atmospheric gases within the pores will also affect the volume determination. If it is desired to obtain a value as close to the true density as possible by liquid displacement, then the sample must be evacuated to promote the greatest degree of penetration into the pores. The greatest difference observed is when the contact angle of the liquid with the solid is greater than 90°. In this *non-wetting* situation, pores will not be penetrated by the liquid unless some external pressure is applied (see Chapters 10 and 11 on mercury porosimetry). See §19.9 for the special application of mercury as the displaced fluid where this behavior is desirable. Careful thermostating, or knowledge of the variation in liquid density as a function of temperature, is also required. Therefore, when reporting apparent density, the liquid used and its temperature should be reported.

19.4 OPEN-CLOSED POROSITY

Certain performance materials rely on their having a cellular structure consisting of cells (bubbles) which might communicate with the exterior (open cells) or which might not (closed cells). Polymer foams (cellular plastics) form the most important class of such materials. Acoustic foams are intentionally created with a very open cell structure in order to attenuate the propagation of sound waves. Buoyancy aids, as one might expect, are substantially composed of closed cells. Packaging and cushioning materials vary according to exact application and method of fabrication. It is possible by comparing the specific volume (reciprocal density) measured by simple geometry, V_g, (see §19.7) with that measured by gas pycnometry V_p, (§19.2) to compute the relative amount of open cell volume and closed cell volume. Note that the latter includes, by definition, the volume of the solid matrix that encloses closed cells and resides between open cells.

$$Open\ cell\% = \frac{\left(V_g - V_p\right)}{V_g} \times 100\% \qquad (19.12)$$

$$Closed\ cell\ \% = \frac{V_p}{V_g} \times 100\% \quad or \quad 100\% - Open\ cell\ \% \quad (19.13)$$

19.5 BULK DENSITY

The volume occupied by the solid (including its internal pores) plus the volume of voids between particles, when divided into powder mass, yields the bulk density. For example, when powder is poured into a graduated container, the bulk density is the mass divided by the volume of the powder bed. Particle shape, particle-particle bond strength and equilibrium moisture content contribute to the *autohesiveness* of a powder. The degree of handling, i.e. the degree to which the powder bed is disturbed, will produce variation in the bulk density observed and can give rise to significant differences between operators and between supplier and customer.

One way to ensure reproducible, minimum bulk density (i.e. maximum bulk volume) is to use a device known as a *Scott volumeter*. Essentially, it consists of a series of glass plates down which the deagglomerated powder cascades into a metal cup of known internal volume. Excess powder is carefully removed with a straight edge and the tared container weighed full of powder.

19.6 TAP DENSITY

The tap density is another expression of bulk density, obtained after tapping (lifting and dropping) the container in a specified manner to achieve more efficient particle packing. The tap density is, therefore, almost without exception higher than the bulk density.

Both active and inactive (excipient) compounds used by the pharmaceutical industry possess many properties or characteristics and exhibit certain behaviors as a result. One very important characteristic is tapped bulk density, that is, the maximum packing density of a powder (or blend of powders) achieved under the influence of well-defined, internally applied forces, i.e. those due to the powder's own weight under the influence of the oscillating acceleration. The tap density of a material can be used to predict both its *flow properties* and its *compressibility*.

The two most commonly used measures of the relative importance of such interparticulate interactions are the *compressibility index* (often referred to as Carr's Index *(CI)* [1] and the *Hausner ratio (HR)* [2] as follows:

$$CI = 100 \times \frac{(V_0 - V_f)}{V_0}\% \quad = 100 \times \frac{(D_f - D_0)}{D_f}\% \qquad (19.14)$$

and

$$HR = \frac{V_0}{V_f} \quad = \frac{D_f}{D_0} \qquad (19.15)$$

where V and D represent powder volume and density respectively, subscript 0 denotes the initial or untapped state and f the final or tapped state. In free-flowing powders the initial bulk and tapped densities will be more similar than in poor flowing powders which yield greater differences between the two values.

To effect the measurement of tap density, powder is poured into a graduated glass cylinder and its so-called loose-poured volume is read. Tapping is performed by lifting the cylinder and dropping it onto a hard surface from a known and reproducible height. This is most conveniently done by a mechanical device that can be pre-programmed to stop after a predetermined number of taps. Such devices, which use an electric motor and cam to do the lifting and dropping, may tap at a rate of hundreds of drops per minute. It may be necessary to revert to a Scott volumeter (see §19.5) to obtain the initial (untapped) density for the most exacting work.

19.7 ENVELOPE OR GEOMETRIC DENSITY

Sometimes also called the particle or pellet density, the envelope density is an expression of the bulk density of an individual particle rather than an assembly of many particles. The required volume is that which represents the sum of the solid mass plus the volume of internal pores.

A simple (cylinder, cube, sphere), sufficiently large, geometric form's volume is readily computed from straightforward measurement of external dimensions by means of suitable calipers or micrometer. See table 19.1. Even the volume of a seemingly complex shape such as a modified ovoid, i.e. a cupped (domed) pharmaceutical tablet with a flat wall (equatorial face), can be computed geometrically. Therefore, any fluid displacement technique should conform to the bounding envelope, which describes the geometric volume, and not penetrate the pore structure. A completely non-wetting fluid, like mercury, in the absence of any applied pressure would suffice. In practice, some intrusion does occur into pores larger than approximately 14μm diameter due to the presence of atmospheric pressure. See also section 19.8.

Gases are obviously inappropriate for this measurement, but free flowing powders can be used [3-5]. Particles that are significantly smaller than the material to be analyzed will more or less fill the voids between the sample particles and conform to the surface of an irregular object.

Table 19.1 Volumes of various geometric shapes.

Shape	Dimensions required	Volume
Sphere	Diameter, d	$\pi\dfrac{d^3}{6}$
Cube	Length of side, l	l^3
Cylinder	Length, l and diameter, d	$\pi\dfrac{d^2}{4}$
Pyramid	Width, w, length, l and height, h	$\dfrac{wlh}{3}$
Cone	Diameter, d and height, h	$\pi\dfrac{d^2h}{12}$

19.8 EFFECTIVE DENSITY

A particle may contain occluded (embedded) foreign bodies, which may increase or decrease its density. It may also contain blind pores, totally encapsulated by the particle, which will effectively reduce the particle's density., e.g. cenospheres, (hollow glass microspheres). In these cases, the measured quantity (by gas, liquid or dry powder displacement) is called the effective density. Monolithic cellular samples were discussed separately in 19.4.

19.9 DENSITY BY MERCURY POROSIMETRY

Mercury porosimetry provides a convenient method for measuring the densities of powders. This technique gives the true density of those powders that do not possess pores or voids smaller than those into which intrusion occurs at the highest pressure attained in the porosimeter. It also provides apparent densities for those powders that have pores smaller than do those corresponding to the highest pressure.

Table 19.2 Density Measurement by mercury displacement and porosimetry – example.

Sample I.D.: Silica Gel
0

Date: _____

Operator: _____
0

Outgassing Conditions: _____ 0

DATA

1.	Weight of empty cell	35.9483	g
2.	Weight of cell filled with mercury	87.1300	g
3.	Weight of cell and sample	37.0205	g
4.	Weight of cell and sample filled with mercury	73.1008	g
5.	Intruded volume at 414 MPa	0.406	cm³

CALCULATIONS (using mercury density table below)

6.	Volume of mercury in cell without sample	3.778	cm³
7.	Volume of mercury in cell with sample	2.663	cm³
8.	Volume of sample and pores smaller than 7.26 µm (6-7)	1.115	cm³
9.	Volume of sample and pores smaller than 18 Å (8-5)	0.709	cm³
10.	Sample density $\dfrac{(3-1)}{9}$	1.52	g/cm³

The worksheet shown in table 19.2 illustrates the calculation of the density of a silica gel sample using the cumulative volumes intruded up to 414 MPa. The volume of the sample, including pores of radii smaller than 7.26 µm, is first determined at ambient pressure (0.1 MPa). This is accomplished by weighing the cell filled with mercury and then the cell, containing sample, filled with mercury. These weighings must be carried out with the penetrometer stem filled to the same level. After converting the weights of mercury to the corresponding volumes using table 19.3, the sample volume can be determined as the difference between the two mercury volumes.

The volume of the sample, and thus the density, including pores smaller than 18 Å, is calculated as the difference between the sample volume and pores smaller than 7.26 µm, shown as entry 8 in table 19.2, and the volume of mercury intruded at 414 MPa. The apparent density, i.e. the volume of a given mass of sample plus voids divided into the sample mass, can be calculated as a function of the void and pore volume from a mercury intrusion curve.

Table 19.3 Density of mercury as a function of temperature.

Temperature (°C)	Density (g/cm³)	Temperature (°C)	Density (g/cm³)
15	13.5585	23	13.5389
16	13.5561	24	13.5364
17	13.5536	25	13.5340
18	13.5512	26	13.5315
19	13.5487	27	13.5291
20	13.5462	28	13.5266
21	13.5438	29	13.5242
22	13.5413	30	13.5217

The ambient to 414 MPa curve for the silica gel sample is illustrated in Fig. 19.2. Using the volume of mercury intruded at various pressures, the volume of the sample including voids and pores, and thus the apparent density, can be obtained as shown in table 19.4. The calculated apparent densities are obtained by subtracting the intruded volume from the initial sample volume and dividing the resulting value into the sample weight.

Fig. 19.3 shows a plot of density versus pore size radius from the data in table 19.4. The horizontal line indicates the true density, obtained by helium pycnometry. The higher density by gas displacement reflects the volume of pores smaller than about 18 Å.

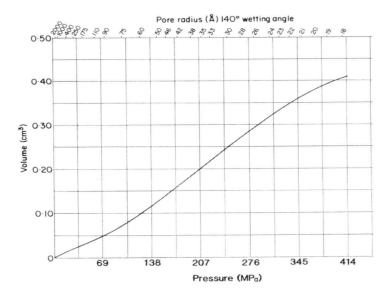

Figure 19.2 High-pressure intrusion curve of silica gel.

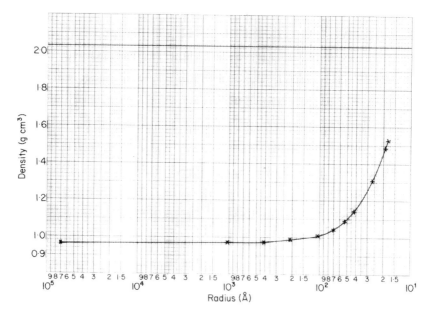

Figure 19.3 Density versus pore radius plot for silica gel.

Table 19.4 Apparent density of silica gel at various pressures.

Pressure (MPa)	Radius* (Å)	Intruded volume† (cm³)	Sample volume (cm³)	Apparent density (g/cm³)
000.10	72600	0	1.115‡	0.967
034	213	0.024	1.091	0.988
069	107	0.050	1.065	1.01
103	71.1	0.080	1.035	1.04
138	53.4	0.117	0.998	1.08
172	42.7	0.160	0.955	1.13
207	35.6	0.203	0.912	1.18
241	30.5	0.245	0.870	1.24
276	26.7	0.285	0.830	1.30
310	23.7	0.323	0.792	1.36
345	21.3	0.357	0.758	1.42
379	19.4	0.385	0.730	1.48
414	17.8	0.406	0.709	1.52

*Calculated from the intrusion pressure, assuming $\theta = 140°$; ‡for a sample weighing 1.078 g; ‡taken from entry 8 of Table 19.2.

19.10 STANDARD METHODS

There exists a multitude of standard methods for the determination of various density values. Most are material specific. The list below is not intended to be exhaustive, but illustrative of the diversity of available methods from the various recognized bodies.

ASTM[1] D2638	Real Density of Calcined Petroleum Coke by Helium Pycnometer
ASTM D5550	Specific Gravity of Soil Solids by Gas Pycnometer
ASTM D4892	Density of Solid Pitch (Helium Pycnometer Method)
ASTM D6093	Percent Volume Nonvolatile Matter in Clear or Pigmented Coatings using a Helium Gas Pycnometer
ASTM D5965	Specific Gravity of Coating Powders
ASTM D6226	Open Cell Content of Rigid Cellular Plastics
ASTM D4781	Mechanically Tapped Packing Density of Fine Catalyst Particles and Catalyst Carrier Particles
ASTM D4164	Mechanically Tapped Packing Density of Formed Catalyst and Catalyst Carriers
ASTM B 527	Determination of Tap Density of Metallic Powders and Compounds
ASTM C 493	Bulk Density and Porosity of Granular Refractory Materials by Mercury Displacement
ASTM D6393	Bulk Solids Characterization by Carr Indices
ISO[2] 787-11	Pigments and Extenders: Determination of Tamped[3] Volume and Apparent Density After Tamping
ISO 10236	Green Coke and Calcined Coke: Determination of Bulk Tapped Density
ISO 5311	Fertilizers: Determination of Bulk Density (tapped))
USP[4] <616>	Bulk Density and Tapped Density
USP <699>	Density of Solids
Ph. Eur.[5] 2.9.15	Apparent Volume
Ph. Eur. 2.2.42	Density of Solids
Ph. Eur. 2.9.23	Pycnometric Density of Solids

1. ASTM International (formerly the American Society for Testing and Materials), West Conshohocken, Pennsylvania, USA.
2. International Organization for Standardization, Geneva, Switzerland.
3 "Tamped" is an unfortunate term since it implies an external force as might be applied from a ram or piston. The referenced method describes *tapping*.
4. US Pharmacopeia (The United States Pharmacopeial Convention, Inc.) Rockville, Maryland, USA.
5. European Pharmacopoeia (European Directorate for the Quality of Medicines), Strasbourg, France.

19.11 REFERENCES

1. Carr R.L. (1965) *Chem. Eng.* **72,** 163.
2. Hausner H.H. (1967) *Int. J. Powder Metall.* **3,** 7.
3. Buczek B. and Geldart D. (1986) *Powder Technol.* **45,** 173.
4. Davis S., Dolle A. and Moreau E. (1988) *Prog. Batteries Sol. Cells* **7,** 342.
5. Medek J. and Weishauptova Z. (1989) *Collect. Czech. Chem. Commun.* **54,** 303.

Index

PARTICLE TECHNOLOGY SERIES

FORMERLY POWDER TECHNOLOGY SERIES

KLUWER ACADEMIC PUBLISHERS – DORDRECHT / BOSTON / LONDON